PROCEEDINGS OF THE 30TH INTERNATIONAL GEOLOGICAL CONGRESS
VOLUME 24

ENVIRONMENTAL GEOLOGY

Proceedings of the 30th International Geological Congress

PROCEEDINGS OF THE
30TH INTERNATIONAL GEOLOGICAL CONGRESS

BEIJING, CHINA, 4 - 14 AUGUST 1996

VOLUME 24

ENVIRONMENTAL GEOLOGY

EDITOR:
YUAN DAOXIAN
INSTITUTE OF KARST GEOLOGY, MGMR, GUILIN, CHINA

CRC Press
Taylor & Francis Group
Boca Raton London New York

CRC Press is an imprint of the
Taylor & Francis Group, an **informa** business

First published 1997 by VSP BV Publishing

Published 2019 by CRC Press
Taylor & Francis Group
6000 Broken Sound Parkway NW, Suite 300
Boca Raton, FL 33487-2742

First issued in paperback 2019

No claim to original U.S. Government works

ISBN 13: 978-0-367-44829-5 (pbk)
ISBN 13: 978-90-6764-239-2 (hbk)

Visit the Taylor & Francis Web site at
http://www.taylorandfrancis.com

and the CRC Press Web site at
http://www.crcpress.com

CONTENTS

GEOLOGY RELATED TO WASTE DISPOSAL

ENVIRONMENTAL GEOLOGY MAPPING

Proc. 30ᵗʰ Int'l.Geol.Congr., Vol. 24, p. 4
Yuan Daoxian (Ed)
© VSP 1997

Preface

This volume includes 26 papers selected from 8 symposia of the 30th IGC, grouped under the general title of Environmental Geology. There were altogether 9 symposia in this direction, but the participants of the Symposium 18-2 "geochemistry of coal and its impact on environments and human health" preferred to publish their articles in a periodical.

The symposia from which papers of this volume are selected include "geomorphology and environmental impact assessment", "trace element and human health", "environmental geology in large river harnessing and development", "volcanic activities and their effects on environments", "formation and evolution of karst and data on environmental change (UNESCO/IUGS IGCP379--karst processes and the carbon cycle)", "regional evaluation and mapping in environmental geology", "environmental geology related to disposal of municipal and industrial waste", and "environmental hydrogeology and engineering geology in aspects of radioactive waste disposal".

Within the 8 symposia, there are 194 papers included in the abstract book volume 3 of 30th IGC. Among them, 79 were arranged for oral presentation, and 64 for poster sessions. The selection of the 26 papers in this volume is based on the recommendations from the convenors of the symposia who were involved in the screening of the relevant abstracts reached at the Secretariat of the 30th IGC before the congress. However, the decision was made after the authors presented their papers and the consultation between convenors.

It is therefore naturally that this volume includes all aspects of environmental geology, i.e. the impact of different geological processes, such as crustal movement, volcanism, climate change, soil erosion, karst and groundwater pressure on human environment; the geological problems in large river development; trace elements and human health; geology of waste disposal; and environmental geology mapping.

The papers reflect the modern trend of environment geological study in using the Earth System Science approach, namely, to assess human environments on the basis of understanding the interaction between the lithosphere, hydrosphere, atmosphere, biosphere and anthroposphere. The evaluation of atmospheric greenhouse gas source and sink in karst processes carried out by IGCP 379; the ideas on the influence of 130,000 year climate changes scenario on a nuclear waste repository; the discussions on the impacts of water, rock, soil, vegetation on trace element transportation and human health; and many practices in comprehensive development of major rivers of the world could be good examples in this direction.

I would like to thank all the authors, symposium convenors and colleagues in the Institute of Karst Geology for their enthusiastic cooperation and contributions in preparing this volume.

Guilin
February 1997

Yuan Daoxian

Proc. 30th Int'l. Geol. Congr., vol. 24, pp. 5-24
Yuan Daoxian (Ed)
© VSP 1997

A Scenario of Climate Change and its Impact on the Future Behaviour of a Deep Geological Repository for Nuclear Spent-Fuel

LOUISA M KING-CLAYTON

QuantiSci Ltd., 47, Burton Street, Melton Mowbray, Leics., UK

FRITZ KAUTSKY

Swedish Nuclear Power Inspectorate, S-106 58, Stockholm, Sweden

NILS-OLOF SVENSSON

Department of Quaternary Geology, Lunds Univ., Tornavägen 13, S-223 63, Lund, Sweden

Abstract

Part of the SITE-94 project recently performed by the Swedish Nuclear Power Inspectorate (SKI) involved the construction of scenarios to assist in the evaluation of the future behaviour of a deep repository for spent-fuel. The project used real geological data from the Äspö site, and assumed that a hypothetical repository (reduced in size to approximately 10%) was situated at about 500m depth in granitic bedrock in this coastal area of SE Sweden. A Central Scenario, involving "prediction" of the climate and consequent surface and subsurface environments at the site for the next c.130,000 years, lies at the heart of the scenario definition work. The Central Climate Change Scenario has been based on the climate models ACLIN (Astronomical climate index) [29], Imbrie & Imbrie [26] and the PCM model [4]. These models suggest glacial maxima at c. 5000, 20,000, 60,000 and 100,000 years from now. The Äspö region is "predicted" to be significantly affected by the latter three glacial episodes, with the ice sheet reaching and covering the area during the latter two episodes (by up to c. 2200 m and 1200m thickness of ice respectively). The scenario can, however, only provide an illustration of future climatic change in SE Sweden. The predictions of how the climate will change over the next 130,000 years is limited by substantial uncertainties. Based upon the climate change scenario timeframe, a further objective of this work has been to provide a first indicator of the physical and hydrogeological conditions below and at the front of the advancing and retreating ice sheets, with the aim of identifying critical aspects for modelling impacts of future glaciations on far-field groundwater flow, rock stress and groundwater chemistry. The output of this study is a time-dependent description of the properties of the site, coupled to predictions of key parameter values at different times in the future, for input into modelling of the hypothetical repository performance. The sensitivity of repository performance to uncertainties in climate predictions and parameter values may be addressed.

Keywords: climate change, glaciation, radioactive waste, Äspö, Sweden, scenario.

INTRODUCTION

Part of the SITE 94 Project [39] involves the construction of scenarios to assist in the evaluation of the future behaviour of a deep repository for spent-fuel [14]. The project uses data from the Äspö Hard Rock Laboratory site and assumes that a hypothetical repository for spent-fuel (reduced in size to c.10%) is situated at about 500m depth in granitic bedrock in this coastal area of SE Sweden (Figure 1).

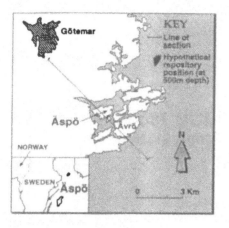

Figure 1. The location of Äspö with the line of section used to construct the Central Scenario.

A Central Scenario, involving prediction of the climate and consequent surface and subsurface environments at the site for the next c. 130,000 years, lies at the heart of the scenario definition work. In the course of time, the climate of this area is likely to change and, on the basis of past climatic variations, periods of permafrost and ice-sheet cover are expected to occur in the area in the future. These changes may significantly alter the pattern of groundwater flow and thus the potential migration of radionuclides released from the hypothetical repository.

The scenario construction process first involves the definition of all the features, events and processes (FEPs) which affect the behaviour of a repository system. The Central Scenario thus requires the following main components: a deterministic description of the most probable future climate states for Sweden, with special reference to the Äspö area; a description of the likely nature of the surface environment in the site area at each stage of the climate sequence selected; quantitative information on how these changes might affect the disposal system for input to performance assessment modelling. As recognised by [14], the Central Scenario is simply a means of *illustrating* possible future behaviour of the system and exploring how such behaviour might arise.

ON THE RELATION BETWEEN PAST AND FUTURE CLIMATES

Attempting to predict future climate changes is a difficult task requiring knowledge and understanding of the earth's past climate changes, as well as requiring a model for predicting the future. It is now known that a period of glaciations of the Northern Hemisphere started some 2.3 million years ago, and since then, during the Quaternary period, a considerable number of glaciations, interrupted by warmer interglacials, have been documented. The periodicity of the glacial/interglacial sequence approaches a 100,000 year cycle during the middle and late Pleistocene. This knowledge of Quaternary glaciations allows us to make a preliminary prognosis of future glaciations. The last interglacial, the Eemian (c. 115,000-130,000 yrs BP), was c.10,000-15,000 years long. As the present warm period, the Holocene, has lasted about this long, and as the Holocene climatic optimum passed several thousand years ago, it is probable that the cooling of the next glaciation has already begun. The great uncertainty in this simple prediction is that it does not consider the causes of the glacial/interglacial cycles. A more elaborate prognosis of future climate would need a climate model based on the variables which regulates the glacial/interglacial cycles.

The present paper aims to present a possible scenario of future climate and glaciations based on a number of published climate models; ACLIN (Astronomical climate index) [29], the Imbrie & Imbrie model [26] and the PCM model [4]. All these models consider the

Milankovitch orbital parameters as the main factor for large scale climate forcing. From these models changes in the global amount of ice and corresponding eustatic sea-level were estimated (Figures 2 and 3). The prediction of coming glaciations in Fennoscandia was based on parallels with the last glaciation. It is important to note that, as a scenario and not a prognosis, the climate scenario outlined here describes a plausible sequence of events, not accounting for alternative paths and parameters. In the following sections important factors contributing to the scenario are discussed, as well as those significant factors not taken into account.

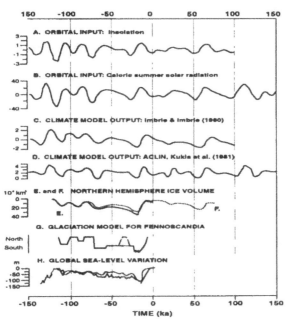

Figure 2. Summary of the past and predicted climate variables used in the development of the Central Scenario. A. Insolation (July 65°N) calculated from the Milankovitch parameters [4] [26]. B. Caloric summer northern hemisphere solar radiation (July 65°N), deviation from year of 1950. C. Climate model by Imbrie & Imbrie [26]. D. The ACLIN Climate model [29] E. Northern hemisphere ice volume model [4] for the past. F. Ice volume model of Berger [4] integrated by Gallée [23] for the next 80,000 years. G. Scandinavian glaciation curve [31]. H. Weichselian sea-level records, from [15], [8], and [38].

A number of variables and assumptions were made in order to construct the Central Climate Change Scenario:

- The scenario is based on climate models which use orbitally induced changes in insolation, Milankovitch forcing, as the primary input. In applying the results from these models it is important to note that they are not appropriate for frequencies more rapid than 20,000 years.
- The ice volume calculated in the Imbrie & Imbrie model and PCM model is used for estimating future fluctuations in global sea-level. For periods not included in these models, glacial volume and global sea-level is subjectively derived from the ACLIN model.

- For future glaciations an ice model with a single Fennoscandian dome and maximum ice thickness of 3000 m is used for a major glaciation.
- The local sea-level is derived directly from the predicted global sea-level by only adding the predicted glacio-isostasy.
- The isostasy during glacial episodes is estimated through comparisons with modelling results for the last Weichselian glaciation given by Fjeldskaar (unpublished data, and [22]). This model applies a low viscosity mantle (1.6×10^{22} poise), a 75-100 km thick asthenosphere (0.7×10^{20} poise) and a lithosphere close to 90 km thick.
- The distribution and extent of ice during forthcoming glaciations is derived by direct parallels with the Weichselian. The timing of future Scandinavian glaciations is assumed to be in phase with the global ice volumes of the models used.

Various uncertainties and alternatives concerning the global climate development are apparent:
- The future climate may not follow the Milankovitch forcing.
- To explain the dominance of the 100,000 year glacial cycles in the past climate record, some other factors such as varying response times, understanding the effects of natural CO_2 effects, glacio-isostatic downwarping, basal sliding of the ice sheets and the feedback between sea-level and ice-sheets, have to be included in the climate models. To some extent this is done in the models used as input to the scenario, but although work is currently being undertaken (*e.g.* [5]), more work is needed in these areas.
- It has recently been shown that during the last glacial, massive iceberg discharges (Heinrich Events) have acted as climate triggers of global extent [13]. Such features may be expected in the future.
- "Flip-flop" variation of climate. Data from central Greenland ice cores (*e.g.* [17]) show that the climate during the last 230,000 years has been unstable, characterised by abrupt climate changes related to changing ocean circulation. These findings further emphasize that the Milankovitch forcing only gives part of the background for climatic changes.
- Human alteration of climate development through effects by human activities (*e.g.* greenhouse warming) have not been considered. Due to the limited amount of fossil fuels, the global warming due to CO_2 could be expected to be short-lived relative to the timescale of the scenario described here. However, a longer-term effect could result if the heating during this short period could melt the Greenland ice or trigger some other major climate regulator of long periodicity. According to ongoing modelling (*e.g.* [5]) the anthropogenic rise in atmospheric CO_2 content may melt the Greenland ice sheet within the next 5,000 years. The resulting albedo effect will cause a delay in the onset of major glaciations and suggests that the Fennoscandian ice would not reappear until 50,000 AP, reaching only half the size expected with a Greenland ice sheet present.

Various uncertainties and alternatives concerning the local conditions at Äspö, primarily isostacy and sea-level variation, are also recognised:

- The global glacial volume as predicted in the global climate models, and thus global eustasy, need not be in the same phase relationship with the Fennoscandian ice volume as during the last glacial. If so, local sea-level changes might reach more extreme values and deviate strongly from this scenario.
- The varying estimates of thickness of the Weichselian ice-sheet (*e.g.* [2], [36], [30]) could give large discrepancies in modelling glacio-isostasy as well as in making parallels for future glaciations. The ice model applied in the scenario, a thick single Fennoscandian ice dome could have alternatives such as a much thinner ice or an ice sheet with separated domes. The last glaciation to affect Scandinavia may have involved the formation of several coalescing domes [3] and the formation of an ice stream following the line of the Baltic [25].
- The scenario applies a simplified picture of local sea-level and isostacy, thereby not accounting for the effects of forebulge and hydro-isostacy. Nor is the continuation of present day land-uplift included. A more thorough way to predict the response of the earth's crust during future glaciations would be to apply detailed geophysical modelling.

THE SITE-94 CENTRAL SCENARIO: THE CLIMATE CHANGE SCENARIO

The present scenario aims to provide an *illustration* of future climatic change in SE Sweden. The predictions of how the climate will change over the next 130,000 years is, however, limited by substantial uncertainties (compared to the actual possibly measurable future evolution of the climate). It is obviously impossible to give true quantitative values for any parameter during the seven different hypothetical evolution stages described below. The presented estimations should only be looked upon as rough measures and should serve as comparisons between the stages.

The climate models used for this study suggest glacial maxima at c. 5000, 20,000, 60,000 and 100,000 years from now. The scenario is based on the climate model output from Imbrie & Imbrie [26] and on the ACLIN model for the period beyond 100,000 AD [29]. The output of these climate models is then compared to the Quaternary geological records especially on glacier and sea-level variations and the scenario is drawn through parallels with the past. The SKB/TVO scenario [1] and the report by Björk & Svensson [6] follows a similar approach and has served as a background for the present scenario. Permafrost development has been modelled [28], elaborating on work by McEwen & de Marsily [32]. There has also been input from Svensson [43, 44]. The proposed evolution of the future climate evolution is illustrated in Figure 3.

0 - 10 000 years. The climate in Sweden will gradually change to cooler conditions with growth of an ice sheet in the Scandinavian Caledonides. Ice sheet thickness in the central mountains will be approximately 1000 m, with no ice in Stockholm or southwards. Crustal downwarping of c. 300 m in the central mountainous part. Sea level will gradually drop to 20 to 40 m below present day sea-level in the Stockholm region as well as in the Äspö region. During the colder parts of this period permafrost will occur in northern Sweden. The water in

Louisa Mking-Clayton

the Baltic will gradually become fresher as the ocean connection decreases.

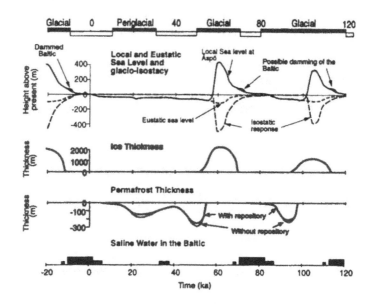

Figure 3. Predictions and observations of global and local (Äspö) eustatic sea-level; sea-level variations, ice cover, glacio-isostasy, permafrost thickness and Baltic salinity. Past eustatic record from Fairbanks [21] and past shore-displacement from Svensson [44]. Permafrost thicknesses calculated from surface temperature predictions [28].

10 000 - 30 000 years. After a minor, somewhat warmer, period the climate will get colder and fully glacial conditions will prevail around 20,000 years from now. The glacial peak will last perhaps 5,000 years. The ice sheet is estimated to reach the Stockholm area, but probably not the Äspö region. Ice thickness in the central part of the ice sheet will be c. 1500 m, while the ice sheet thickness in the Stockholm region will be c. 800 m. Crustal downwarping of about 500 m in the central part of the ice sheet and c. 60 m in the Stockholm latitude will occur. During deglaciation, when the ice front is located at the Stockholm region, the sea-level is estimated to be c. 25 m below present day sea-level. At Äspö sea-level will drop to c. 50 m below present day sea-level. The total effect of the glacial loading at Äspö is difficult to evaluate, but will probably not be large. Permafrost conditions will exist in southern Sweden, including Äspö. The water in the Baltic will be fresh, but probably in level with the oceans, due to erosion at its outlets.

30 000 - 50 000 years. Interstadial with a dry and cold climate (similar to the present climate on Greenland). Glaciers will be present in the Swedish mountains and permafrost will be present in northern Sweden. In the Stockholm region the eustatic rise of sea-level and decreasing isostatic uplift will result in a sea-level c. 30-40 m below present day sea-level. At Äspö there will not be much isostatic uplift, but a eustatic sea-level rise of maybe 10-20 m may occur. Thus the sea-level will be c. 30-40 m below present day level. The Baltic will mainly be fresh but some saline water may enter it.

50 000 - 70 000 years. Full glacial conditions will prevail. Due to the previous cold conditions the ice sheet will respond more rapidly. The glacial culmination will be around 60,000 years. The ice sheet will cover the whole of Sweden down to north Germany, comparable to the maximum of the Weichselian glaciation. Ice sheet thickness in the central part will be c.3000 m. In the Stockholm region the ice thickness will be c.2500 m. The Stockholm region will probably be covered by the ice sheet for at least 10 000 years, and possibly longer. Downwarping of c.700 m will occur in the central part of the ice sheet, c.600 m in the Stockholm area and c.500 m at Äspö. During deglaciation, when the ice front is located at the Stockholm region, the sea-level is estimated to be c.150 m at Stockholm and c.80 m above present day coastline at Äspö. There could also be a damming of the Baltic to 10-30 m above ocean level. The Baltic will be mainly fresh, but during deglaciation saline water may enter the Baltic through isostatically depressed areas. Permafrost will be present in large areas of Europe.

70 000 - 80 000 years. A rapid deglaciation will lead to and culminate in interglacial conditions at 75,000 years. Total crustal uplift is estimated to be c.700 m in the central parts of the previous ice sheet, c.600 m at Stockholm and c. 500 m at Äspö. This will be a relatively "warm" period with a climate in the Stockholm region similar to the present climate in northern Sweden. Small mountain glaciers and permafrost will occur in the very north. Parts of southern Sweden will be resettled and farming might be possible. Sea-level and salinity will be similar to the present day all over Sweden and the isostatic rebound will restore the land-surface to approximately its present state. Permafrost will only be present in the very north of Scandinavia.

80 000 - 120 000 years. The climate will gradually become colder with maximum glacial conditions at 100,000 years. The ice sheet will be extensive, covering large parts of Fennoscandia. Permafrost will occur in large areas outside the ice-margin. In the Stockholm region maximum ice thickness is estimated to be c.1500 m, and at Äspö c.1000 m. Downwarping of c.500 m in the Stockholm area is predicted during the main phase of the glaciation. At Äspö the maximum downwarping will be a little less than at Stockholm, perhaps 400 m. The relative sea-level at Stockholm at deglaciation will be c.100 m above present day coastline. The sea-level at Äspö at deglaciation may be c.80 m above present day sea-level, but damming of the Baltic to 10-30 m above ocean-level is possible. The Baltic will be mainly fresh but during deglaciation saline water may enter the Baltic through isostatically depressed areas.

120 000 - 130 000 years. Interglacial. The next warm period with a climate similar to the present in the whole of Scandinavia. Sea-level and salinity in the Baltic will be similar to the present.

GLACIAL CYCLE SCENARIO CONCEPTUALISATION

The next step in the scenario definition process was to create a sequence of pictorial scenarios describing the potential evolution of the hypothetical repository site over a single glacial cycle (advance and retreat of an ice sheet) (Scenarios 1-7; Figure 4). This scenario sequence is based on the glaciation which is predicted to climax at c. 60,000 years from now. Extrapolation of this sequence could then be made to the cover the entire period of the Central Scenario.

The intention is that the scenarios represent the most significant features, events and processes for analysis of the future safety of a repository for spent fuel. Thus, the objective of the scenario development was to provide a first indicator of the physical, hydrogeological and hydrogeochemical conditions at the front of and beneath the advancing and retreating ice sheets. This has involved features such as the hydraulic connection between free water in, on and under the ice with groundwater in the underlying rock, the rate of ice build-up and decay at a site-scale, and evidence for the geometry of the ice front during both advance and retreat. These features, in addition to other related features, such as the development of permafrost and sea level changes, are considered in terms of their impact on hydrogeology, groundwater chemistry, rock stress and surface environments for input into performance assessment modelling.

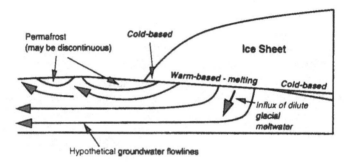

Figure 4. Schematic conceptual model of the thermal and hydraulic conditions which may exist beneath a mid-latitude ice sheet. Based on various sources [e.g. 41, 1, 12, 19, 24, 18].

Many simplifying assumptions have been made in the development of the scenario sequence, in addition to those outlined in earlier sections. The topography of the Äspö region and area is assumed to be unchanging throughout the glacial cycle. Only three fault zones (EW-1, NE-1 and the Ävrö Zone in Scenario 1 of Figure 5) are considered in detail, although the fracture network in the crystalline bedrock is known to be far more complicated. The flow paths shown within the bedrock are therefore only intended to display approximate flow directions. The tectonic setting of Äspö is considered to remain constant through the glacial cycle, such that glacial loading or glacio-isostatic effects are the only factors affecting bedrock stresses. The morphology of the ice sheet assumes that it develops over a predominantly non-deformable substrate (crystalline bedrock) during the early and later periods of the glacial cycle, over the majority of Norway and Sweden, but that it may advance and retreat over peripheral more deformable substrate, in inland Europe and the Gulf of Bothnia, during the glacial maxima and perhaps retreat (*e.g.* [10], [25]). Surface profiles have been determined by consideration of maximum ice thickness estimates and the horizontal extent of the ice sheet, modelling by Boulton & Payne [11] and by comparison with the Antarctic and Greenland ice sheets (*e.g.* [41], [19]). Repetition of phases, such as multiple advance and retreat due to ice surging or climatic changes within the glacial cycle, in addition to the formation of ice lobes and interlobate complexes, have not been considered. The temperature and basal melting regime of the ice sheet has been based on models of mid-latitude ice sheets (*e.g.* [41], [19]) in addition to reconstructions of the Weichselian Ice Sheet and predictions of the next ice sheet to affect Scandinavia [12].

The conceptual model used as a basis for the scenario development discussed in the following sections is shown in Figure 4.

Present day / Interglacial (Scenario Stage 1)

The base case scenario is that of an interglacial period with relatively warm conditions free from the influence of permafrost or ice, akin to those at the present day at Äspö. The section is depicted in Scenario 1 (Figure 5) and has been modified from Smellie & Laaksoharju [42]. The line of section is orientated approximately NW-SE (Figure 1). It is assumed that the hypothetical repository is situated at a depth of 500 m, in the SE of Äspö and that it has been closed (stopped operating) just prior to this scenario. At present the regional maximum principal stress is known to be oriented in a predominantly NW-SE direction. Glacio-isostatic stresses are considered to be zero for this base case although isostatic rebound may still be occurring over Scandinavia (e.g. [3]).

The nature of present day groundwater flow in the Äspö area and region has been simplified from Smellie & Laaksoharju [42] and SITE 94 work [40] in order to accommodate modelling. Regional groundwater flow, reflecting the regional topographic gradient, is taken to be approximately WNW to ESE although more localised deep groundwater flow occurs perpendicular to the line of section. The various fracture zones act as zones of recharge and discharge.

The base case groundwater chemistry assumes the presence of five groups of water classified according to ^{18}O, D and Cl contents (P. Glynn & C. Voss, pers. comm., 1994): recent Na-HCO_3-rich waters present to a few hundred metres depth; old, dilute glacial meltwater group present in isolated shallow sections in the northern part of Äspö; highly saline deep water group found at depths greater than about 500 m; intermediate 4000 to 6000 mg/l Cl group found at c. 200 to 500 m depths; group with a seawater or Baltic signature present only in very isolated areas below 300 m. It has been assumed that the combination of these chemistries resulted solely from the previous glacial cycle, although this is unlikely.

Periglacial: c.48,000 years (Scenario Stage 2)

Permafrost is considered to be present in the Äspö region to a maximum depth of about 250m (Figure 5) although it is punctuated by taliks. Fracture zone NE-1 has been assumed to form a talik (unfrozen opening in the permafrost), whereas both EW-1 and the Ävrö Zone are assumed to be sealed by the permafrost to a depth of 250m. There is obviously some uncertainty as to the continuity of permafrost and the spatial distribution of taliks. Isostatic uplift is considered to be negligible during this period, although development of a forebulge in front of the advancing ice sheet situated further north may affect Äspö. The forebulge may reach a maximum height of tens of metres at Äspö when it is a hundred kilometres or so in front of the ice sheet margin, although the magnitude of forebulge development is debatable (*e.g.* [45, 33]). Sea level is approximately 50 m below the present day level at Äspö for this scenario. Lakes may develop on the periglacial surface within topographic depressions, perhaps where the sea once was present in the base case. If a lake existed before development of the permafrost, an open talik could develop [32]. Vegetation will be scarce in these tundra conditions and it is likely that the ground surface will be characterised by bare rock with only local organic accumulations as tundra marshes in isolated poorly-drained topographic lows which may develop along outcropping fracture zones.

During this period there will be no direct glacial (ice) effects on the bedrock at Äspö. However, as the ice sheet advances towards Äspö, development of a flexural forebulge will affect stress patterns within the bedrock (*e.g.* [35, 39]). Tensional opening or reactivation of fractures may occur during forebulge uplift prior to subsequent compressive loading. However, there is apparently no evidence for reactivation of fractures during glacial advances and fracture reactivation is therefore not considered in this scenario. However, as permafrost develops, the formation of ice within any fractures in the bedrock, and the resulting expansion of the fracture water volume, will exert stresses that may propagate the fractures although they will remain sealed within the permafrost zone.

Infiltration and recharge will be severely restricted, particulary where the zones of base case flow are sealed by permafrost (fracture zone EW-1 and the Ävrö Zone). If taliks completely penetrate the permafrost, pathways will be available either for recharge or discharge of sub-permafrost water. The concentration of recharge in smaller areas and fewer points can lead to increased recharge and discharge rates at these points. Although permafrost is not considered to form at repository depths, it may well have an indirect effect on the repository by altering the volume of groundwater flow, as well as local flow directions, through the repository. Forebulge uplift is likely to cause a re-orientation of regional flow directions, assumed here to be predominantly a function of topographic gradient, until the forebulge has passed away from the Äspö region.

The formation of very saline waters may occur at the base of the permafrost zone due to the longer residence times of groundwaters trapped beneath the permafrost [32]. If radionuclides are present in the groundwaters, they may become concentrated amongst these saline waters beneath, or within, the permafrost zone. Alternatively, they may be concentrated at the lower levels of chemically stratified lakes acting as lacustrine taliks. Subsequently, the radionuclides may be released as a concentrated pulse when thawing takes place (see below).

Glacial advance: 50,000 - 60,000 years (Scenario Stages 3a-c)
During the first c. 5,000 years of this phase the sea level at Äspö is predicted to be below present day sea level, such that the advancing ice sheet will be completely landbased and will initially reach Äspö on dry land (3a). The margin of the ice sheet is expected to have an approximately WSW-ENE orientation as it passes across Äspö (although this does not account for the possibility of a more lobate development of the ice sheet, *e.g.* [3]). The ice sheet will advance across permafrosted bedrock, possibly with permafrost reaching depths up to 250 m. The ice sheet will therefore be cold-based until temperatures at the base of the ice sheet reach the pressure melting point and the permafrost melts [24, 18] (3c).

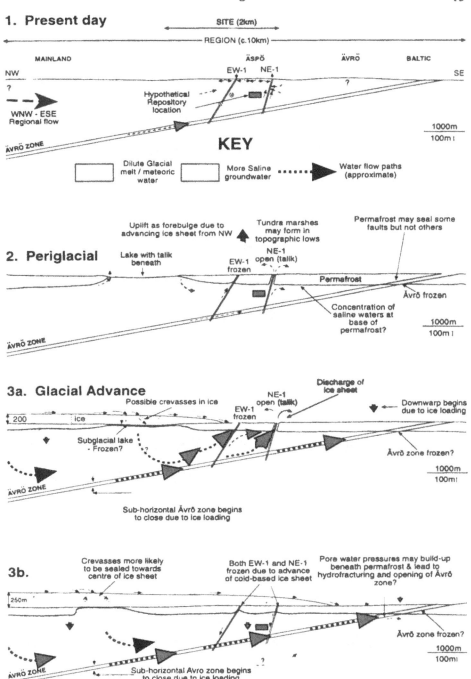

Figure 5. Sequence of scenarios for the Äspö area and region through the single glacial cycle predicted to occur from 50,000 to 70,000 years from now. This page: present day, periglacial and glacial advance stages.

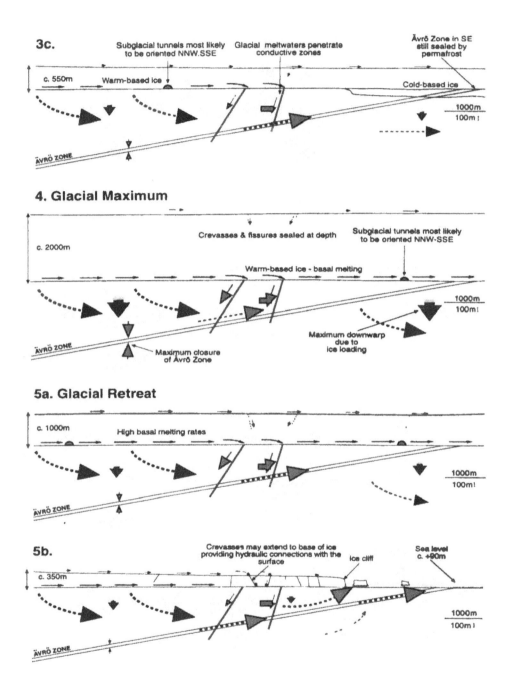

Figure 5. (Continued). This page: glacial advance, glacial maximum and glacial retreat stages.

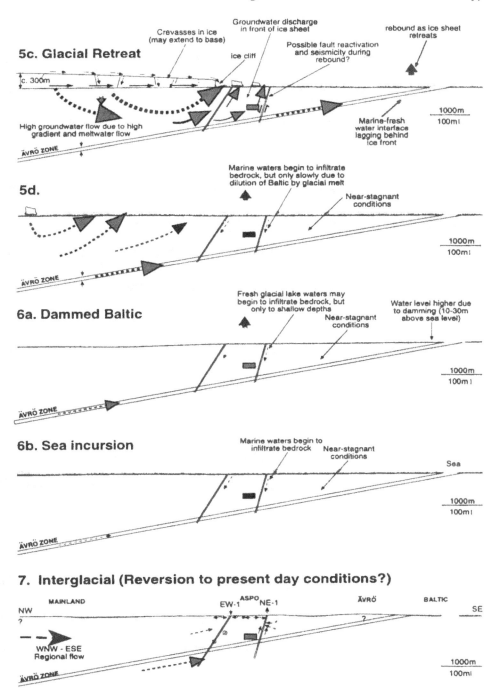

Figure 5. (Continued). This page: glacial retreat, dammed Baltic, sea incursion and interglacial stages.

The flexural forebulge will prograde further ahead of the ice front and away from the Äspö region. As the ice front comes within several tens of kilometres of Äspö the bedrock will begin to downwarp due to the loading effect of the ice sheet to the northwest (3a). Downwarping will continue as the ice sheet itself advances onto the Äspö bedrock (3b & c). As downwarping begins to take place, it is likely that subhorizontal fractures (such as the Ävrö Zone) will begin to close due to the increasing vertical stresses. Near-vertical fractures (such as NE-1, EW-1) may be expected to close due to increases in horizontal stresses, or may undergo shearing, which, in combination with increased pore water pressures, may act to enhance the conductivity of the fractures. In bulk, the bedrock will tend to experience overall compression during loading [35]. However, the response of fractures to glacially induced stresses is very difficult to predict and requires an understanding of the pre-glacial nature and history of the fracture system.

During initial advance into the Äspö region, the ice sheet will probably be cold-based and overlie permafrost, so that little or no subglacial meltwater will be generated [24, 18]. Groundwater recharge and discharge patterns may be similar to those under periglacial conditions (3a). However, any groundwater present within the bedrock beneath the permafrost zone underlying the ice sheet will be forced outwards and subsequently upwards in front of the ice margin due to the excess hydraulic gradients considered, as a conservative first approximation, to be controlled by the thickness profile of the ice sheet. In addition, groundwater generated from subglacial melting further towards the centre of the ice sheet is also expected to make its way to the ice sheet margins. Regional flow beneath the ice sheet will occur in a direction approximately perpendicular to the ice margin, almost parallel to the line of section in this case. Upward flow would therefore be possible along the fracture zones, although only NE-1 is considered to allow discharge at the surface in the scenario. However, the duration of such discharge is likely to be short, since the ice sheet is expected to pass across the site within a few hundred years at most, based on average rates of advance (*e.g.* 50m/a for the Weichselian Ice Sheet; [12]) although the possibility of a stillstand cannot be ruled out. Insulation by the permafrost is likely to induce the build up of high pore water pressures and hence may cause fracture reactivation and/or hydrofracturing in the outer subglacial or proglacial permafrost zones. Supraglacial meltwater will provide some surface run-off and potential recharge. As the ice sheet covers the Äspö area, recharge and discharge will be severely inhibited. Flow will be directed downwards, but may remain upwards locally along the Ävrö Zone, although closure of the zone may start to prevent flow directly along it. As the ice sheet advances and thickens, the base of the ice is expected to reach pressure melting point and hence melting will occur of both the permafrost and basal ice (*e.g.* [24, 18]) (3c). Subglacial water pressures will build up leading to subglacial water discharge, either into available transmissive zones in the bedrock, and/or along tunnels at the base of the ice where the underlying bedrock is of low transmissivity (*e.g.* [16]). The hydraulic pressure gradient imposed by the ice sheet will tend to force the flow downwards and towards the ice front. Subglacial tunnels may feed other transmissive fractures or will channel subglacial flow to the margin of the ice sheet to be discharged as glaciofluvial outwash. Groundwater flows may be many times that of the present day case (e.g. [11]).

For the early advance of the ice sheet (3a & b), groundwater chemistry is expected to be similar to that of the periglacial scenario (2) although glacial runoff may locally infiltrate the uppermost parts of the bedrock. As the ice advances and basal meltwater is produced (3c), any meltwater infiltrating the bedrock will tend to dilute the existing groundwaters, particularly any previous deep brines or saline waters concentrated beneath the recently melted permafrost. Dilute, oxidising glacial waters will penetrate to increasing depth as the ice sheet grows and hydraulic pressures increase. Any radionuclides stored beneath the melting permafrost will tend to be flushed out to either re-enter the groundwater system or be discharged as glacial outwash SE of Äspö.

Glacial Maximum: c.60,000 years (Scenario Stage 4)

At the glacial culmination the ice sheet is expected to reach a thickness of 2,200 m at Äspö. Downwarping due to ice loading may be c.500 m in the Äspö area at this time. Confinement by the large thickness of ice may prevent tectonic stress release during the entire period of glacial loading [27]. Loading stresses are expected to cause maximum closure of the sub-horizontal Ävrö Zone. The effect of stresses on the near-vertical fracture zones is unclear due to the play-off between horizontal stresses acing to close, and high pore water pressures acting to open, the fractures.

It is considered that Äspö will still lie beneath the melting zone of the ice sheet at this time and hence the bedrock will continue to receive subglacial discharge [11]. High infiltration rates are therefore expected to continue into localised transmissive zones. Estimates of mean erosion of only a few tens of metres for past glaciations in Sweden and Finland (*e.g.* [37]) suggest that erosion is unlikely to affect significantly the subglacial topography and hence the hydrogeology of the Äspö area, although localised erosive channels may form. Hydraulic connections between the surface of the ice sheet and its base are considered unlikely due to the sealing of crevasses and fissures by ice flow, such that the entire meltwater supply will arise from basal melting of the ice sheet. The transmissivity of the Ävrö Zone is expected to be reduced due to the high vertical stress. Therefore, groundwater flow may be localised above and not within this zone. NE-1 and EW-1 are still considered to be transmissive due to the play-off between increased pore water pressures and moderately high horizontal stresses.

The chemistry of the groundwater will reflect the continued high input of dilute, oxygenating glacial meltwater into the fracture systems and perhaps penetration of such waters to repository depths, which scoping calculations suggest is a possibility [40].

Glacial retreat: 60,000 - 70,000 years (Scenario Stages 5a-d)

The ice sheet is considered to retreat in seawater across Äspö, and hence the ice margin is likely to be in the form of an ice cliff of the order of a hundred metres high (*e.g.* [41]). Permafrost will not become re-established since the area will be beneath the sea. When the ice front has retreated to the Stockholm area, the sea level is predicted to be 80-100 m above the present day coastline at Äspö. Rapid retreat may be facilitated by the process of frontal calving at the seaward margin. Glacial deposition is considered to be more significant than at other times in the glacial cycle, although it is not considered to modify topography significantly.

The bedrock will begin to uplift and undergo extension, releasing the stresses which had built up during loading by the ice sheet (*e.g.* [35]). Certain major fracture zones may be reactivated, although it is not possible to predict which ones are likely to be affected. For the purposes of the scenario exercise it is assumed that NE-1 undergoes reactivation with a normal sense of displacement (5c). This displacement would be consistent with greater uplift to the SE (due to rapid retreat of the ice sheet across the Baltic area) and overall extension which is likely to accompany rebound. Normal fault reactivation and seismic activity is likely to involve coseismic compression of the local bedrock [35]. The Ävrö Zone is expected to become more transmissive as the loading stresses decrease.

High rates of basal melting will continue as the warm-based ice sheet retreats across Äspö. Groundwater flow will continue to be directed downwards and outwards (5a,b). The hydraulic gradient across the ice cliff margin will be very high, causing strong upward flow and discharge of groundwater in front of the margin (5c) as the ice cliff passes over Äspö. The duration of such discharge is likely to be short, within decades to a few hundred years at most, since the ice sheet is expected to retreat across the site at even greater rates than its initial advance (*e.g.* 150-200 m/a, even up to 1000 m/a; [11, 25]). However, this does not account for the possibility of stillstands in retreat or ice surging (such as the Younger Dryas event; [25, 10]). Possible hydraulic connections between the surface of the ice sheet in relation to the unstable calving of the ice sheet into water or at the margins of localised ice streams may increase the potential subglacial discharge of meltwater by up to two orders of magnitude [11] by introducing surface meltwater to the base of the ice sheet. As the ice sheet retreats further, the sea will cover the Äspö area. The subglacial water input into the groundwater system will be dramatically reduced as the ice front retreats. Reactivation of NE-1 during deglaciation and the associated crustal rebound, may increase its transmissivity. The overall state of extension of the bedrock as it rebounds may tend to increase its porosity and hence induce recharge (5d). However, in the vicinity of NE-1, coseismic compression could induce short-term (days to months) discharge of groundwater [34].

As the ice sheet retreats the marine-fresh water interface may be considered to occur in a landward thinning wedge modified by the transmissive fracture system, which lags behind the ice margin (5c,d). Once the ice sheet has moved well away from Äspö, sea water may begin to slowly infiltrate the Äspö bedrock. Glacial melt water will contribute less to the groundwater system as the ice sheet retreats away.

Damming of the Baltic: c.70,000 - 71,000 years (Scenarios 6a-b)

There is a possibility that the Baltic will become dammed during the later stages of deglaciation to 10-30 m above sea level (*e.g.* [20]) (6a). This will cause the waters of the Baltic to become increasingly dilute with time as the meltwater input increases. Incursion of the sea is then predicted (6b), possibly by a dramatic jökulhlaup event [7], or several events, where water escapes rapidly from the dammed lake after breakdown of the ice barrier or depositional topography causing damming.

Major fracture zones may continue to be reactivated, although no further fault movements are considered in these scenarios. The Ävrö Zone is expected to continue to open-up as rebound continues. A jökulhlaup event may induce seismic activity and reactivation of fracture zones, although such activity is not included in this scenario step. The dammed lake level will exert a constant hydraulic head over the region and hence groundwater flow will be low. By comparison with the development of the Baltic Ice Lake during the last glaciation [20] the path of water escape may be to the north of Äspö such that the Äspö region is not directly affected. However, there is no reason to believe that breaching of the ice dam could not occur in the Äspö vicinity. Damming of the Baltic will cause the Baltic's waters to become increasingly dilute with time as the meltwater input increases. Hence, the low density lake water will not infiltrate to great depths. Re-establishment of the sea across the area would lead to more saline waters gradually infiltrating into the bedrock, encouraged by continued expansion of the bedrock during glacial rebound.

Interglacial; 71,000 -75,000 years (Scenario 1 or 7)

By about 75,000 years from now, permafrost may only exist in northern Sweden and the climate at Äspö may be similar to that present in northern Scandinavia. This scenario assumes that hydrogeological, hydrochemical and stress conditions fully revert to those of the present-day base case before the following glacial period starts. However, at this time it is still expected that sea level will be higher than at the present day by 10-20 m, such that conditions at Äspö will not revert fully to those of the present day. An alternative scenario may be needed to describe this period. The predicted high sea level during this interglacial implies that flushing of groundwater by meteoric waters and deep saline waters during post-glacial uplift and subaerial exposure will not occur to the extent experienced by the Äspö area prior to the present day. Marine and glacial waters could therefore provide a more major component of the groundwater beneath Äspö at the start of the next glacial cycle in comparison to the present situation.

IMPACTS ON THE DISPOSAL SYSTEM

The scenario sequence outlined above for the single glacial cycle has been extended and modified in order to represent the glacial evolution of the Äspö area over the next 120,000 years. Estimates of key parameter values and potential relative changes in magnitude at different times in the future, have been compiled for preliminary input into performance assessment modelling (Figure 6). Values are approximate and are intended only to give an insight into potential relative temporal changes.

SUMMARY AND CONCLUSIONS

It has been the intention that the Central Scenario illustrates potentially significant aspects of climate change for analysis of the future safety of a hypothetical repository for spent fuel. A number of glacial cycles are considered to affect the Äspö region during the next 120,000 years, during which the mechanical and hydrogeological stability of a potential repository would be affected by such factors as ice loading, permafrost development, temperature changes, biosphere changes and sea level changes. The main periods of groundwater discharge and enhanced groundwater flow, which are of considerable significance to the performance

assessment of a potential repository, are considered to occur during the early advance and late
retreat phases of a glaciation. The latter discharge phase is considered to be more significant
than the former phase which is inhibited by the development of permafrost in the proglacial
zone. In addition, hydrochemical changes are of great significance to the performance of a
repository. During ice sheet development dilute, oxygenated waters are forced into the bedrock,
particularly via more transmissive fracture zones, by a combination of head increases and
permafrost sealing. Flushing by dilute meltwaters continues until the ice sheet retreats from
the Äspö region. Rock mechanical effects are also likely to be significant. During ice
loading the bedrock undergoes bulk compression whereas subsequent glacial unloading causes
expansion of the rock volume, which may affect the hydraulic properties of the bedrock. The
possibility of seismic reactivation of fracture zones during isostatic rebound is also of
importance. The various estimates and assumptions made during the development of the Äspö
scenarios have been outlined.

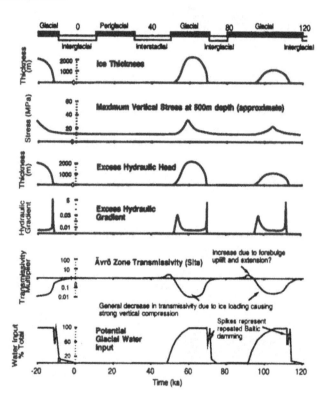

Figure 6. Graphical representation of potential changes in some climatic, stress, hydraulic and hydrochemical
parameters over the next 120,000 years at the Äspö site. Vertical axes are only approximate. Stress data have been
taken from modelling undertaken for the SITE 94 project [39].

The scenario development has been reliant on future predictions of climate change which are
known to involve many uncertainties. In addition, there appears to be a general lack of
observational data on the direct effects of an ice sheet on the hydrogeology, stress state and
hydrochemistry of the underlying bedrock. It is suggested that future research should focus on

the collection and analysis of palaeohydrogeological data at previously glaciated sites and bedrock data beneath present day ice sheets. However, despite all the uncertainties involved, we believe that this study has made a step towards an understanding of the way in which a hypothetical deep geological repository and its environment could be affected by future changes in climate.

Acknowledgements

SKI are thanked for their financial support. Ghislain de Marsily and Edward Ledoux are thanked for their input to the permafrost calculations which could unfortunately not be discussed in detail in this paper. Ove Stephansson, Matti Eronen, Jan Lundqvist and Svante Björk are thanked for their helpful reviews of the work in general.

REFERENCES

K. Ahlbom, T. Äikäs, & L.O. Ericsson, SKB/TVO Ice Age Scenario. *SKB Technical Report*, 91-32. Swedish Nulear Fuel and Waste Management Company, Stockholm (1991).

B.G. Andersen. In. Denton G.H. and Hughes T.J. *The last great ice sheets*. John Wiley & Sons. (1981).

K. Anundsen. Crustal Movements and the Late Weichselian ice sheet in Norway. *SKI Technical Report* KAN 3 (93) 12, Swedish Nuclear Power Inspectorate, Stockholm. (1993).

A. Berger, H. Gallée, T. Fichefet, I. Marsiat and C. Tricot,. Testing the astronomical theory with a coupled climate-ice sheet model. In: Geochemical Variability in the Oceans, Ice and Sediments. Paaeogr., Palaeoclimatol., Palaeoecol., 89(1/2), Ed.s Labeyrie, L. & Jeandel, C, *Global and Planetary Change Sect.*, 3(1/2), 125-141. (1990).

A. Berger, M.F. Loutre and H. Gallee. Sensitivity of the LLN 2-D climate model to the astronomical and CO_2 forcings (from 200 kyr BP to 130 ky AP). Universite Catholique de Louvain, Faculté des Sciences, Department de Physique, *Unité ASTR, Scientific Report*, 1996/1. (1996)

S. Björck & N-O. Svensson. Climatic change and uplift patterns - past, present and future. *SKB Technical Report*, 92-38. Swedish Nulear Fuel and Waste Management Company, Stockholm. (1992).

H. Björnsson. Explanation of jökulhlaups from Grimsvotn, Vatnajökull, Iceland. *Jökull*, 24, 1-26. (1974)

A.L. Bloom & N. Yonekura. Coastal terraces generated by sea level change and tectonic uplift. In: *Models in geomorphology*. Binghampton symposia in geomorphology; international series, 14. (1985).

M. Böse. Reconstruction of ice flow directions south of the Baltic Sea during the Weichselian glaciations. *Boreas*, 19, 217-226. (1990)

G.S. Boulton,& A.S. Jones. Stability of temperate ice sheets resting on beds of deformable sediment. *Journal of Glaciology*, 24, 29-43. (1979).

G.S. Boulton,& A. Payne. Simulation of the European ice sheet through the last glacial cycle and prediction of future glaciation. *SKB Technical Report*, 93-14. Swedish Nulear Fuel and Waste Magaement Company, Stockholm. (1993).

G.S. Boulton, G.D. Smith, A.S. Jones & J. Newsome. Glacial geology and glaciology of the last mid-latitude ice sheets. *Journal of the Geological Society of London*, 142, 447-474. (1985).

W.S. Broecker. Massive iceberg discharges as triggers for global climate change. *Nature*, 372, 421-424. (1994).

N.A. Chapman, J. Andersson, P. Robinson, K. Skagius, C-O. Wene, M. Wiborgh & S. Wingefors. Systems Analysis, scenario construction and consequence analysis definition for SITE-94, *SKI Report* 95:26, Swedish Nuclear Power Inspectorate, Stockholm. (1995).

J. Chappel, and N.J. Shackleton. Oxygen isotopes and sea level. *Nature*, 324, 137-140. (1986)

P.U. Clark & J.S. Walder. Subglacial drainage, eskers, and deforming beds beneath the Laurentide and Eurasian ice sheets. *Geological Society of America Bulletin*, 106, 304-314. (1994).

W. Dansgaard, S.J. Johnsen, H.B. Clausen, D. Dahl-Jensen, N.S. Gundestrup, C.U. Hammer, C.S. Hvidberg, J.P. Steffensen, ZA.E.Scveinbjörnsdottir, J. Jouzel & G. Bond. Evidence for generally instability of past climate froim a 250-kyr ice-core record. *Nature*, 364, 218-220. (1993).

D. Drewry. *Glacial Geologic Processes*. Edward Arnold. 276 pp. (1986).

C. Embleton & C.A.M. King. *Glacial Geomorphology*, Arnold Publishers Ltd. (1975).

20. M. Eronen. A scrutiny of the late Quaternary history of the Baltic Sea. *Geological Survey of Finland, Spe* *Paper*, **6**, 11-18. (1988).

21. R.G. Fairbanks. A 17,000-year glacio-eustatic sea level record: influence of glacial melting rates on the Youn Dryas event and deep-ocean circulation. *Nature*, **342**, 637-642. (1989).

22. W. Fjeldskaar & L. Cathles,. Rheology of mantle and lithosphere inferred from postglacial uplift in Fennoscan In *Glacial isostasy, sea-level and mantle rheology*. Sabadini, R., Lambeck, K. and Boschi, E. (Eds).Nato ASI Se C., **334**, 1-19. (1991).

23. H. Gallée. Conséquences pour la prochaine glaciation de la disparition eventuelle de la calotte glaciaire recouv la Groenland. *Scientific Report*, **1989/7**, Institut d'Astronomie et de Géophysique G. Lemaitre, Université Catholi de Louvain-la-Neuve, Belgium. (1989).

24. R.C.A. Hindmarsh, G.S. Boulton & K. Hutter. Modes of operation of thermomechanically coupled ice she *Annals of Glaciology*, **12**, 57-69. (1989).

25. P. Holmlund & J. Fastook,. Numerical modelling provides evidence of a Baltic Ice Stream during the Your Dryas. *Boreas*, **22**, 77-86. (1993).

26. J. Imbrie & J.Z. Imbrie. Modelling the climatic response to orbital variations. *Science*, **207**, 943-953. (1980).

27. A. Johnston. Supression of earthquakes by large continental ice sheets. *Nature*, **330**, 467-69. (1987).

28. L.M. King-Clayton, N.A. Chapman, F. Kautsky, N-O. Svensson, G. de Marsily, & E. Ledoux, (in press). ' Central Scenario for SITE 94. *SKI Report*, **94:16**, Swedish Nuclear Power Inspectorate, Stockholm,. (1996).

29. G. Kukla, A. Berger, R. Lotti, and J. Brown,. Orbital signature of interglacials. *Nature*, **290**, 295-300. (1981).

30. E. Lagerlund. An alternative Weichselian glaciation model with special reference to tE. he glacial history Skåne, South Sweden. *Boreas*, **16**, 433-459. (1987).

31. Mangerud & Anderson, 1990. *** TO BE COMPLETED***

32. T.J. McEwen & G. de Marsily,. The Potential Significance of Permafrost to the Behaviour of a Deep Radioact Waste Repository. *SKI Technical Report*, **91:8**. Swedish Nuclear Power Inspectorate, Stockholm. (1991).

33. N.A. Mörner,. The Fennoscandian uplift and Late Cenozoic geodynamics: geological evidence. *Geojournal*, 287-318. (1989).

34. R. Muir Wood. Earthquakes, water and underground waste disposal. In: Waste Disposal and Geology; Scient Perspectives. *Proceedings of Workshop WC-1 of the 29th International Geological Congress*, Tokyo, 169-1 (1992).

35. R. Muir Wood. A review of the seismotectonics of Sweden. *SKB Technical Report*, **93-13**. Swedish Nulear Fuel Waste Management Company, Stockholm. (1993).

36. A. Nesje & S.O. Dahl, S.O. Autochtonous block fields in southern Norway: implications for the geome thickness, and isostatic loading of the Late Weichselian Scandinavian ice sheet. *Journal of Quaternary Science* 225-234. (1990).

37. V. Okko. Maaperä. In: Rankama, K (ed.). *Suomen geologia*. Kirjayhtymä, Helsinki. (1964).

38. N.J. Shackleton,. Oxygen isotopes, ice volume and sea level. *Quaternary Science Review*, **6**, 183-190. (1987).

39. B. Shen, & O. Stephansson. Near-Field Rock Mechanics Modelling for Nuclear Waste Disposal (SITE-94). *Technical Report*, **95:17**, Swedish Nuclear Power Inspectorate. (1995).

40. SKI SITE 94. Deep repository performance assessment project, *SKI Report*, **96:36**, Swedish Nuclear Pov Inspectorate, Stockholm, Sweden. (1996).

41. B.J. Skinner & S.C. Porter. *Physical geology*. 750 pp. John Wiley & Sons. (1987).

42. J. Smellie & M. Laaksoharju,.The Äspö Hard Rock Laboratory: Final evaluation of the hydrogeochemical p investigations in relation to existing geologic and hydraulic conditions. *SKB Technical Report*, **92-31**. Swec Nulear Fuel and Waste Management Company, Stockholm. (1992).

43. N-O. Svensson. Lat e Weichselian and early Holocene shore displacement in the central Baltic, based dtratigraphical and morphological records from eastern Småland and Gotland, Sweden. *Lundqua Thesis* **25**. (198

44. N-O. Svensson. Late Weichselian and early Holocene shore displacement in the central Baltic sea. *Quatern International*, **9**, 7-26. (1991).

45. R.I. Walcott,. Isostatic response to loading of the crust in Canada. *Canadian Journal of Earth Sciences*, **7**, 716-7 (1970).

Proc.30ᵗʰ Int'l.Geol.Congr.,vol.24,pp.25-30
Yuan Daoxian (Ed)
© VSP 1997

Study of the Crustal Stability of the Beishan Area, Gansu Province, China--the Preselected Area for China's High Level Radioactive Waste Repository

GUOQING XU, JU WANG, YUANXIN JIN AND WEIMING CHEN
Beijing Research Institute of Geology, China National Nuclear Corporation, Beijing 100029, P. R. China

Abstract

The Beishan area, Gansu province, China, is the preselected area for China's high level radioactive waste repository, and is located in the Erdaojin-Hongqishan compound anticline of the Tianshan-Beishan folded belt. The candidate host rock for the repository is granite. The regional brittle faults are nearly EW- striking, shallow and non-active faults. The crust in the area is of block structure, with the crust thickness of 47 through 50 km. The depth contour of the crust is nearly EW striking, with very little variation. The gravity gradient is less than 0.6 mGal/km. The seismic intensity of the area is less than 6, and no earthquakes with Ms -4¾ have been happened. Since Tertiary, it is a slowly uplifting area, and the velocity of uplifting is 0.6--0.8mm per year. The direction of the modern principal compressive stress is 40°, and the superimposed fault angle ranges between 55° and 80°, which is in the range of stable superimposed fault angles. The geological characteristics of the Beishan area shows that the crust in the area is stable, and it has a great potential for the construction of the high level radioactive waste repository.

Keywords: Crustal Stability, Beishan Area, High Level Radioactive Waste, Repositiry, Site Selection

INTRODUCTION

A high level radioactive waste (HLW) repository is a kind of permanent large scale facility used to dispose of high level radioactive waste. The facility of this kind should have a safety period of 100,000 years, during which the disposed radioactive waste should not be leached out and should not harm the human beings. This shows that the siting for a HLW repository is very strict.

During siting process, two factors should be considered: social economic conditions and natural conditions. Among the natural conditions the crustal stability of an area is the first decisive factor to decide whether the area is suitable for the construction of a HLW repository or not.

The site selection for China's HLW repository began in 1985, and has undergone the stages of national screening and regional screening. At present, the Beishan area, Gansu province, northwest China, is preliminarily selected as a key potential area for the repository. The crustal stability of the Beishan area is the decisive factor for the potential of the area.

The scope of crustal stability studies includes the study of the evolution process and modern dynamic conditions of the crust, the analysis of the relationship between geological features and engineering buildings. This paper will discuss the regional geological features and the crustal stability of the Beishan area, and furthermore, discuss the possibility to select favourable districts for the construction of a HLW repository.

REGIONAL GEOLOGY

Regional Tectonics

Geographically, the Beishan area is located to the north of Yumen town, northwest of Gansu province, China. Tectonically, it is located in the Erdaojin-Hongqishan compound anticline of the Tianshan-Beishan folded belt (Fig. 1). The anticline is nearly EW striking with its core composed of schist, laminated migmatite of Laojunmiao Group and Dakouzhi Group of pre-Changchengian age, and its limbs composed of schist, marble, quartzite and migmatite of Yujishan Group, Huayaoshan magmatite (plagioclase granite), Baiyuantoushan Caledonian magmatite (quartz diorite) and Qianhongquan Hercynian magmatite (monzonitic granite and orthoclase granite).Those magmatite rocks are the main candidate host rocks for the repository.

The border between the Beishan area and the Hexi Corridor Transitional Zone is the blind Sulehe fault. Within the Beishan area there are three EW-striking large scale ductile shear zones (DSZ), namely, South Erduanjin DSZ, Zhongqiujin-Jinmiaogou DSZ, and Erdaojin-Hongqishan DSZ. The activity of the brittle faults developed within the shear zones is a key factor affecting the crustal stability of the area.

The Zhongqiujin-Jinmiaogou DSZ is EW-striking with a length of 90km and a width ranging between 50 and 2000m. It was developed in minonlitic gneiss, granite and sericite slate during Hercynian period. The largest depth of the shear zone is about 10km. Several EW-striking brittle faults are developed in the shear zone. However, those faults do not cut the Quaternary sediments, and no earthquakes with $Ms>4\frac{3}{4}$ have been happened along them, which show that they are not active faults.

The Erdaojin-Hongqishan DSZ is about 130km long and has a largest width of 7km. The southern and northern borders of the shear zone are two EW-striking brittle faults which are not active fault either.

Crustal Structure

According to the main tectonic units, deep geophysical features and the crustal thickness, the crustal structure of the western Gansu province can be divided into Beishan-Alashan area, North Qilian Mt.-Hexi Corridor area and Qilian area. The selected area for the HLW repository is located within the Beishan-Alashan area. The depth contour of the crust in the area is NWW-EW striking. The thickness of the crust is 47 ~ 50km, and it gradually increases from north to south, but with very little variation. The gravity anormaly is $150 \times 10^{-5} \sim 225 \times 10^{-5}$ m/s^2, the gravity gradient is less than 0.6 mGal/km. On the gravity anormaly map, the gravity anormaly contour

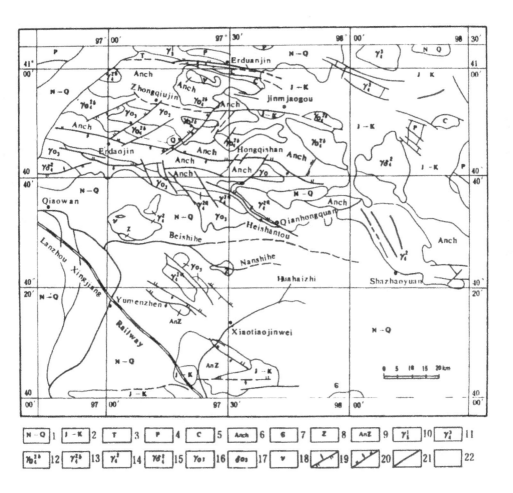

Figure 1. Geological sketch map of the Beishan area, Gansu province, NW China — the preselected area for China's high level radioactive waste repository

1-Tertiary and Quaternary sediment; 2-Cretaceous and Jurassic sandstone, shale and mudstone; 3-Triassic conglomerate and pebbly sandstone; 4-Permian System; 5-Carboniferous System; 6-Pre-Changchengian schist, gneiss, marble and migmatite; 7-Cambrian System; 8-Sinian System; 9-Pre-Sinian System; 10-Yanshanian granite; 11-Hercynian granite; 12-Hercynian plagioclase granite; 13-Hercynian orthoclase granite; 14-Hercynian plagioclase granite and two-mica granite; 15-Hercynian granite diorite; 16-Caledonian plagioclase granite; 17-Caledonian quartz diorite; 18-gabbro vein; 19-normal fault; 20-reverse fault; 21-fault is distributed very sparse without obvious step zones, indicating that there are no great faults extending to the depth of the crust.

Based on the above mentioned characteristics, the crust in the Beishan area possesses block structure with good integrity.

Earthquake Activities
Earthquakes are the demonstration of modern crustal movement, and have close relation with tectonic movement, especially with the intensive movement of great deep-rooted faults. The Beishan area is located north to the Hexi Corridor earthquake zone. In the area there are lack of active great deep-rooted faults and strong earthquakes. According to the data provided by the National Seismological Bureau of China, no earthquakes with $Ms>4 3/4$, have been happened in the Beishan area. On the "Seismic Intensity Regionalization Map of China (1:4,000,000)", the Beishan area is within a VI seismic intensity region.

With a very sharp comparison to the Beishan area, the Hexi Corridor earthquake zone has several active NWW deep-rooted faults. Along the North Hexi Corridor Great Fault several intensive earthquakes took place, e.g. the $M=7 1/4$ earthquake in Mingle county in 1790; the $M=7 1/2$ earthquake in Shandan county in 1954; the $M=6$ earthquake in Gaotai county in 756.

Neotectonism and Tectonic Stress Field
Neotectonism is referred to the crust tectonic movement since Tertiary, including the horizontal and vertical movement of the crust, volcanism, earthquake and slide etc. According to the neotectonic characteristics, the western Gansu province can be divided into 3 parts, namely, the Qilianshan blocking and intensive uplifting region, the Corridor depression region and the Beishan weakly uplifting region. The landform of the area is characterized by flatter gobi and small hills with elevations above sea level ranging between 1000m and 2000m. The height deviation is usually several ten meters. Since Tertiary it is a slowly uplifting area without obvious differential movement. The uplifting velocity of the crust in the area is about $0.6 \sim 0.8$ mm/a, much lower than that of the Qilian region ($1.5\sim1.8$ mm/a).

Comprehensive analysis of structural deformation of the Cenozic faults and folds indicate that the area is undergone horizontal compression at present, and the principal compression stress is between 30° and 60°. The data provided by the mechanism at the source of earthquakes show that the direction of the principal compress stress is about 40°.

The strike of the main faults in Beishan area is between 95° and 120°. The angle between the direction of the principal compress stress and the striking of the main faults, which is also called superimposed fault angle, ranges between 55° and 80°, which are within the ranges of stable superimposed fault angles, suggesting that the main faults are stable and will not have strike-slip displacement.

THE REGIONALIZATION AND EVALUATION OF THE CRUSTAL STABILITY

There are 9 indices to evaluate the crustal stability of an area, namely, (1)crustal structure and deep-rooted faults; (2) Cenozoic crust deformation; (3) Quaternary block and faults; (4) Quaternary and modern tectonic stress field; (5) Quaternary volcanos and geothermal fields; (6) ground deformation and displacement; (7) gravity field; (8) earthquake strain energy; and (9) earthquakes. According to the indices crustal stability can be divided into 4 classes: (1) stable region; (2) basically stable region; (3) sub-stable region; and (4) unstable region.

The western Gansu province can be divided into 3 regions: (1) Beishan stable region; (2) Yumenzhen-Huahai sub-stable region, and (3) Hexi Corridor unstable region. The characteristics of those regions are summarized in table 1. From the table we can see that the Beishan region has a block crustal structure with good integrity and without regional active faults, and it is a slowly uplifting region. The direction of the principal compression stress in the area is about 40°, and the superimposed fault angles are between 55° and 80°, which is in the range of stable superimposed fault angles, indicating that the main faults will not have strike-slip displacement. There is no gravity steps in the area, and no records of intensive earthquakes either. The earthquake intensity of the area is less than 6. These features show that the crust in Beishan area is more stable than its southern regions, and is a suitable candidate region for China's HLW repository, from which some districts can be further selected for detailed site characterization.

REFERENCES

1. Li, Xingtang et al., Research theory and method of regional crust stability, Geological Press, Beijing, (1987).
2. China National Seismological Bureau, The geophysical exploration achievement of the crust and upper mantle in China, Seismological Press, Beijing, (1986).
3. China National Seismological Bureau, Atlas of seismic activity in China, China Science and Technology Press, Beijing, (1991).
4. Geological Bureau of Gansu Province, The geological map and its explaination of Yumenzhen Quadrangle (1:200,000), (1972).
5. Geological Bureau of Gansu Province, The geological map and its explaination of Houhongquan Quadrangle (1:200,000), (1969).

Table 1. The Crust Characteristics of Western Gansu Province, China

Characteristics	Beishan Stable Region	Yumenzhen-Huahai Substable Region	Hexi Corridor Unstable Region
crust structure and deep fault	the crust is block structure with good integrity, NW-striking basement faults are distributed	the crust is mosaic structure with NE-striking basement faults	the crust is crushed structure controlled by the north Qilian deep fault and Aljin active fault
Active fault and quaternary crust movement	with few active faults, it's a slowly uplifting region, with precipitation velocity less than 0.1 mm/a	with Sanweishan and Aljin active faults, the precipitation velocity is between 0.1 ~ 3.5 mm/a	with Aljin active fault, the largest precipitation velocity is 4.7 mm/a
Superimposed fault angle	55° ~ 80°	45° ~ 70°	80°
Gravity field	the gravity anormaly is smooth, the gradient is less than 0.5 mGal/km	with regional positive and negative anormalies, the gradient is 1 ~ 2 mGal/km	located in gravity anormaly step zone, with negative gravity anormaly
largest earthquake(Ms)	3.0	5.0	73/4
Energy $\sqrt{E}$ ($10\sqrt{erg}$)	0.14	10.59	25.54
Frequency (Ms ·3.0)	1	4	2
possibility for the construction of HLW repository	possible	not possible	not possible

Proc.30th Int'l.Geol.Congr.,vol.24,pp.31-41
Yuan Daoxian (Ed)
© VSP 1997

Groundwater pressure changes associated with earthquakes at the Kamaishi Mine, Japan
- A study for stability of geological environment in Japan -

KOHSON ISHIMARU[1] AND ISAO SHIMIZU[2],*

1 Tono Geoscience Center, Power Reactor and Nuclear Fuel Development Corporation (PNC), Toki, Gifu, Japan
2 Kamaishi Site Office, Power Reactor and Nuclear Fuel Development Corporation (PNC), Kamaishi, Iwate, Japan.
*Present address: Electric Power Development Co. Ltd., Tokyo, Japan

Abstract

Seven seismographs have been installed at four different levels of the Kamaishi Mine, northeast of Japan, and groundwater pressure in three broholes have been also monitored. During the period from November 1991 to December 1995, twenty one cases of earthquake-related changes in groundwater pressure have been observed. The range of groundwater pressure changes are generally less than 0.1 kgf/cm^2 with the maximum of 0.35 kgf/cm^2. These changes tend to recover to the original state within several days. Since an annual groundwater pressure changes due to the rainfall was less than 1.0 kgf/cm^2, groundwater pressure changes induced by seismic impact were within the range of annual hydraulic variation.

In our observations, the groundwater pressure change related to earthquake possibly depend on principally static crust strain rather than maximum acceleration, because the siginificant correlation was observed between groundwater pressure changes before and after earthquakes and theoretical strain calculated from the magnitude and epicentral distance.

keywords: earthquake, groundwater pressure, Kamaishi Mine, geological environment

INTRODUCTION

The PNC is conducting a study for stability of geological environment as a basis for the performance assesment study of geological disposal of High-level radioactive waste(HLW) in Japan. The Japanese islands are located along the Circum-Pacific Mobile Belt, where natural processes and events such as seismicity, volcanism and other tectonic movements occur over a long period of time. Such natural processes and events may perturb the stability of geological environment in Japan, and are regarded as the key issues for HLW.

Among other geoscientific studies, seismological study at the Kamaishi Mine has been carried out by PNC since 1990, as a part of study for stability of geological environment in Japan. The main objective of this seismological study is to evaluate the influence of earthquake on groundwater. Some results of the seismological study at the Kamaishi Mine during the period from November 1991 to December 1994 were already reported by Shimizu et al.[1996].

In this paper, groundwater pressure changes associated with earthquakes, which is one of most concerned issue for HLW, will be discussed based on seismological observation at the Kamaishi Mine.

SEISMOLOGICAL STUDY AT THE KAMAISHI MINE

The Kamaishi Mine is located approximately 600 km north of Tokyo. The geology in this area is mainly of Paleozoic to Mesozoic sedimentary formations, and early Cretaceous Ganidake igneous complex and Kurihashi granodiorite (Fig.1). The Ganidake igneous complex is mainly composed of granodiorite with minor amount of diorites and monzonites. The Kurihashi granodiorite is mainly composed of granodiorite and is distributed near the Ganidake igneous complexes. The Fe-Cu ore bodies of the Kamaishi Mine are associated with the Ganidake granodiorite along its contact with the older sedimentary rocks.

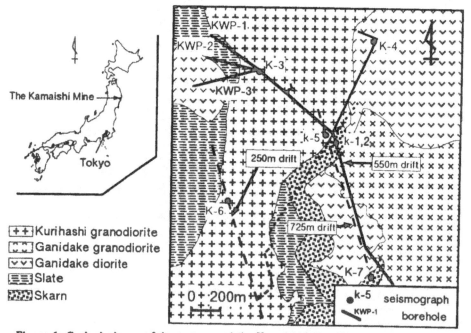

Figure 1. Geological map of the area around the Kamaishi Mine. The map shows locations of seismographs and boreholes for monitoring of water pressure.

Seismological observations have been conducted since February 1990 to measure the attenuation characteristics of the ground motion at the different levels in the Kamaishi Mine drifts (Fig.2). Seismographs K-1,K-2,K-3 and K-4 were installed on February 1990, K-5 and K-6 were installed on January 1991. K-1,K-5,K-2 and K-6 seismographs are set on a vertical line. K-6 seismograph is set at the deepest drift, 615m below the ground surface. Groundwater pressure is monitored in three boreholes (KWP-1,2,3) which are located in the elevation level (EL.) of 550m drift, 315m below the ground surface.

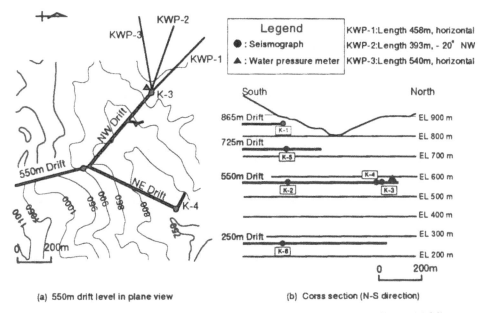

(a) 550m drift level in plane view   (b) Corss section (N-S direction)

Figure 2. Allocation of seismographs and water pressure meters in the Kamaishi Mine.

a) Seismological observations

Two hundred and forty-nine earthquakes with an acceleration of more than 0.5 gal were recorded from February 1990 to December 1995. The epicenters of these earthquakes, according to the Japan Meteorological Agency, were mostly plotted on the Pacific Ocean side of north-eastern part of Japanese island, and epicentral distance were mostly from 20 km to 250 km (for example, Figure 3 shows the distribution of epicenters for the 27 earthquakes which were observed during the one-year period from January to December 1995). The earthquakes were mostly of a magnitude smaller than 6.0. The maximum magnitude was 8.1 in October 1994, and it's epicentral distance was about 675 km. Maximum accelerations observed at the surface level of the Kamaishi Mine (K-1) were mostly below 4 gal. The maximum value of maximum acceleration obtained is 53 gal (Magnitude 5.9) in November 1993, and it's epicentral distance was about 85 km.

The ratios of maximum acceleration observed at seismographs K-1, K-5, K-2 and K-6 are shown in Figure 4. Figure 4 includes 41 ground motion data observed simultaneously by the four seismographs. The maximum accelerations at K-6 are about 1 to 1/2 times as large as than those at K-2 and K-5, while acceleration at K-6 are only 1/2 to 1/4 times as large as those at the K-1.

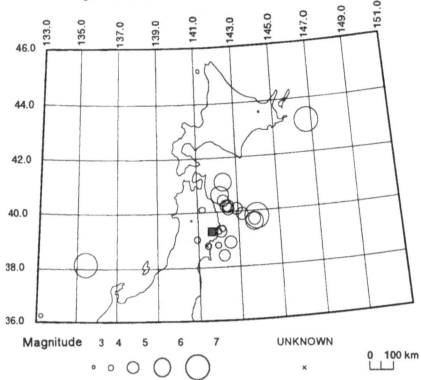

Magnitude 3 4 5 6 7   UNKNOWN

Figure 3. Epicentral distribution of 27 earthquakes observed from January 1995 to December 1995. The earthquakes larger than 0.5 gal on the K-2 seismograph are plotted.

b) Monitoring of groundwater pressure

Groundwater pressure is monitored in three boreholes (KWP-1,2,3), which were drilled for mineral exploration during 1971 to 1973 (Fig.2). These boreholes, located in the EL. 550m drift of 315m below the ground surface, encounter sedimentary rocks at about 120m from the drift wall. The monitoring was started in February 1990 by installing water pressure meters (resolution of approximating 0.001 kgf/cm^2) of a strain guage type with measurement range of 0 to 50 kgf/cm^2. Measurement interval is set at 10 seconds.

Figure 5 shows an example of the long-term variation of groundwater pressure for the KWP-1 borehole from November 1991 to December 1995. The water pressure for the KWP-1 borehole is consistently low in winter (November to February) while it tends to rise rapidly in spring (April to May). The monthly rainfall are also shown in Figure 5. It is cosidered that the variation of groundwater pressure is mainly controlled by the rainfall. The annual groundwater pressure changes is mostly less than 1.0 kgf/cm^2 in three boreholes.

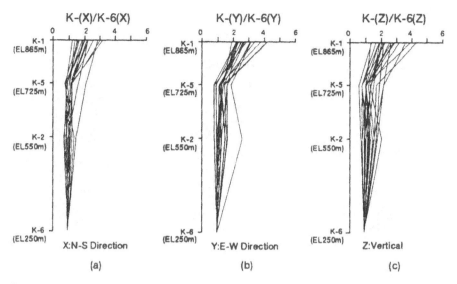

Figure 4. Ratio of acceleration of earthquakes between the K-6 seismograph and another seismographs. (a) N-S direction, (b) E-W direction, (c) vertical direction. This figure includes 41 earthquakes motions observed during the period from January 1991 to December 1994 [Shimizu et al.,1996].

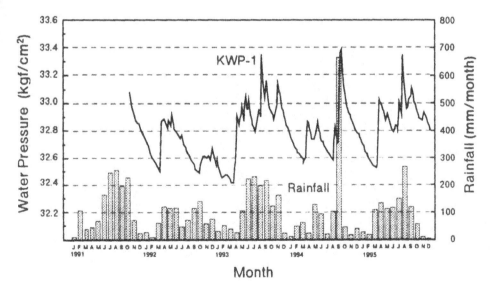

Figure 5. Long-term variations of groundwater pressure and rainfall during the period from November 1991 to December 1995. The groundwater pressure was monitored for the KWP-1 borehole and the rainfall at Aonoki station which is located approximately 2km north-east of the Kamaishi Mine.

c) Groundwater pressure changes associated with earthquakes

Figure 6 shows the short-term periodic fluctuation of groundwater pressure for the KWP-1 borehole which is considered to shows an influence of the earth tide. Figure 6 also shows the change in groundwater pressure related to an earthquake which occurred at 22:19, July 12, 1993. As coseismic variation, the water pressure increased step-wise by about $0.02 kgf/cm^2$ immediately after the earthquake, and tended to recover to the original trend within several days. Such a change in groundwater pressure was observed for 21 examples from November 1991 to December 1995. The epicentral distribution of 21 earthquakes which induced changes in groundwater pressure are shown in Figure 7. The maximum change in groundwater pressure is the drop of $0.35 kgf/cm^2$, while the others are mostly within the range of approximately $0.1 kgf/cm^2$. The periods of these changes were mostly less than a week.

No groundwater pressure change was observed for some earthquakes with accelerations as high as 18 gal, while the groundwater pressure changed in response to the earthquakes with accelerations less than 2 gal. These results imply that the maximum acceleration is not the critical factor in triggering a change in groundwater pressure.

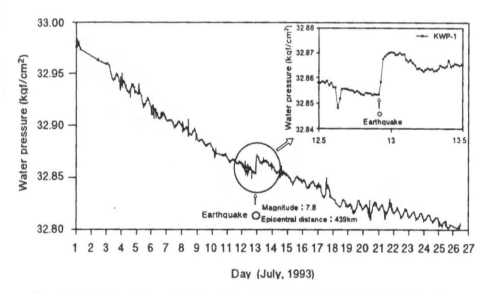

Figure 6. Example of groundwater pressure record for the KWP-1 borehole during the period from July 1 to July 26, 1993. Open circle indicates occurrence of earthquake. The groundwater pressure record shows periodic fluctuation and a step-change related to an earthquake.

d) Relation between earthquake-related changes in groundwater pressure and crustal strain level

According to Shimizu et al. [1996], the earthquakes, related to changes in groundwater pressure, are plotted on an area of larger than 10^{-8} crustal strains, as calculated by using the soft inclusion model of Dobrovolsky et al. [1979] (Fig. 8).

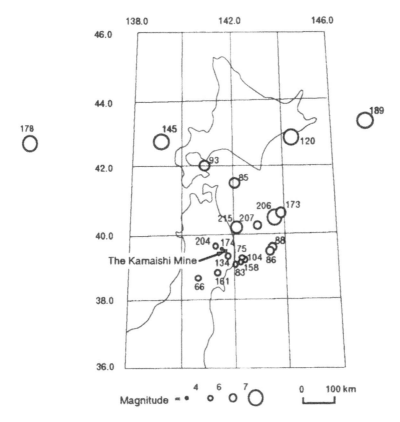

Figure 7. Epicentral distribution of 21 earthquakes which induced changes in groundwater pressure during the period from January 1991 to December 1995. The numbers in the figure indicate the earthquake number observed in the Kamaishi Mine.

The major earthquake-related changes in groundwater pressure observed for the KWP-2 borehole are shown in Table 1. The groundwater pressure for the KWP-2 borehole responded sensitively to these earthquakes among three boreholes. The groundwater pressure changes associated with earthquakes were estimated from coseismic change including precursory change before earthquake (Fig. 9). The relationship between the groundwater level(pressure) changes on Table 1 and the static crust strain calculated after Dobrovolsky et al.[1979] from the magnitude and epicentral distance is shown in Figure 10. The significant correlation was observed in Figure 10. This result indicates that the range of groundwater pressure changes depends on the static crust strain. As further investigation, monitoring of the in-situ change of crustal strain related to earthquake is necessary for verifying these relationship.

Kohson Ishimaru et al

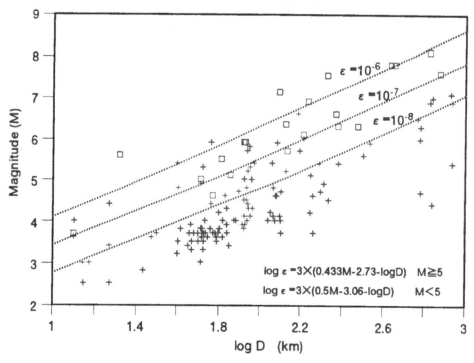

$$\log \varepsilon = 3 \times (0.433M - 2.73 - \log D) \quad M \gecolon 5$$
$$\log \varepsilon = 3 \times (0.5M - 3.06 - \log D) \quad M < 5$$

Figure 8. Relationship between magnitude and epicentral distance (January 1991 to December 1994)[Shimizu et al.,1996]. D, epicentral distance from the mine. ------ , theoretical strain by Dobrovolsky et al.[1979].   , changes in groundwater pressure observed. + , no changes in groundwater pressure observed. The formula by Dobrovolsky et al.[1979] is for the maximum deformation not the exact deformation in the mine.

Table 1. Major earthquakes and water level changes observed at the Kamaishi Mine.

No	Date	Latitude	Longitude	Magnitude	Depth (km)	Epicentral Distance (km)	Maximum Acceleration (gal)	Water level change in KWP-2 (cm)
215	7/ 1/95	40° 14'	142° 23'	7.0	50	120.7	34.68	216.7
207	29/12/94	40° 06'	143° 00'	6.3	—	145.0	13.41	97.2
206	28/12/94	40° 24'	143° 42'	7.5	—	212.6	31.25	344.4
204	21/12/94	39° 28'	141° 15'	5.0	89	40.4	13.10	69.4
189	4/10/94	43° 22'	147° 40'	8.1	30	675.1	36.94	208.3
145	12/07/93	42° 47'	139° 12'	7.8	34	438.9	1.37	255.6
120	15/01/93	42° 51'	144° 23'	7.8	107	455.7	25.37	22.2
88	18/07/92	39° 23'	143° 39'	6.9	30	174.0	4.98	120.0
86	16/07/92	39° 20'	143° 35'	6.1	—	165.4	1.91	19.4
83	16/05/92	38° 53'	142° 06'	4.6	51	59.6	12.34	22.2
75	13/04/92	39° 07'	142° 25'	5.1	58	72.6	21.33	30.6

Earthquake-related change in groundwater pressure larger than 0.01 kgf/cm^2 are shown. The change of groudwater pressure are converted into the change of water level.

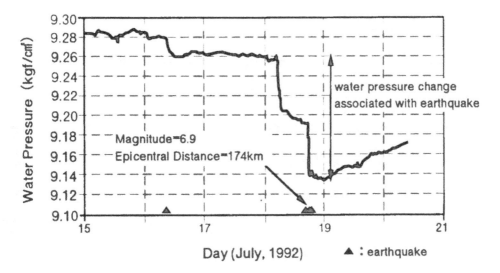

Figure 9. Groundwater pressure change associated with the earthquake far east off Sanriku. Groundwater pressure change before and after the earthquake is the drop of 0.12kgf/cm², while the coseismic change is the drop of 0.05kgf/cm² .

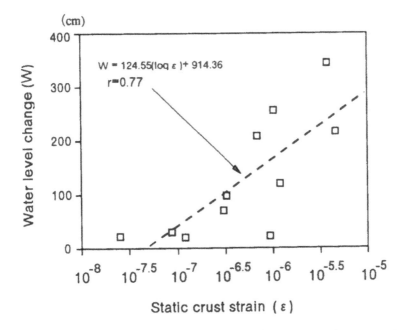

Figure 10. Relationship between the water level change and static crust strain at the Kamaishi Mine. The data of water level change are shown in Table 1. The static crust strain is calculated after Dobrovolsky et al.[1979] from the magnitude and epicentral distance.

DISCUSSION

Papers concerning coseismic changes of groundwater from the view of the prediction of earthquake are not few. Igarashi and Wakita[1991] reported the coseismic changes of water level in the observation well at Kamakura city, the southern Kanto district, Japan. They inferred that the range of the coseismic changes was smaller than 4.0 cm while the range of the changes caused by earth tide was around 10 cm.

Besides, many informations about coseismic behavior of groundwater by the 1995 Hyogoken-nanbu earthquake are available [e.g., Koizumi et al.,1996, Sato et al.,1996, Toda et al.,1996]. The following results were made through by literature survy. The range of water level changes was from a few centimeters to several meters, and continuous periods of these changes were mostly several weeks around the Kinki district, central Japan, more than several kilometer distance from the epicenter. It was also recognized that continuous reliable observation data about behavior of groundwater before and after the earthquake were limited.

At the Kamaishi Mine, twenty one cases of earthquake-related changes in groundwater pressure were observed during the period from November 1991 to December 1995. The range of groundwater pressure changes are generally smallerl than 0.1 kgf/cm^2 , with the maximum of 0.35 kgf/cm^2. These changes tend to recover to the original state within several days. Since an annual groundwater pressure change due to the rainfall has been observed as large as 1.0 kgf/cm^2, groundwater pressure changes induced by seismic impact were within the range of annual variation of groundwater pressure.

Thus, the results obtained during the seismological study at the Kamaishi Mine and the other informations indicate a possibility that groundwater pressure (or level) changes related to earthquakes may be impermanent and within the range of long-term hydraulic variation due to the rainfall. Consequently, it can be considered that the effect of earthquake on groundwater is insignificant from the viewpoint of stability of geological environment.

CONCLUSIONS

The results of the seismological study at the Kamaishi Mine are summarized as follows:

(1) Accelerations vary with depth, where accelerations at 615 m and 140 m below ground surface are 1/2 - 1/4 times and 1 - 1/2 times, respectively, the surface values.

(2) From the monitoring data of groundwater pressure at depths below 315 m, groundwater pressure changes about 1.0 kgf/cm^2 due to seasonal variation of rainfall.

(3) The changes in groundwater pressure associated with earthquakes follow a step-wise change and these changes recover to their original state within several days. These changes were mainly around 0.1 kgf/cm^2,with the maximum recording of 0.35 kgf/cm^2.

(4) The significant correlation was observed between groundwater pressure changes before and after earthquakes and theoretical crustal strain levels larger than 10^{-8}.

(5) Groundwater pressure changes induced by earthquakes are generally within the range of annual hydraulic variations due to rainfall. Seismic effects on stability of geological environment, especially groundwater, are possibly insignificant.

Acknowledgements

This study has been partially conducted under the guidance of an Advisory Committee of "The study on earthquakes in the in-situ experiments at the Kamaishi Mine". We thank members of the Committee. The seismic information was kindly provided by the Japan Meteorologlcal Agency.

REFERENCES

Dobrovolsky, I.P., Zubkov, S.I. and Miachin,V.I. Estimation of the size of earthquake preparation zones, *Pageoph*, **117**, 1025-1044 (1979).

Igarashi, G.and Wakita,H. Tidal responses and earthquake-related changes in the water level of deep wells, *Journal of Geophysical Research*, **96**, 4269-4278 (1991).

Koizumi, N., Kano, Y., Kitagawa, Y., Nishimura, S. and Nishida, R. Groundwater anomalies associated with the 1995 Hyogo-ken Nanbu earthquake, *Bulletin of Laboratory for Earthquake Chemistry, University of Tokyo*, **7**, 16-17 (1996).

Sato, T., Matsumoto, N., Takahashi, M and Tsukuda, E. Anomalous groundwater discharges and coseismic groundwater level changes in relation to the 1995 Kobe (Hyogo-ken-nanbu) earthquake, *Bulletin of Laboratory for Earthquake Chemistry, University of Tokyo*, **7**, 43-44 (1996).

Shimizu, I., Osawa, H., Seo, T., Yasuike, S. and Sasaki, S. Earthquake-related ground motion and groundwater pressure change at the Kamaishi Mine, *Eng. Geol.*, **43**, 107-118 (1996)

Toda, S., Tanaka, K., Chigira, M., Miyagawa, K. and Hasegawa, T. Coseismic Behavior of Groundwater by the 1995 Hyogo-ken Nanbu Earthquake, *Journal of the seismological society of Japan* (in Japanese), **48**, 547-553 (1996).

Proc.30th Int'l.Geol.Congr.,vol.24,pp.42-61
Yuan Daoxian (Ed)
© VSP 1997

Study on Tianchi volcano, Changbaishan, China

LIU RUOXIN

Institute of Geology, State Seismology Bureau, Beijing 100029, P.R. China

WEI HAIQUAN

China University of Geosciences, Beijing 100083, P.R. China

LI XIAODONG

Department of Geophysics, Beijing University, Beijing 100871, P.R. China

Abstract

Many research workers pointed out that Changbaishan Tianchi volcano is the most dangerous one of the world now. Recent studies present new evidence of several historic large eruption. Tianchi volcano is a composite central one with some successive eruption in Holocene. Time dating research shows the calendar age of the last large eruption of Tianchi volcano: 1215±15 AD, which is consistent with the acid peak at 1227-1229 AD from the ice core sampling and the climate event happened roughly in 1230 AD. That large eruption produces various genetic types of accumulation due to the different dynamic processes. Magnetotelluric sounding and CT prospecting reveal the crust magma chamber and the mantle reservior under Tianchi volcano, meaning that Tianchi volcano is still active, and has the potential danger of eruption. The volcanic effects on global climate of the last large eruption are simulated with a two dimensional energy balance model by introducing the volcanic radiative forcing that is calculated based on a diagnostic diffusion-decay model. The numerical results indicate that the stratospheric aerosols formed by the last large eruption may last for 3 years, and the climatic impact of the last large eruption is very significant and may last more than six years.

Keywords: Eruption, Tianchi Volcano, Genetic nomenclature, Dynamical Processes, Triggering Mechanism, Climate Model, Radiative Forcing, Numerical Simulation

INTRODUCTION

The study on volcano eruption has been one of the hotest point in the geoscience studies because of several famous large volcano eruptions in recent years. Some of those large eruption cost tens of thousands of lives and caused intensive destruction, as in the tragic 1985 eruption of Colombia's Nevado del Ruiz. Some of them had significant impacts on global climate, as the 1991 large eruption of Mount Pinatubo did. The studies on volcanology, environmental effect of volcanic eruption, volcanic impact on global climate, volcanic hazard reduction, etc., have been concerned and deemed highly and broadly.

The Changbaishan-Tianchi volcano, located at the Sino-Korea border in eastern part of Jilin province, is one of the most dangerous volcanoes in the world[11]. New evidences reveal that the historical large eruption of Tianchi volcano, which happened in 1215±15 AD, is one of the largest and most violent eruption in recent 2000 years over the world[3, 11, 12, 13, 14]. Some recent geological investigations indicate that Tianchi volcano had several large eruptions before that large one, such as the eruption happened in about 4105 a BP. The potential danger of large eruption of it has been stressed by a lot of studies [11, 12].

Many volcanologists have paid great attention to Tianchi volcano for the reason of its peculiarity in academic research, such as the magma composition variety, the tremendous polygenetic volcanism phenomena, the successive eruptions in Holocene, and so on. Many foreign volcanologists visited Tianchi volcano and did a lot research works in the last decade. Although many papers that deal mainly with the petrology and geology relating to Tianchi volcano have been published, a few of them were published in English for many reasons. In Part 2 of this paper, general background about Tianchi volcano, including some basic concepts and the geological setting, will be introduced. Part 3 is a systematic review of the volocanological studies about Tianchi volcano, it consists of the following several sections: airfall deposit, ignimbrite, plinian column, eruption history, and so on. Part 4 discusses the triggering mechanism of the last large eruption from these aspects: the crust magama chamber, the mantle magma reservior, and magma supply and transport, etc.

It is generally acknowledged that volcanic stratosphere aerosols scatter incoming solar radiation to space, increasing planetary albedo and cooling the earth's surface and troposphere, and absorb solar and terrestrial radiation, warming the stratosphere [8, 9, 15]. Although a lot of papers about volcanic impact on climate have been published after Mt. Pinatubo large eruption (1991) and El Chichon eruption (1982), some details of climate effects of volcanoes have not yet been demonstrated. Because there have been very few large eruptions in the past century, it is very difficult to discuss the climate effects of volcanoes in detail with statistics method. In Part 5 of this paper, the climate effects of the last large eruption of Tianchi volcano are simulated with a two dimension energy model (2D-EBM). The last part is the conclusion.

GEOLOGICAL SETTING

Tianchi (sky-lake in Chinese) volcano located at China-Korea border which has a height of 2751 m above sea level. The main part of it is in the Changbaishan (ever-white Mountain) area, NE China. This is a polygenetic central volcano, with successive eruptions during the human history. Its magma composition has a great variety, with a tremendous volcanological feature change, from a lava shield made of alkali basalt and tholeiite to trachyte and comendite cone and comenditic pumice, as shown in Figure1.

The base of the volcano consists of augite-olivine basalts (K/Ar age of 1.2-2.6 Ma) with a high value of alkalinity, locally named as Junjianshan (Battle-Ship Mountain) Formation and Baishan Formation covered an area of several thousands square kilometers. This lava

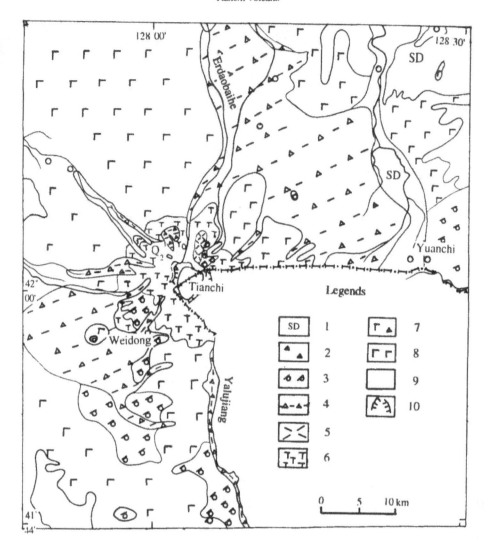

Fig.1 Simplified geological map of Tianchi volcano, Changbaishan, China.

1.secondary deposits; 2. lahar; 3. Holocene airfalls; 4. Holocene ignimbrite; 5. Holocene comendite; 6. Pleistocene trachyte; 7. Pleistocene basaltin scoria; 8. neogen basalts; 9. pre-tertiary basement; 10. caldera cliff.

shield has a gentle dip (smaller than 4°) in the height of 1000 to 1800 m above sea level. The cone of Tianchi volcano consists of trachytes, comendites, and their equivalent pyroclastic rocks. Covering an area of about 70 km², it situates at 1800-2744 m elevation interval with a local name of Baitoushan (White-Head Mountain). The lower part of the cone has a gentle dip of 5-10°,while the upper part has a dip increased to 30°, inner walls of the crater dip 60-80°. The upper part of the cone consists of comendite, obsidian, trachyte, and pyroclastic rocks with a total thickness of about 100 m. K/Ar dating gives ages of 0.55, 0.44, 0.2, and 0.1 Ma for deposits from the base to the top of the cone. The upmost part of the cone is usually covered by some plinian pumice falls and ignimbrites with ^{14}C ages of >2024 BP, and 1215±15 AD.

Tianchi is one of the most beautiful crater lake in China. Its water surface elevation is about 2190m. It presents an irregular ellipse shape with an area of 9.2km^2, a major axis (from north to south) of 4.4km and a minor axis (from east to west) of 3.4km long respectively. The average depth of the lake water is 204 m with the deepest value of 373m at the lake center. It stores 20 billion cubic meters of water. The summit caldera of the north rim has a V-shape gully with the depth of several hundred meters. Tianchi volcano is the only head waters for the three major rivers (Songhua, Yalu, and Tumen) in the China-Korea-Russia border.

LAST LARGE ERUPTION

Tianchi Volcano, the same as the other modern volcanoes, produced various genetic types of pyroclastic deposits including airfalls, ignimbrites, and surges. In a general review for the last deposits from Tianchi Volcano, Wei and Liu gave a highlight description on the deposits [19]. The main features of the different tephra will be discussed in this section.

Airfall deposits
The airfall deposits fallen from the plinian column above Tianchi Volcano to the land surface around it are preserved very well. This makes it possible to reconstruct the dynamical processes in the plinian column [11]. The typical falls are described as follows:

Airfalls on the crater rim: These light yellowish colored falls are of a feature of partially welded pumices in which accident lithics are contained. The pumice size often varies from 0.5 to 6cm with an averaged maximum pumice (Mp) of 44 cm. Its averaged maximum lithics diameter (Ml) is 177 cm while the fines fragments are strongly depleted. There are always some changes in the deposit profiles on the grainsize of the tephra, content of the lithic fragments, welding, and the coloration. All of these account for the changes of the dynamical parameters during eruption, which will be discussed later.

Airfalls on the lower flank of the cone: These pumiceous deposits, showing white gray color, are featured by the good sorting, horizontal layering, and vertical grainsize syclotheming of the deposits. These plinian falls have a thickness of usually between 10 and 20m at different places of the cone. Ml for these falls is often in an interval of 30--40 cm while their Mp falls between 10 to 20cm. No welding has been found on these loose pumiceous fragments for a lateral transporting distance more than 2.5 km from the vent.

Airfalls on the basaltic plateau far away from the cone: The distal facies of the plinian airfall deposits from Tianchi Volcano is of a typical feature of its horizontal layering and homogeneous fragments composition. These deposits have a thickness of the order of 10cm at a distance of the order of 10km, and further more, several centimeters at a distance of about one thousand kilometers east of the volcano. Mp for these distal falls is always less than one centimeter. Some papers indicated Mp of 0.5cm at the location of 60km east of the volcano, for instance, and 0.2mm at Tomakomai one thousand kilometer away from the volcano [14].

Ignimbrite deposits

Ignimbrite deposits also well developed around Tianchi Volcano. Two types of ignimbrite deposits, that is, ignimbrite sheet and ignimbrite flow, can be distinguished according to their vertical deposition texture and horizontal geometrics occurrence, as well as the composition analysis. The ignimbrite flows usually distributed in some valleys and canyons nearby, but the sheets distributed over broad regions on the basaltic plateau [11]. The fragments of the co-ignimbrite deposits, or its equivalents with different facies, scattered regularly at different places away from the caldera.

Ignimbrite sheets: These loose ignimbrite sheets often appear in a light gray or light yellow color. Their pumice content usually exceeded 30 percent of the total tephra. Pumices greater than 20 cm in diameter are very popular in the upper part of the deposition profile. Reverse grading of the pumices and a pumice concentration zone (2bp) is a general feature of the sheet while the grainsize lamination appears very often in the deposit sections. Thickness of these deposits is usually between 5 to 10m while a brown leaching zone at its base can be found some times. The facies characteristics of the deposit vary along with the distance away from the volcano. A dark colored strongly welded ignimbrite sheet, for example, overlaps on the top of the caldera and the upper part of the cone. Its columnar joints, strongly plastic deformation of the fiamme, and the recrystalization of the vitrics show evidence of high temperature setting regime. At the distal part of the sheet, a fine depleted layer (FDI) formed due to the strong fine depletation. This will be discussed in the following section.

Ignimbrites flows: These dense ignimbrites flows located at the radial valleys and ponds around the cone. A dark-gray color and thermal cooling joints are general features of the deposits. Partially or initially welding at the lower part of the profile makes the deposits petrified. Lithic content of the deposits may be as high as 20 percent with a normal grading of the lithic fragments. Flowing segregation layer of the pumices and fines can be found at the upper part of the deposit profile, which is a symbol of partially fluidization during its flowage. The typical thickness of these flows is about 20m at most places. A maximum thickness of more than 100 m in the canyon made the deposits totally lithified where VPI (valley and pond ignimbrites) formed. Different flow units originated from the partial collapsing of the plinian column left a grand sight of the gully. These accumulative materials represent often the intermediate facies of the ignimbrites where has usually had some evidences of flowing segregation structure formed by fluidization.

Co-ignimbrites accumulative materials: These deposits accompany always by the main body of the ignimbrites, including ground surges, ash cloud surges, lag breccia, veneer deposits as well as the co-ignimbrites falls. The ground surges formed in the head of the flow due to the turbulent torrential agency. For instance, a ground layer of breccia with thickness of 1m should be 9km away from the volcano, and exist under the main body of the ignimbrite flow unit. Most of the composition of the ground layer is the partially rounded lithic. Very few fine matrics exist in the ground layer. The ash cloud surges that were elutriated out from the fast moving flow body settled down onto the top of the flow unit.

Very few ash cloud surges preserved in the area out of the ignimbrite distribution because of the secondary transportation. In other hand, the ash cloud surges interbedded with the pumice flow deposits in the ignimbrite succession. It is easily to recognize that no matter its fines grainsize population or the composition components, these ash cloud surges have a strict clue to their parent flows. It was suggested that lag breccia can be taken as a symbol of climax of the eruption [18]. One can see such deposits on the north and west flanks of the volcano. A laboratory analysis shows that lag breccia is produced mainly by air fall agency. Veneer deposits were the tail deposits of the flows. A 20cm-thick, dark-gray lenticular layer overlaid its former flow unit.

Reconstruction of the column height

A light gray-yellow pumice fall layer is well preserved around Tianchi Volcano. It is no difficult to get the averaged maximum lithics diameter at individual outcrops around the volcano. Figure 2 is the dispersal pattern for the isopleth maximum lithics (MI) of Tianchi Volcano. This makes it possible to reconstruct the plinian column height by different dynamic models. Carey and Sparks proposed a dynamical model for the plume height controlled by clast trajectory pattern [1]. The column height of that large eruption of Tianchi volcano was determined to be between 20 and 30km by comparing the MI of Tianchi Volcano to the calculated by that model, as shown in their Fig 12 and 13 for Hb 20km and 30km respectively. It is estimated that the speed of the wind from NW to SE is 30m/s.

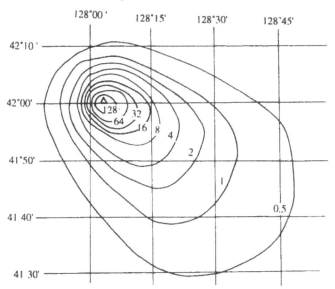

Fig.2 MI isopleth disperssion of Tianchi volcano airfall deposits (cm).

MI-distance plot diagram can be easily obtained from the MI isoplith pattern. As shown in Fig 3, MI-distance plots for up-, cross-, and down-wind are well consistent with the model values. From the cross-wind plot it can be seen that Hb and Ht arrive their tops of 25 and 35km high during the eruption climax. Because the hight of the tropopause at this latitude is about 10km, the main part of the erupted materials must have been injected into the stratosphere in that large eruption. In addition, the down-wind Ml-distance plot roughly coincides with the 30km plot model calculated. This means there should be NW wind with the speed higher than 30m/s. For the lithics Ml>8cm, the plot shows a sharper slope. This has a good agreement with the blistic trajectory of the lithics. For any given range of clast grainsize corresponding to the given column height and wind speed, the distance interval varies in different directions. The major axis of the trajectory, both for the down-wind and the up-wind directions, shows a greater distance interval. This means that the wind speed varies in time. When the wind speed is smaller than 10m/s, the given size clast travells an additional distance up-wind relative to the normal wind speed case.

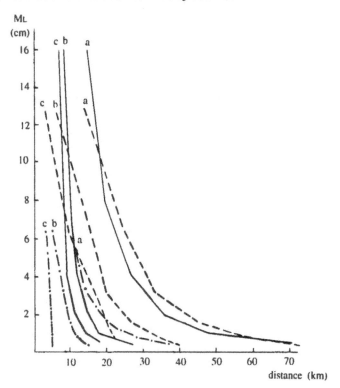

Fig.3 Down(a), cross(b), and up(c)-wind Ml-distance plots.

Solid, dotted and dashed line, for Tianchi volcano, Hb=30km and 20km model (from Carey, 1986), respectively.

The pumiceous plinian fall layer often shows a sharp change on its profile grainsize. The lower part of the deposits is always coarser than the upper part. As the deposit grainsize is determined by the column height, there should be obvious variations in column height during eruption. The grainsize of the lower deposits, corresponding to the early period of the eruption, reveals that the eruptions with the column height, Hb, about 20km last for several days. On the contrary, in the later period of the plinian fall eruption, the column hight is lower, about 10km as indicated by the grainsize of the upper deposits. Fig. 4, the isotherm of the fall deposits, clearly indicates that. There may be a secondary thickening as shown in the

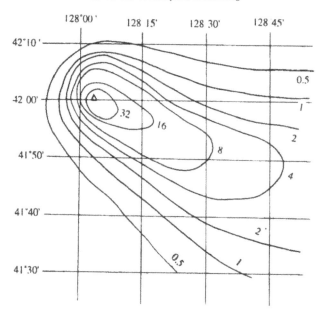

Fig. 4 Thickness isotherm of Tianchi airfall deposits (m).

major axis of the isotherm, as what can been seen easily, a light gray colour pumice fall layer overlays on the top of some hills in North Korea. Although Fig. 4 does not show the secondary thickening in SE direction and the thickness shortage in NW direction directly, those can be infered from the relatively lower plinian column and the stronger wind. In short words, in the large eruption of Tianchi volcano, the typical value of the plinian column height is between 10 and 20km, and the maximum value exceeds 25km during the climax of the eruption. The results of the isopleth area/clast size/column height relation given by Carey [1] surport these conclusions above mentioned.

Eruption history

The volcanic cone of Tianchi volcano consists of some trachyte, comendite and a few basaltic rock layers, whose hight is about 1000m. The volcanoclastic rock is seldom seen in the volcanic cone. From K-Ar chronology measurement, the age of the trachyte and comendite rock in middle-upper part of the volcanic cone is about 0.55-0.10Ma [11, 12]. The K-Ar ages are 1.12Ma, 0.04(±0.01)Ma, and 0.06(±0.02)Ma for the trachyte rock in the bottom of the cone, the upper trachyte in Haifengkou and the interbed of the upper basaltic rock, respectively. All of these indicate that the cone of Tianchi volcano was gradually formed by several effusions from the early to late Pleistocene .

A lot of pumiceous airfall deposits, pumice flow (ignimbrites) deposits, lahars and secondary accumulative meterials broadly distributed in the hills at the crater rim, the slopes of the cone and the valley plateau around the cone of Tianchi volcano. Dig-hole survey in Yuanchi, about 30 km east of Tianchi caldera, and the surface mapping in the east of Tianwenfeng clearly reveals two events of pumiceous airfall deposits of broad scope, as shown in Figure 5. The airfall deposit layer of comendite formed by the earlier eruption presents yellowish. According to the analysis of the ^{14}C age of the covered organic muddy layer, it is sure that the age of the earlier large eruption was before 2024 a BP. In the other hand, the comenditic pumice deposit layer produced by the later large eruption shows light gray, the ^{14}C age of its bottom is about 1065±95 a BP. In other words, the later plinian falls was deposited about 1000 years ago, and the earlier plinian falls was deposited about 2000 years ago.

According to the systematic ^{14}C time dating results for a large gagatite of a 300-years-old tree found in the ignimbrites nearby, the averaged apparent age of ten gagatite samples is 1000±16 a BP, and it varies between 1170 and 877±52 a BP from the centre to the edge of the gagatite. By fitting this age series with the high precise calibrated curve of tree ring (Fig.6) [17], it is concluded that the calendar age of the most outward section of the gagatite is 735±15 a BP, that is 1215±15 AD. Based on the above analysis, the date of the last large eruption of Tianchi volcano should be 1215±15 AD. It is necessary to point out that this date coincides with the acid peak at 1227-1229 AD from the ice core sampling [3] and the climate event happened roughly in 1230 AD revealed by the paleoclimate study [20]. From the Tianwenfeng profile (Fig.7), some eruption materials of several middle scale eruption also can be found above the light gray pumice deposit produced by that large eruption. These materials had been regarded as the results of the eruptions in the period from 1668 to 1702 AD in historic records [12]. In addition, another earlier large eruption that happened in 4105±90 a BP had been reported also [12].

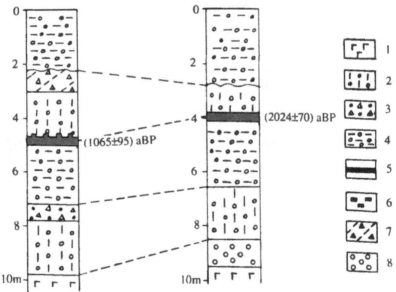

Fig.5 Drill core profile of the eastern slope of Tianchi volcano

1. basalt; 2. light brown-yellow pumice; 3 pyroclastic flow deposit; 4. secondary deposit; 5. puddly soil layer; 6. carbonized wood; 7. surge deposit; 8. puoluvium.

Based on above disscusion, the eruption history of Tianchi volcano can be divided into two periods: one is the period from earlier to later Pleistocene (1.12-0.04Ma), the other is the

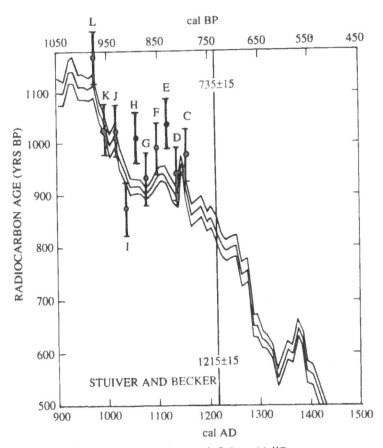

Fig. 6 Calibrated curve of high precise tree ring match-fitting with ¹⁴C age.

period since 4105(±90) a BP. The first period is the cone-forming period, which can be characterized by trachyte and comendite eruption. In the second period, several pyroclastic eruptions of comendite happened, and this period can be characterized by ignimbrite-form-ing eruptions. One of the largest eruption in the second period hapened in 1215(±15) AD, i.e. 791 years ago. The last group of eruptions, with middle scale, happened during the period from 1668 to 1702, i.e. 294 years ago.

It is necessary to point out that, during the cone-forming period of Tianchi volcano, there were not only alkalic basalt eruption as indicated by the alkalic basalt interbed in the middle-upper cone body, whose K-Ar age is about 0.06(±0.02)Ma, but also basaltic lava eruption in the same time as shown in the basalt layer in Haishi river and Dongfanghong forest field (about 30km north-east of Tianchi volcano), whose K-Ar age is between 0.18-0.19Ma. It is important to point out that, the comendite magma and the basalt magma erupted in the same time in early stage of that largest eruption (1215±15 AD) [12]. This will be disscussed in the next section.

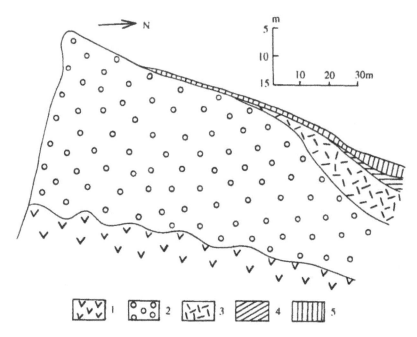

Fig. 7 Profile map for recent erupted materials east of Tianwenfang.

1. trachytes and comendites; 2. brown-yellow pumice (>2024 a BP); 3. light gray pumice (1215±15AD); 4. red-brown clastogenetic lava (historically recorded eruption); 5. dark gray pumice (historically recorded eruption).

TRIGGER MECHANISM OF MAGMA ERUPTION

In order to investigate the magma activities of Tianchi volcano, the magnetotelluric sounding had been taken in two profiles with the depth of 40km which consist of 15 sites. The magnetotelluric sounding results clearly show a high electric conductive body with the value from 1 to 15Ω/m, which locates in the depth of 12-25 km under Tianchi volcano, as shown in Fig. 8. That high conductive body was interpreted as the geophysical proof that there must be a crustal magma chamber [11, 12]. The two dimensional horizontal range of that high conductive body was estimated at 10×13km. In addition, there is another small centre with high conductivity in the upper part of the vertical profile, that may indicate the warm water under surface.

From the global seismic network digital data [4], the CT results of S wave in three profiles, all of them cross Tianchi volcano, reveal a low speed section which exists in the depth of 40-65 km and with a width of 100-200 km as shown in Fig. 2 of Guo [4]. Although the resolution of CT data is not very high to describe the structure of the profile in detail, this low speed section in CT profile doubtlessly means that there is a source of the basaltic magma --a magma reservoir in the upper part of the upper mantle under Tianchi volcano.

In Cenozoic era, the volcanic activities is very popular in the broad area around Tianchi volcano [6]. As disscussed in the previous section, while the cone of Tianchi volcano was

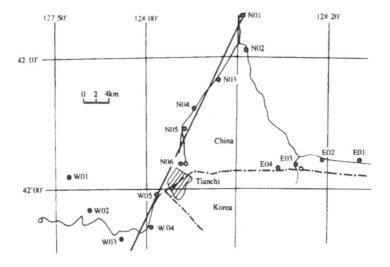

Fig. 8a Location of MT sites of Tianch volcano region

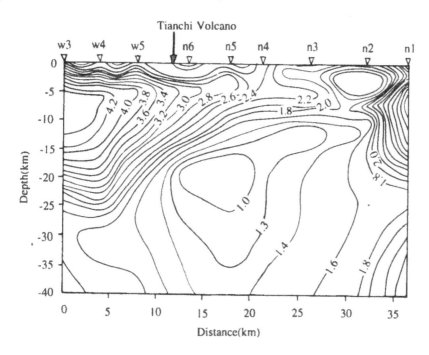

Fig. 8b Structure of resistivity obtained by 2D inversion, the value is resistivity (log ohm/m)

forming in Pleistocene epoch, the basaltic cinder cone and the basaltic lava flow which include xenolith of the mantle peridotite exist around Tianchi volcano, and there are some interbeds of alkali basalt even in the cone. What is more interesting is that, in the low part of the layer formed by the last largest eruption (1215±15 AD), the comenditic pumice and the basaltic pumice exist in the same layer of the pumice airfall deposit , i.e., the comenditic pumice and the basaltic pumice erupted and accumulated simultaneously in the early stage of that largest eruption [12]. These facts strongly prove that there must be a mantle magma reservior in the upper mantle under Tianchi volcano in which the basaltic magma exist. In that largest eruption process, the trigger factor was very possibly the upward magma injection from the mantle magma reservoir to the crust magma chamber that was uptight and ready to erupt. Because of sudden triggering, the mantle magma and the crust magma erupted out of the surface before they mingled with each other sufficiently, and the result is what can be seen in the profile of the pumice deposit: the white comenditic pumice and the black basaltic pumice coexist, and it is very easy to distinguish them.

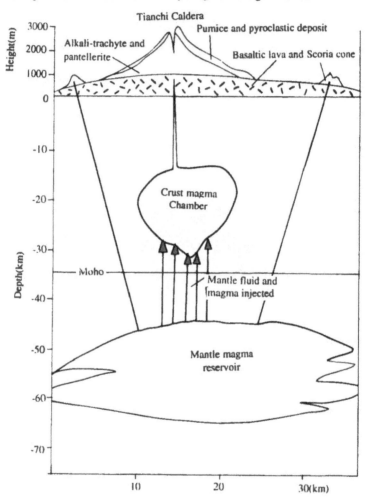

Fig.9 Magma forming-mixing-erupting model in 1215±15AD eruption of Tianchi volcano.

As disscussed above, the geophysical sounding results are consistent with the actual data of the activities of Tianchi volcano in Cenozoic era, especially in the period from Pleistocene epoch to Holocene epoch, as well as the characteristics of the last largest eruption (1215±15

AD): the comenditic pumice and the basaltic pumice erupted and accumulated simultane-ously in the early stage of that largest eruption as revealed by the profile of the pumice deposit. In order to summarize the dynamical mechanism of that largest eruption, Fig. 9 gives the dynamical model of the magma activity [12].

CLIMATE AND ENVIRONMENTAL EFFECTS

The last large eruption of Changbaishan-Tianchi volcano is a special candidate for the cli-mate effect simulation because it is not only the most dangerous active volcano over the world now but the unique large vocano in middle-high latitude in north hemisphere since several centuries ago. The estimated volcanic eruption index (VEI) of that large eruption arrives 7, in other word, this historic eruption is the largest one in the world in recent 1000 years, with the same VEI as the great eruption of Tambora in 1815, which caused the famous climate event—the year without a summer" [16]. Therefore, that large eruption should have significant impacts on global climate.

Model and Simulation Scheme

In spite of the increasing complexity of coupled general circulation models(CGCMs), they do not yet include certain physical processes that may be important. Even if the "perfect" CGCMs were developed, they can not predict climate change correctly and can not give the satisfactory simulation on some special time scales. For the volcano-climate studies, it is difficult to identify the volcanic signals from the CGCM's simulation results. In the other hand, energy balance models(EBMs) have been used widely for theoretical studies of the large scale climate. While typically limited to few spatial dimensions and employing parameterized dynamics and physics, they offer several advantages over more complex models. In special, their mathematical and physical simplicity makes them amenable to ana-lytical methods, which can contribute to the understanding of cause and effect relationship. They provide a framework for studying climate change that is comprehensive and economi-cal. The remarkable thing about the EBMs is that although they employ a very simple for-mation of the climate problems, they seem to be obtain very good agreement with the ob-served global climate [9].

The model, used for simulating the climate effects of the last large eruption of Tianchi vol-cano, is a two dimensional one level energy balance model (2D-EBM) described in detail in Li [8, 9]. The model parameters and the resolving method have been given in previous papers [8, 9]. It is necessary to point out that these model parameters are based on the ob-served present climate. In the 2D-EBM simulation, the radiative forcing of the stratospheric volcanic aerosols is determined according the amount of erution materials and the eruption hight discussed in Part 3, and the eruption date, in January, is specified arbitrarily.

Volcanic Aerosols and Radiative Forcing

In order to account for the stratospheric aerosol thickness (AOT), a diagnostic model of the spatial and temporal distribution of AOT is developed. The model is based on the simplify-ing assumption that the dominate processes governing aerosol dispersion are diffusion transport in latitude and exponential decay in time. This approach to calculate the space and time distribution of aerosol (or AOT) was developed by King [7], and proved to be highly effective and reasonable. This diagnostic model has three parameters: time constant, diffusion coefficient and initial global mean AOT (at typical wave length). These parameters can be determined by a best fit to the actual AOT data. It was found that these parameters were very similar for different eruption, such as Mt. St. Helens (May 1981), El Chichon (March 1982), Mt. Pinatubo (June 1991), etc. [9]. In other words, these parameters vary within a small range for the present climate. Figure 10a is the spatial and temporal distribution of the relative AOT from Tianchi eruption simulated by the diagnostic model.

According to the theoretical study on the radiative forcing of AOT [6], the radiative forcing varies in latitude and season even for the given constant AOT. The maximum perturbation in radiative balance occurs at high latitudes in spring and fall, whereas polar winter shows a net gain in the radiative balance due to the increased greenhouse effect. Latitudes equatorward of 50 degree show a nearly uniform reduction in the radiative balance. In special, the typical value of the radiative forcing for AOT=0.1 in the middle and low latitudes is about -3 W/m^2. As Robock did in his radiative forcing calculation for St. Helens eruption, the thin approximation theory was used to calculate the radiative forcing for the climate model simulation, i.e., the AOT values were multiplied by the normal values given by Harshvardhan to provide the radiative forcing for any given latitude and time. Figure 10b is the calculated spatial and temporal distribution of the radiative forcing based on Figure 10a for the simulation of the two dimensional energy balance model.

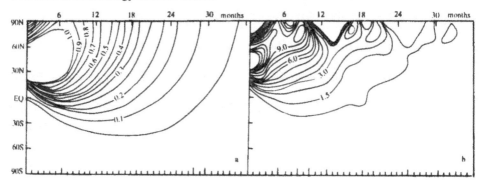

Fig.10 Simulated relative AOT (a) and its radiative forcing (b) (W/m^2)

Figure 10 indicates that there are obvious differences in space and time for AOT and its radiative forcing. Within several months after the eruption, volcanic aerosols mainly distributed in the high latitude near the eruption latitude, and they cause significant reduction in radiation balance. For example, the radiative forcing at 60°N exceeds 9W/m^2 until 9 months after the eruption. The period from the initial time to 9 months after the eruption is characterized by the remarkable diffusion process of the volcanic aerosols. In this period both AOT and its radiative forcing show noticeable differences in different latitude. From several months to 20 months after the eruption, the volcanic aerosols spread into lower latitudes and decay in the meantime. After then, the decay process dominates the distribution of the volcanic aerosols. Corresponding to AOT variation, the radiative forcing becomes smaller and smaller, more uniform and more uniform in space. For example, the values of the forcing are lower than $0.75W/m^2$ in the whole North Hemisphere after 30 months.

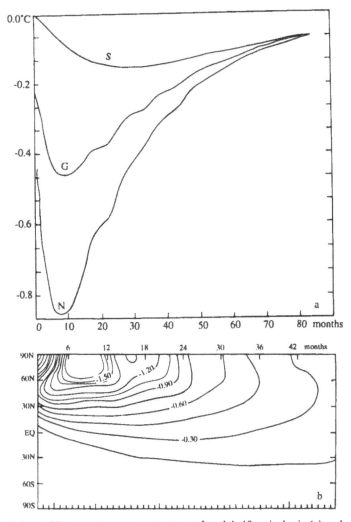

Fig. 11 Simulated monthly mean temperature response for global/hemispheric (a) and zonally averaged (b).

Global/Hemispheric and Zonally Averaged Temperature Response

The radiative forcing as described in above section is applied to the two dimension energy balance model (2D-EBM). Figure 11 to Figure 12 show the 2D-EBM simulated temperature response, that is, the temperature difference between the control run for the present climate and the volcanic experiment for the Tianchi eruption. Figure 11a and Figure 11b are the global/hemispheric and the zonally averaged temperature response respectively. From Fig-

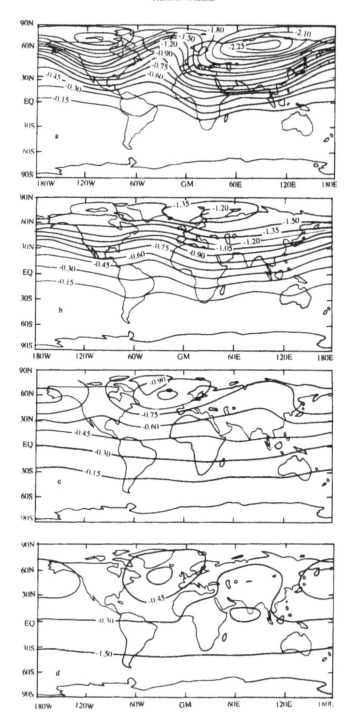

Fig.12 Spatial distribution of simulated transient temperature anomaly in 6(a) 12(b) 24(c) and 36 months (d) after the eruption, respectively.

ure 11a it can be drawn that the impact of that large eruption on global climate last over 6 years. The amplitudes of the maximum temperature decrease are 0.85, 0.16 and 0.47°C for the North Hemisphere (NH), the South Hemisphere (SH) and the global, which occur in 7, 28 and 8 months after the eruption respectively. As shown in Figure 11a, the amplitude of the temperature decrease is larger in NH than that in SH. In special, within a few months after the eruption, for example, the diference of temperature response between two hemispheres exceeds 0.6°C in the period from 6 to 9 months after the eruption. This is what would be expected from the forcing characteristics and the thermodynamic difference between two hemispheres.

Figure 11b indicates that the amplitude of the maximum effect of that eruption exceeds 1.5°C, which occurs in the high latitudes near the eruption latitude in 6 months after the eruption. This value is comparable with the interannual standard deviation of the monthly mean zonally averaged temperature for the same latitudes, which varies from 1°C in the summer to 2.5°C in the winter. For this reason, the volcanic signals of that large eruption can be detected from the historical climate records. As noted in the radiative forcing section, the first period (several months after the eruption) is characterized by the diffusion process. In that period, the aerosols mainly distribute in the latitudes near the eruption location, and the forcing varies significantly in latitude. Therefore, the dominant feature of the zonelly averaged temperature drop in the first period, i.e., the remarkable difference between different latitudes, can been interpreted mainly by the forcing. After then, the meridional differences of temperature response become smaller and smaller. The isoline tends to be uniform in 3 years after the eruption, and the temperature drops vary between 0.3 and 0.4°C in NH. The amplitude of the uniform temperature decrease after 3 years decays in time, this can be drawn from the oceanic lag effect due to the high thermal inertia.

Geographic Simulation

Figure 12 is the spatial distribution of the 2D-EBM simulated monthly temperature anomaly for 3, 6, 12, 18, 24 and 36 months after eruption respectively. From these maps it can be seen that, within a few months after the eruption, two strong cooling centers appear in the land areas near the eruption latitude. For example, the temperature drops in the central part of the Europe-Asia continent increase rapidly, reaching a peak cooling over 2.25°C. The cooling is weak in NH oceans, and there is no cooling tendency in the most part of SH. It is noticeable that the strong cooling centers move to the high latitudes in time. These features are doubtlessly related to the thermodynamic difference between ocean and continent and the radiation forcing described in above section.

After 6 months, the cooling in the continents gradually becomes less obvious, on contrast, the cooling in oceans becomes more and more intense. The difference of the cooling between the continent and ocean becomes smaller and smaller, and the isolines distribute roughly along the latitude line. For example, in 18 months after the eruption, the enclosed centers disappear, and the typical value of cooling in the middle latitudes is about 0.75°C. In additional, the cooling spreads to the middle latitude of SH. In 24 months after the eruption, the cooling in ocean is more obvious than that in continent, the enclosed centers with the temperature anomaly of -0.9°C appear in the north part of the Pacific and the Atlantic. This is also true for 36 months after the eruption, but with a weak maximum cooling of 0.45°C. That means the lag effect of the ocean dominates the cooling situation. In 5 years after the eruption, the cooling is very uniform in space, and the temperature anomaly is about -0.2°C.

In summary, the transient response of the eruption of Tianchi volcano can be divided into three categories. The first is the strong cooling occurring in the continents of NH within a few months after the eruption. This results from the rapid response to the radiation forcing due to the small inertia of continent. The second is the relatively latitudinal-uniform cooling occurring in the period from 18 months to 3 years after the eruption. The last is the weak cooling after the above period, which is caused by the oceanic lag effect.

CONCLUSION

1. Geological mapping reveal that, Tianchi volcano is a polygenetic central volcano with successive eruptions during the human history. The last large eruption of Tianchi volcano, happened in 1215±15 AD, produced various types of pyroclastic deposits.

2. The genetic classfication studies indicate that the ignimbrite deposits around Tianchi volcano can be divided into two types according to their field occurrence. The ignimbrite sheets distributed over a broad region than the ignimbrite flows did.

3. The reconstruction results from dynamical model show that, during the last large eruption of Tianchi volcano, the height of plinian column varies from 10 to 20km (Hb), and the Hb value reaches its peak 25km at the climax of the eruption.

4. From Magnetotelluric sounding and CT results, there should be a crust magma chamber and a mantle reservior under Tianchi volcano. This means Tianchi volcano is still active, and has the potential danger of eruption.

5. The spatial and temporal distribution of the Tianchi volcano aerosols calculated by the diffusion-decay diagnostic model shows that, the stratospheric aerosols formed by this eruption may last for 3 years, which had a significant impact on solar radiation, the maximum net radiation deficit exceeds 9W/m^2.

6. The climate response simulated by 2D-EBM indicates that the last large eruption may produce a maximum monthly averaged local temperature decrease over 2.25°C, and a maximum monthly averaged hemispheric temperature decrease of 0.85°C. The climatic impact may last more than six years.

ACKNOWLEDGEMENT

We thank Prof. Wang Shaowu, Dr. A. Robock, Prof. G.P.L. Walker, Prof. Song Shengrong,

Prof. Liu Jiaqi for their contribution to this paper. This research was supported by NSF of China.

REFERENCES

1. S. Carey and R. S. J. Sparks. Quantitative models of the fallout and dispersal of tephra from volcanic eruption column. *Bull Volcano*, **48**, 109-125 (1986).

2. V. B. Chiiakov, et al. Age of the vocanic ash of Baitoushan volcano in North Korea. *DANSSR*, **306:1**, 169-172 (1989)(in Russian).

3. J. Gill, et al. Large volume, mid-latitude, Cl-rich volcanic eruption during 600-1000 AD, Baitoushan, China. *AGU Chapman*, 91-99 (1992) .

4. Guo Lucan, Ma Shizhuang, and Zhang Yushen. Research on: "magma chamber" of Changbai Mountain Volcanoes by means of seismic tomography. *Computerized Tomography Theory And Applications* **5(1)**, 47-52. (1996) (in Chinese).

5. Harshvardhan. Perturbation of the zonal radiation balance by a stratospheric aerosol layer. *J. Atmos. Sci.* **34**, 1274-1285 (1979).

6. Jin Bolu and Zhang Xiyou. Researching Volcanic Geology In Mount Changbai. *NE national educational press* (1994) (in chinese).

7. M. D. King, Harshvardhan, and K. Arking. A model of radiative properties of the El Chichon stratospheric aerosol layer. *J. Climate and Applied Meteorology*, **23**, 1121-1137 (1984).

8. Li Xiaodong, Wang Shaowu and Liu Ruoxin. The Advances in volcano-climatology. *Seismological and Geomagnetic Observation and Research* **17:4**, 74-80 (1996) (in Chinese)

9. Li Xiaodong. The volcanic impact on global climate. *Chinese science and technology press*, Beijing (1995) (in Chinese).

10. Liu Jiaqi. The chronology research on Cenozoic volcanic rocks from NE China. *Acta Petrology Sinica*, **4:4**, 21-31 (1987).

11. Liu Ruoxin, Recent eruptions of Tianchi Volcano, Changbaishan. *Science publish house.* (in press).

12. Liu Ruoxin. et al, Progress of the study on Tianchi volcano, Changbaishan, China, *Seismological and Geomagnetic observation and research.* **17:4**, 2-11 (1996)(in Chinese).

13. H. Machida, et al. The recent major eruption of Changbaishan volcano and its environmental effects. *Geographical reports of Tokyo Metroplitan University*, **25**, 262-269 (1990).

14. H. Machida, et al. The temperate forest ecosystem, ITE Symposium No 20, Institute of Terrestrial Ecology. *The Lavenham Press*, UK, 23-26 (1987).

15. A. Robock. The Mt. St. Helens volcanic eruption of 18 May 1980: minimal climate effect. *Science*, **212**, 1383-1384 (1981).

16. A. Robock. The volcanic contribution to climate change of the past 100 years. In: *Green-House-Gas-Induced Climate Change, a Critical Appraisal of Simulation and Observations.* M. E. Schlesinger (Ed.). pp. 429-443. Elservier, Amsterdam (1991).

17. B. Stuiver, and Becker. B. High-precision decadal calibration of the radiocarbon time scale, AD 1950-2500 BC. *Radiocarbon*, **28(2B)**, 805-1030 (1986)

18. G.P.L. Walker. Origin of coarse lithics breccias near ignimbrite souce vent. *J. Geotherm. Volcan. Res*, **25**, 157-71 (1985).

19. Wei Haiquan and Liu Ruoxin. The genetic nomenclature and hazards analysis on the recent pyroclastic accumulations from Tianchi Volcano. In: *Volcanism and Human Environment.* Liu Rouxin et al (Eds). pp. 21-27. Seismic publishing house, (1995) (in Chinese).

20. Zhang Peiyuan, Wang Zheng, and Liu Xiaolei. The climate evolution stages in China during the last 2000 years. *Science in China (series B)*, **9**, 998-1008, (1994).

Proc. 30th Int'l. Geol. Congr., vol. 24, pp. 62-73
Yuan Daoxian (Ed)
© VSP 1997

Recent Soil Erosion on the Loess Plateau through Geochemical and Palaeomagnetic Studies on the Huanghe Estuarine Sediment in the Bohai Sea, China

HAO CHEN and GRAHAM B. SHIMMIELD
Dept. of Geology and Geophysics, University of Edinburgh, Edinburgh EH9 3JW, UK

Abstract

An estuarine sediment core, JX91-3B, was recovered about 30 km off the Huanghe River delta in the Bohai Sea. Multidisciplinary studies, including geochemistry and palaeomagnetism, have been carried out on the core. Principal component analysis (PCA) of XRF major and trace elements has revealed four groups of elements relating to two main processes involved in the sedimentation. Palaeomagnetic analysis supports the idea that *in situ* depositional environment in the Bohai Sea and sediment supply from the Huanghe River are the two controlling factors. Lithogenic elements, represented by Si and Al, frequency-dependent susceptibility χ_{fd} (%) and particle size distribution are controlled by the former, while magnetic susceptibility, phosphorus, REEs and titanium are determined by the later. On the Loess Plateau, biogeochemical elements, such as P and Mo, are generally enriched and magnetic susceptibility is enhanced in topsoil in contrast to loess, therefore topsoil erosion, once occurs, can produce a distinct signal in the sediment. As topsoil on the plateau is normally protected by vegetation, topsoil erosion, unlike loess erosion, is less likely to occur except during heavy rainstorms. A comparison is made between the magnetic susceptibility and the hydraulic records of the Huanghe River over the past 60 years, which shows a sound coincidence of magnetic susceptibility spikes in flooding years. This confirms magnetic susceptibility, P, REEs and Ti in the Huanghe estuarine sediment are good indicators of soil erosion on the Loess Plateau, and consequently reveals a longer yet unknown history of topsoil erosion on the plateau.

Keyword: soil erosion, estuarine sediment, geochemistry, magnetic susceptibility, Huanghe River, Loess Plateau, Bohai Sea

INTRODUCTION

The Huanghe (Yellow) River traverses the vast Loess Plateau in North China and transports an average of 1.2 billion tons of sediment annually to the Bohai Sea [9, 15, 20]. As about 90% of the sediment is from the Loess Plateau [20], the Huanghe River has practically served as a strong land-sea connection which makes it possible to study soil erosion on the Loess Plateau by examining Huanghe estuarine sediment in the Bohai Sea. Sedimentary records of recent soil erosion on the plateau could be preserved straightaway in the rapidly accumulated estuarine sediment [5, 9], however, as sedimentation processes in the Bohai Sea would inevitably impose their impact on the sediment, it will be crucial to identify as many indicators as possible in the sediment regarding all the main processes so that the two sets of records can be separated and recovered.

To tackle this problem effectively, detailed multi-disciplinary studies, together with statistical analysis, are essential. While palaeomagnetic studies on the Chinese loess have been so successful concerning palaeoclimate and magnetostratigraphy [8, 10, 11, 26], much less work of this kind has been done in the Bohai Sea [9]; on the other hand, more systematic geochemical work has been carried out at the sea than on the plateau [25, 27, 31]. Besides these difficulties caused by incomplete data sets, another problem for making a direct land-sea correlation was the lack of a high resolution chronostratigraphy for the estuarine sediment, which is recently solved by Chen and Shimmield [5] using historical records of channel switchings on the Huanghe River delta and ^{210}Pb dating. This paper will focus on geochemical and palaeomagnetic studies, which prove to be two most successful analytical methods, on the Huanghe estuarine sediment concerning the soil erosion on the Loess Plateau.

Geological setting and core description

The Bohai Sea is a semi-isolated inlet in northeast China with an average water depth of 18 m [9]. The rather quiet depositional environment in the sea helps the Huanghe River develop a huge modern delta, which is still growing at 23 km^2/a [16]. The Huanghe sediment dominates most part of the sea and the sediment around the delta is almost exclusively from the Huanghe River. Because of the enormous sediment input from the river, the accumulation rate in the estuarine area is pretty high [9], ensuring the sedimentation record a high resolution.

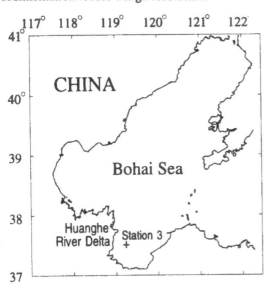

Figure 1. Location of JX91-3B in the Bohai Sea. The 3.2 m long core was recovered at Station 3 in cruise JX91 about 30 km off the Huanghe River Delta using a pneumatic Mackereth type corer at a water depth of 10 m.

A sediment core, JX91-3B, was recovered during cruise JX91 using a pneumatic Mackereth type corer [12] at about 30 km off the Huanghe River delta within the Huanghe estuarine area (Fig. 1). Fig. 2 shows the lithology profile of the core. The normal Huanghe sediment is fine-grained mud, hardly coarser than 20 µm [9], which is seen to comprise the main part of the core. It is extraordinary to find three sandy layers interspersed in the mud with means exceeding 60 µm (ref. to Fig. 4), as the typical mode of loess only falls in the range of 20 - 40 µm [18]. In fact, the sandy layers were generated under a much stronger hydraulic condition when the river discharged towards the coresite where the fine-grained fraction of the original Huanghe sediment could not get deposited [5]. The three sandy layers are found to be formed during 1897-1904, 1929-1934 and 1934-1953 AD respectively, and the shell layers were associated with initial flushing at the beginning of channel switching [5, 16].

GEOCHEMICAL AND PALAEOMAGNETIC ANALYSES

Geochemistry

Not all the XRF major and trace elements of JX91-3B vary in a same pattern down the length of core (ref. to Fig. 5). In order to analyse these multivariate data, the widely used principal component analysis (PCA), is recommended [17], which has already proved to be successful on geochemical analysis of marine sediments [e.g. 22]. PCA is a data reduction technique used to identify a small set of variables that account for a large proportion of the total variance in the original variables.

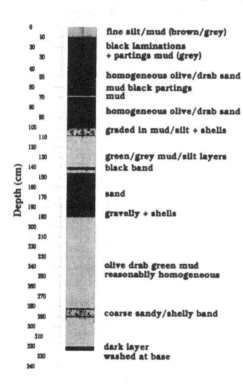

Table 1 shows the PCA results for JX91-3B. As the third component pc3 contributes less than 5% to the total variance, the first three components, pc1, pc2 and pc3, combined to account for over 80% of the total variance, are thought to have enclosed all the major controlling processes present. Here each principal component is regarded as one particular process, and each value for an element shows the extent of its impact on this elements. Negative value simply indicates an opposite impact on the element. Therefore an element that possesses the greatest value on one component but smallest on the other two should best represent that particular process; while those with modest values may be influenced more or less by other process(es). Fig. 3 illustrates the PCA results in 3D diagram and in biplots, for all the principal components are orthogonal. Four elements, namely Si, Al, P and Sr, are identified as representatives of their groups relating to three independent processes.

Figure 2. Lithology profile of JX91-3B. Note three homogeneous sandy layers are interspersed in muddy sediment and the upper two are separated by a thin mud layer. Two obvious shell layers locate at the bottoms of two sandy layers.

The major rock-forming minerals in loess are quartz (SiO_2) and feldspar (up to 90%), especially in the coarser fractions [2]. It can be therefore expected that Si content will be higher in sand but lower in mud. In Fig. 4, which shows the four element representatives in comparison with particle size data, Si does vary very much in line with the mean of particle size distribution, indicating the Si group (also including Na and Zr) is controlled by lithology and hence the *in situ* depositional environment. It is quite understandable that the Al group, including the majority of the rest elements and representing finer fractions of the Huanghe sediment, varies opposite to the Si group.

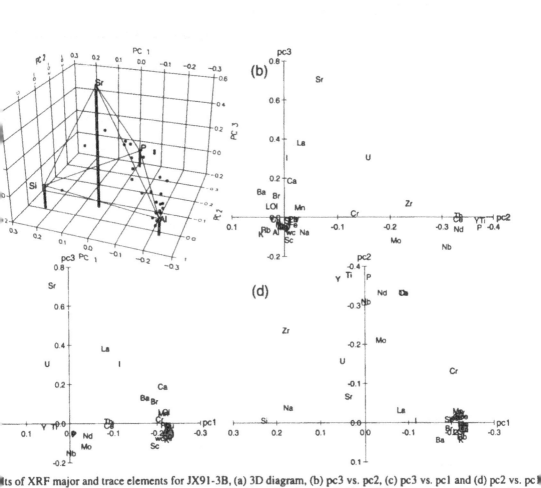

ts of XRF major and trace elements for JX91-3B, (a) 3D diagram, (b) pc3 vs. pc2, (c) pc3 vs. pc1 and (d) pc2 vs. pc1
elements is determined by three processes quantified in pc1, pc2 and pc3. Two of the components, pc1 and pc3 are
dependent. Si, Al, P and Sr are four representatives of four groups of elements controlled by these processes.

Table 1. Eigenanalysis of the Correlation Matrix of JX91-3B

Eigenvalue	18.874	6.256	1.462				
Proportion	0.572	0.190	0.044				
Cumulative	0.572	0.762	0.806				
Variable	PC1	PC2	PC3	Variable	PC1	PC2	PC3
Si	0.229	-0.006	-0.017	Ni	-0.227	0.010	-0.037
Al	-0.227	0.016	-0.076	Cu	-0.228	0.017	-0.011
Fe	-0.228	-0.016	-0.037	Zn	-0.226	0.005	-0.045
Mg	-0.227	0.003	-0.050	Pb	-0.219	-0.016	-0.009
Ca	-0.214	-0.014	0.188	Th	-0.089	-0.332	0.009
Na	0.174	-0.038	-0.076	U	0.051	-0.158	0.306
K	-0.222	0.045	-0.086	Rb	-0.224	0.036	-0.063
Ti	0.036	-0.378	-0.017	Sr	0.038	-0.067	0.706
Mn	-0.215	-0.029	0.050	Y	0.060	-0.367	-0.018
P	-0.010	-0.372	-0.054	Zr	0.178	-0.236	0.070
Sc	-0.194	-0.008	-0.115	Nb	-0.003	-0.309	-0.151
Ba	-0.173	0.044	0.130	Mo	-0.038	-0.211	-0.118
V	-0.225	-0.023	-0.011	I	-0.116	-0.005	0.304
La	-0.083	-0.032	0.382	Br	-0.194	0.015	0.111
Ce	-0.090	-0.331	-0.011	LOI	-0.217	0.024	0.057
Nd	-0.041	-0.332	-0.062	wc	-0.206	-0.012	-0.076
Cr	-0.206	-0.132	0.020				

LOI: Loss Of Ignition; wc: water content.

As for Sr, its association with Ca in shells is well known [31], but because of the high content of calcium carbonate in Huanghe sediment [9], Ca failed to turn up in this group. Nevertheless, Sr has already been confirmed to be almost exclusively of shell origin in the Bohai Sea sediment [31], which can also be found in Figs. 2 and 4. It is highly suggested these shells were washed to the coresite by initial flushing of the riverwater via tidal flat at the very beginning of a new channel switching that later formed the sandy layers [5]. The Sr spike at 205 cm down core is boosted by shells too, actually many tiny shell fragments, which were not quite distinguishable to naked eyes. So the Sr group is also related to depositional environment, or more accurately, the change in depositional environment.

It is interesting to note that unlike the Si, Al and Sr groups, the P group shows nothing comparable to the lithology and hence *in situ* depositional environment (Fig. 4). As the second largest process.(pc2), accounting for nearly 20% of the total variance (Table 1), the P group controlling process is not at all negligible. Since the P group receives little influence from the prevailing sedimentation processes, nor is it likely to be affected by minor processes even collectively in the Bohai Sea, it must have been associated with the original sediment supply from the Huanghe River.

Palaeomagnetism

Magnetic susceptibility (S.I.) and frequency-dependent susceptibility $\chi_{fd}(\%)$ are two commonly used palaeomagnetic parameters for loess study [e.g. 7, 13, 14, 32]. Generally speaking, magnetic susceptibility reflects the overall concentration of magnetic minerals, while the frequency-dependent susceptibility targets the ferromagnetic grain size distribution, the ultra-fine fraction in particular [23, 24, 28, 32].

Fig. 4 shows clearly that the magnetic susceptibility agrees with P pretty well, though in more detail due to a smaller sampling interval (~ 2.37 cm) for palaeomagnetic samples.

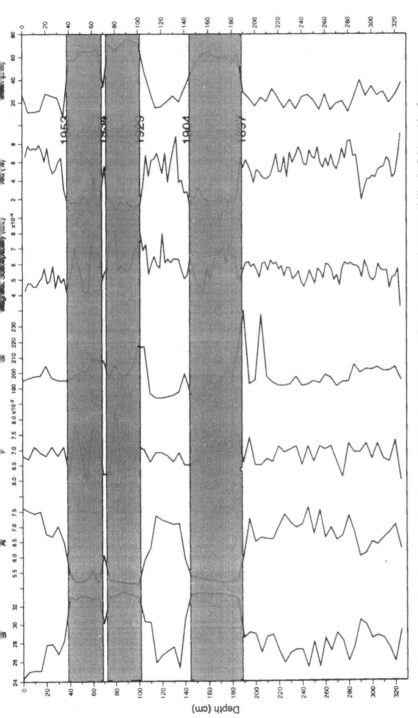

Figure 4. Profiles of representative elements Si(%), Al(%), P(%) and Sr(ppm), magnetic susceptibility X(S.I.) and its frequency dependence Xfd(%), and mean(μm) of particle size distribution for JX91-3B. It can be seen that Si, Al, Xfd and mean are associated with lithology, while Sr is related to shell layers, and P and X are in good agreement but independent from the others. The three sandy layers indicated by hatched areas are dated by Chen and Shimmield [5].

Nevertheless, χ_{fd} follows the pattern of Al and opposite to that of Si and particle size mean, indicating the grain size of magnetic minerals is in line with the bulk sediment. It is noticed that while χ_{fd} fluctuates very little within the rather homogeneous layers, the susceptibility varies significantly. This means the magnetic susceptibility variation in the sediment is not resulted from grain size change of magnetic minerals, but simply the effect of their variable input with grain sizes relevant to the bulk sediment. Apparently sedimentation processes have effectively controlled the magnetic minerals in terms of grain size, but failed to influence the magnetic susceptibility, which is actually a matter of mineral quantities pre-set by sediment supply.

It has been recognised, from both geochemical and palaeomagnetic studies, the sedimentation process in the Bohai Sea is the foremost one of the two main processes controlling the estuarine sediment. Meanwhile, all the evidence shows the other main process is associated with the sediment supply from the Huanghe River, which is directly linked to soil erosion on the Loess Plateau.

INDICATORS OF SOILS

Soil erosion on the Loess Plateau has two aspects: one is topsoil erosion concerning surficial soils, such as farmland and contemporary soils which are normally covered with vegetation; the other is loess erosion relating to pristine loess, which may also include fossil soils (palaeosols) interbedded in loess profiles. Change in relative proportion of these two erosions is the true determinant to the compositional variation in the Huanghe sediment. Soils, in contrast to loess, is believed to form under more humid and warmer conditions, and as a result of pedogenesis, its geochemical and magnetic properties become rather different from those of loess [4, 10, 11, 14, 25]. As suggested from the above analyses, it is possible to identify the various indicators of soils on the plateau.

Geochemical indicators
Phosphorus, the representative element in P group, is a typical biochemical element essential to plants. It gets enriched in topsoil through various mechanisms. Once released from weathered phosphate, P is easily precipitated with poorly crystalline forms of Fe and Al, which is rich in soils, and accumulates in soils [21]. In the meanwhile P in loess, as compared with topsoil, suffers more from leaching and frequent mechanical and chemical erosion owing to lack of protection from vegetation. Organic P, largely held by biota, accounts for a considerable proportion of total P in soil in the long term [21], which can not be matched by the P content in loess. Moreover, P can be enriched absolutely in soils due to the great deal of phosphorus added to the farmland as traditional (mainly organic) fertilisers in the early days and synthetic fertiliser after the 50's. Another biochemical element in P group, molybdenum, is a constituent of several enzymes, including nitrogenase [21]. Mo is thus essential for the biological activities [6], and apparently should present a higher content in soils than in loess.

The other elements in P group are mainly rare earth elements (REEs) and those connected with them. The REEs in this study include La, Ce and Nd, but as yttrium and thorium are frequently found in association with the lanthanide elements, they are often regarded as

REEs [6]. As titanium is usually found to coexist with REEs in heavy minerals [31], it may also be included in REEs in the following discussion. The main connection of REEs with soils is found in the ores containing REEs, mostly phosphates [6], which tend to undergo rapid weathering involving chemical as well as biological processes in soils. While the phosphorus is taken up by biota, the REEs would be likely attracted by organic matter, iron oxide minerals and clay minerals, and retain in soils. The enrichment of these elements in soils is achieved to a large extent by contrast of the depletion of elements in loess, especially at the top part of profile.

Magnetic susceptibility

It has long been noticed that magnetic susceptibility is enhanced in contemporary soils as well as in palaeosols [25]. This phenomenon is well known as magnetic enhancement [28]. Throughout the Loess Plateau, a higher iron-oxide content and a much stronger magnetic susceptibility signal in farm soils and contemporary soils have been confirmed [1, 25]. It is generally agreed that a higher detrital fraction of total magnetic minerals is present in soils, and that the pedogenetic and biogenetic fraction plays an important part in the enhancement as well [7, 13]. In addition, the traditional Chinese farming on the plateau also contributes to the magnetic enhancement in farm soils by burning the remains of crops after harvest, which can convert non-magnetic iron-bearing minerals to ferrimagnetic minerals, which is known as the burning effect [28].

Both the P group elements and magnetic susceptibility have presented a closer relationship with soils rather than loess and demonstrated certain enrichment or enhancement in soils. As a matter of fact, as a group these elements and minerals also maintain close connections among themselves. It is well known that iron-oxides, the main bearers of magnetic susceptibility, can absorb P very effectively; and that Ti often associates with Fe in their solid solution, ferrimagnetic titanomagnetites. In soils, iron oxide minerals, such as iron oxyhydroxide, are acting as hosts for many cations released from phosphate weathering, including REEs, because hydroxide radicals (-OH) are often negatively charged [21]. Organic matter contains organic P and also attracts many other elements, such as REEs. After all, it is biogeochemical processes in soils that make the soils so different from loess geochemically and magnetically, and that is how the P group elements and magnetic susceptibility are made indicators of soils.

TOPSOIL EROSION ON THE LOESS PLATEAU

As shown in Fig. 5, the Loess Plateau geographically consists of two types of land, the vegetation-covered soils and the barren loess. The typical landform of the plateau, erosion gully including various rills and gullies, is carved by water erosion, although loess landforms of gravitational erosion and wind erosion may also be present. Being the main form of water erosion, the gully erosion overwhelms on the plateau, while other water erosions, such as sheetwash, are generally dwarfed [20, 26].

Topsoil erosion

Because most soils on the plateau are found in somewhat flat area, often on the top of *yuan*, *liang* and *mao* [20] where the slope gradient is relatively small, the topsoil erosion can be provoked mainly by sheetwash, which is much less erosive than the gully erosion in a slope system [30]. On the other hand, the loess, without protection from vegetation but with greater porous space and looser intergrain contact, is much more vulnerable to water erosion, mainly in the form of gully erosion and pipe erosion [26, 30].

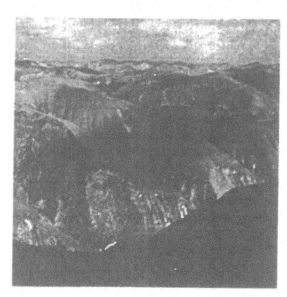

Figure 5. The highly eroded terrains of the Loess Plateau [20]. It shows grassland and farmland at foreground and almost barren loess at distance. Gully erosion is seen to be most important in creating the loess landforms, though severe sheetwash erosion may also occur during heavy rain storms.

Topsoil erosion is less likely to occur than loess erosion during modest rainfall owing to the protection from the vegetation and the small slope, however, when the rain intensity exceeds a threshold and overland flow is generated as excess rainfall over infiltration and as a result of soil saturation, it can also take place on a large scale. The water flow carries elements in suspension as well as in solution. The later is viewed as a more important erosion procedure to the elements either soluble or, more commonly, in occluded form, such as phosphorous. This solution and suspended matter with higher content of soil-related elements and minerals will definitely produce a strong signal in the total sediment eroded from the plateau, and will be eventually recorded in the Huanghe estuarine sediment in the Bohai Sea.

Huanghe hydraulic records and topsoil erosion indicators

Although the relationship between topsoil erosion and sediment records has been decided in principle, it needs to be confirmed that other processes do not interfere the relationship too much and the topsoil records are well preserved. In order to test the validity of the relation, a comparison between the magnetic susceptibility, a typical indicator of soils, and the hydraulic records of the Huanghe River [19, 29] is made (Fig. 6). The magnetic susceptibility is selected as a representative of the soil indicators simply because its profile has a higher resolution to present the record in more detail. As the overall accumulation rate at the coresite is over 1 cm/yr [5], the magnetic susceptibility should be able to show every recent topsoil erosion event. The time-scale for the magnetic susceptibility profile in Fig. 6 is modified slightly between the fixed dates at the lithological boundaries by bringing the susceptibility peaks into phase with flooding years with consideration of the actual situations described by Pang and Si [16].

A pretty good agreement is seen between the magnetic susceptibility and the hydraulic records, despite that the relative magnitudes of water discharge, sediment load and magnetic susceptibility are not perfectly matched. This merely reflects the complexity involved in the relationship among the three as regarding to the actual precipitation procedure, e.g. short but heavy rainstorm would be possible to bring about relatively more topsoil erosion than a long but modest rainfall. This agreement confirms that magnetic susceptibility is an indicator of topsoil erosion induced by flooding on the plateau and the estuarine sediment can preserve records of soil erosion on the Loess Plateau. Obviously, just as magnetic susceptibility, the P group elements are also indicators of topsoil erosion on the plateau, though some of them may be subject to other influences and not be so typical.

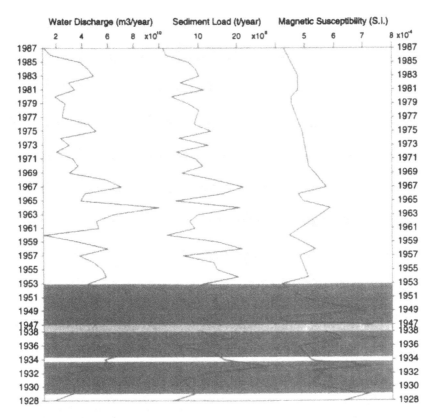

Figure 6. Comparison between hydraulic records of the Huanghe River over the past 60 years [19, 29] and magnetic susceptibility in the top part of JX91-3B. N.B. there is a gap between 1938-1947 AD when the levee of the river was deliberately breached at Huayuankou during W.W.II and no Huanghe sediment was supplied to the Bohai Sea during that period.

DISCUSSIONS AND CONCLUSION

Firstly, the PCA grouping of elements is mainly based on the most significant three processes, however, other processes, though not revealed, may also play a role in the distribution of elements in Fig. 3 and may result in different groups especially to some sensitive elements. Moreover, some elements may be prone to be influenced by more than one processes simultaneously, or have several different origins and in different forms. That is why some elements are found scattered and away from the three axes in Fig. 3. The P group, for instance, includes elements (REEs mainly) that are not only determined by sediment supply, i.e. topsoil erosion on the plateau, but also controlled by other processes, such as coprecipitation of REEs with P at estuary [3]. Therefore, only magnetic susceptibility, phosphorous and titanium are regarded here as trusty indicators of topsoil erosion in the Huanghe sediment in the Bohai Sea.

Secondly, although sediment supply seems to be independent from sedimentation processes, all the sediment is deposited through these processes anyway. A good example is the magnetic susceptibility. The original grain size distribution of magnetic minerals is different in soils and loess as a result of pedogenesis and bacterium activities [7, 32], so the susceptibility spike in sediment should demonstrate a higher fraction of ultrafine-grained magnetic minerals as expected, but did not. All the sediment has actually been reworked by the sedimentation processes before deposited at the coresite. It is only because of an absolute increase of magnetic mineral input in the Huanghe sediment that the topsoil erosion signals have survived the reworking of sedimentation processes.

At last, it should be pointed out that the relationships among the elements and minerals as discussed previously should not be over-extended. Even at the central part of the Bohai Sea, they may not be valid any more, since the depositional environment and biogeochemical conditions have changed from the Huanghe estuary. Moreover, in the whole Huanghe River catchment, erosion of other areas covered by vegetation, but not exactly on the Loess Plateau, may also contribute to some extent to the topsoil erosion signals in the estuarine sediment records in the Bohai Sea.

Geochemistry and palaeomagnetism have been jointly applied to study Huanghe estuarine sediment in the Bohai Sea, providing a new insight into the relationship between the Huanghe estuarine sediment records and the soil erosion on the Loess Plateau. With the help of principal component analysis of the XRF major and trace elements, together with detailed magnetic susceptibility measurements, two kinds of main processes controlling the sediment have been revealed. The dominant one is sedimentation processes in the Bohai Sea and the other is soil erosion processes on the Loess Plateau. Relationships between soils on the plateau and magnetic susceptibility and the P group elements are established, making them indicators of topsoil erosion. Soil erosion on the plateau is discussed and the variation of sediment supply owing to different erosions is linked with precipitation on the plateau. A comparison made between magnetic susceptibility and Huanghe hydraulic records over the past 60 years confirms the relationship between the topsoil erosion and estuarine sediment records as indicated by magnetic susceptibility, phosphorous, titanium and REEs.

Acknowledgements

We thank the British Council in Beijing, CAS (IOCAS) and NERC for their financial and material support for the sampling and measurements of JX91-3B. HC would thank Prof. K. M. Creer, Dr L. Zhou and Dr B. Maher for their helpful discussions and comments on the palaeomagnetic work.

REFERENCES

1. Zhisheng An, S. Porter, G. Kukla and Jule Xiao. Evidence of magnetic susceptibility to the change of monsoon on the Loess Plateau over the last 130, 000 years. *Chin. Sci. Bull.* **7**, 529-532 (1990).
2. A. Bronger and T. Heinkele. Mineralogy and clay mineralogical aspects of loess research, *Quat. Int.* **7:8**, 37-51 (1990).
3. R.H. Bryne, Xuewu, Liu and J. Schijf. The influence of phosphate coprecipitation on rare earth distributions in natural waters, *Geochim. Cosmochim. Acta* **60:17**, 3341-3346 (1996).
4. J.A. Catt and A.H. Weir. Soils. In: *The envolving Earth.* L.R.M. Cocks (Ed). pp. 63-85. Cambridge University Press, New York (1981).
5. Hao Chen and G.B. Shimmield. High resolution chronostratigraphy of the Huanghe estuarine sediment in the Bohai Sea, China (in preparation) (1997).
6. P.A. Cox. *The Elements on Earth, Inorganic chemistry in the environment.* Oxford University Press (1995).
7. F. Heller, X.M. Liu, T.S. Liu and T.C. Xu. Magnetic susceptibility of loess in China, *Earth Planet. Sci. Lett.* **103**, 301-310 (1991)
8. F. Heller and M.E. Evans. Loess magnetism, *Review of Geophysics* **33:2**, 211-240 (1995).
9. IOAS (Institute of Oceanology, Academia Sinica). *Geology of the Bohai Sea.* Science Press, Beijing (1985).
10. T.S. Liu *et al.* (Eds). *Loess and the environment.* China Ocean Press, Beijing (1985).
11. T.S. Liu, Z.S. An, B.Y. Yuan and J.M. Han. The loess-palaeosol sequence in China and climatic history, *Episodes* **8**, 21-28 (1985).
12. F.J.H. Mackereth. A portable core sampler for lake deposits, *Limnol. Oceanogr.* **3**, 181-191 (1958).
13. B.A. Maher and R. Thompson. Mineral magnetic record of the Chinese loess and palaeosols, *Geology* **19**, 3-6 (1991).
14. B.A. Maher and R. Thompson. Palaeoclimatic significance of the mineral magnetic record of the Chinese loess and palaeosols, *Quat. Res.* **37**, 155-170 (1992)
15. J.D. Milliman and R.H. Meade. World-wide delivery of river sediment to the ocean, *J. Geol.* **91**, 1-21 (1983).
16. Jiazhen Pang and Shuheng Si. The estuary changes of the Huanghe River I: changes in modern time, *Oceanologia et Limnologia Sinica* **10:2**, 136-141 (1979).
17. I.C. Prentice. Multivariate methods for data analysis. In: *Handbook of Holocene palaeoecology and palaeohydrology.* B.E. Berglund (Ed). pp. 775-797. John Wiley & Sons, Great Britain (1991).
18. K. Pye. *Aeolian dust and dust deposits.* Academic, San Diego, Calif. (1987).
19. Yunshan Qin and Fan Li. Study of influence of sediment loads discharged from Huanghe River on sedimentation in Bohai Sea and Huanghai Sea, *Studia Marina Sinica* **27**, 14-27 (1986).
20. Mei'e Ren, Renzhang Yang and Haosheng Bao (Eds). *An outline of China's physical geography.* Foreign Languages Press, Beijing (1985).
21. W.H. Schlesinger. *Biogeochemistry, An analysis of global change.* Academic Press, London (1991).
22. G.B. Shimmield and S.R. Mowbray. The inorganic geochemical record of the Northwest Arabian Sea: A history of productivity variation over the last 400 k.y. from Sites 722 and 724. In: *Proceedings of the Ocean Drilling Program, Scientific Results, 117.* W.L. Prell and N. Niitsuma *et al.* (Eds). pp. 409-429. College Station, TX (1991).
23. A. Stephenson. Single domain distributions, I, A method for the determination of single domain grain distributions, *Phys. Earth Planet Inter.* **4**, 353-360 (1971).
24. A. Stephenson. Single domain distributions, II, The distribution of single domain iron grains in Apollo 11 lunar dust, *Phys. Earth Planet Inter.* **4**, 361-369 (1971).
25. Jianzhong Sun and Mingjian Wei. Iron oxide in loess and its climatic significance. In: *Quaternary of the Loess Plateau.* Jianzhong Sun and Jingbo Zhao (Eds). pp. 113-123. Science Press, Beijing (1991).
26. Jianzhong Sun and Jingbo Zhao (Eds). *Quaternary of the Loess Plateau.* Science Press, Beijing (1991).
27. S.R. Taylor, S.M. McLennan and M.T. McCulloch. Geochemistry of loess, continental crust composition and crustal model ages, *Geochim. Cosmochim. Acta* **47**, 1897-1905 (1983).
28. R. Thompson and F. Oldfield. *Environmental magnetism.* Allen & Unwin, London (1986).
29. Kaichen Wang. *Relationship between the Huanghe estuary and the lower reach channels and their management.* Institute of Hydrology, Huanghe Committee (1980).
30. I.D. White, D.N. Mottershead and S.J. Harrison. *Environmental systems, second edition.* Chapman & Hall, Oxford (1993).
31. Y. Zhao and M. Yan. *Geochemistry of sediments of the China shelf sea.* Science Press, Beijing (1994).
32. L.P. Zhou, F. Oldfield, A.G. Wintle, S.G. Robinson and J.T. Wang. Partly pedogenic origin of magnetic variations in Chinese loess, *Nature* **346**, 737-739 (1990).

*Proc.30^(th) Int'l.Geol.Congr.,vol.24,*pp.74-85
Yuan Daoxian (Ed)
© VSP 1997

Effects of Runoff Characteristics on the Difference in Drainage Density in Soya Hills, Northern Japan

YUKIYA TANAKA

Department of Geography, Faculty of Education, Fukui University, Fukui 910, JAPAN

and

YASUSHI AGATA

Department of Geography, Graduate School of Science, University of Tokyo, Tokyo 113, JAPAN

Abstract

Soya Hills, which are composed of Neogene or lower Pleistocene sedimentary rocks, show the clear differential erosional topography; i.e., Miocene hard shale and mudstone hills have low drainage density, whereas those of Pliocene unconsolidated siltstone and Pleistocene unconsolidated sandstone have high drainage density. These differences in drainage density were examined from the hydrogeomorphological point of view. The analysis of geomorphological and hydrological measurement was performed in four drainage basins underlain by hardshale, mudstone, unconsolidated siltstone and unconsolidated sandstone, respectively. The results of hydrological analysis show the clear differences in runoff characteristics between muddy rocks and unconsolidated sandy rocks; i.e., in hardshale and mudstone drainage basins, interflow component was dominant, in contrast, in unconsolidated siltstone and unconsolidated sandstone ones, quickflow component was dominant. The differences in runoff characteristics caused the differences in erosional process. The optimal environment for erosional process of muddy rock basins is thought to be different from the case of unconsolidated sandy rocks. Consequently, the variety of drainage density in hills would be generated by the combinations of differences in runoff characteristics and environmental changes.

Keywords: Soya Hills, Drainage Density, Runoff Characteristics, Hydrological Analysis.

INTRODUCTION

The relationships between topography and rock types have been examined qualitatively (e.g., Nakagawa,1960; Howard,1967). In the past two decades, quantitative researches on the differential erosional topography have been done with measurement of rock properties in the field or laboratory (e.g., Suzuki *et al.*, 1985; Tanaka, 1990).

On the other hand, the differences in hydrological characteristics associated with rock types were examined recently (e.g., Freeze, 1972; Sudarmadji *et al.*, 1990; Hirose *et al.*, 1994; Onda, 1992 and 1994; Tanaka *et al.*, 1996).

However, the relationships between rock types and erosional processes has not been well resolved from hillslope-hydrological point of view. Therefore, the analysis of geomorphology and hydrology was performed in Soya hills, northern Hokkaido, Japan, where clear differential erosional topography is found.

STUDY AREA

Geology

Soya hills are located at the northern part of Hokkaido, northern Japan(Fig.1).

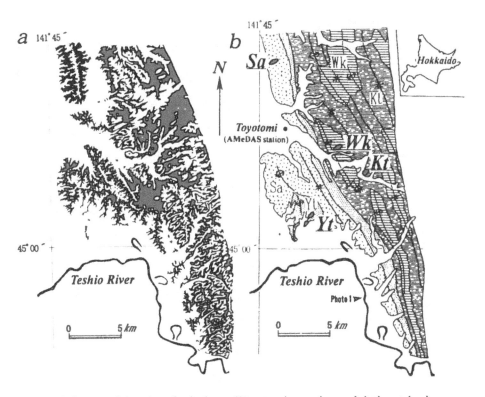

Figure 1. Zebra map(*a*) and geological map(*b*) around experimental drainage basins.
a: This map was drawn by painting every alternate belts of 50m in altitude jet-black on 1:25,000 topographic maps issued by Geographical Survey Institute of Japan(after Suzuki *et al.*, 1985).
b: This map was compiled from the geological maps(after Nagao, 1960; Hata and Ueda, 1969).
Sa: Sarabetsu formation (Lower Pleistocene unconsolidated medium and coarse grained sandstone), Yt: Yuchi formation (Upper Pliocene unconsolidated fine grained sandstone and siltstone), Kt: Koetoi formation (Lower Pliocene mudstone) and Wk: Wakkanai formation (Upper Miocene hard shale). *Sa*, *Yt*, *Kt* and *Wk* mean Sa, Yt, Kt and Wk experimental drainage basins, respectively.

The study areas are underlain by Neogene or Lower Pleistocene sedimentary rocks, which are divided into four rock types; i.e., Miocene hard shale(Wk) with many   joints of tectonic and weathering origins, Miocene-Pliocene mudstone(Kt) with less joints than **Wk**, Pliocene unconsolidated fine-grained sandstone(Yt) and Pleistocene unconsolidated medium-grained sandstone(**Sa**). These rocks are folded with dips of about 20° into north to northwest trending fold axes.

As shown in Fig.2, the rocks of this area have different values of cone-penetration hardness (P) and permeability(K):i.e., P values (kgf/cm^2) of **Wk, Kt, Yt** and **Sa** are 1,100, 300, 56 and 31, respectively, while K values (cm/s) of **Wk, Kt, Yt** and **Sa** are 2×10^{-3}, 1×10^{-7}, 1×10^{-5} and 1×10^{-3}, respectively(Suzuki *et al.*, 1985).

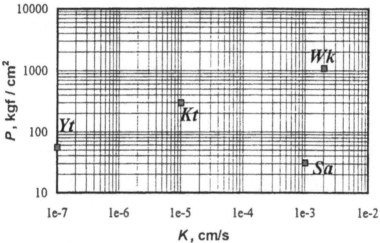

Figure 2. Physical properties of fresh rocks of study area (after Suzuki *et al.*, 1985). P: Cone-penetration hardness, K: Permeability.

Hill Morphology

Differential erosional topography, with less than 250 meter height above sea level, is clearly found in Soya hills(Figs.1, 3). Topographical characteristics, such as drainage basin relief, stream length and drainage density are controlled by rock types; i.e., **Wk**-hills have maximum basin relief and very few long and deep valleys, which results in very coarse topographical texture, compared with other rock hills. **Kt**-hills are similar with **Wk**-ones, but have lower basin relief than **Wk**-ones. **Sa**- and **Yt**-hills have very rugged topographical texture with low basin relief and a lot of short well branching streams. Drainage density of second or third order stream of **Wk**- and **Kt**- hills shows the small value(about 9km/km^2), whereas that of **Sa** and **Yt** shows the large one(about 15km/km^2)(Suzuki *et al.*, 1985).

Figure 3. Rock controlled denudational hill morphology of study area.
The location taken this picture is shown in Fig.1.

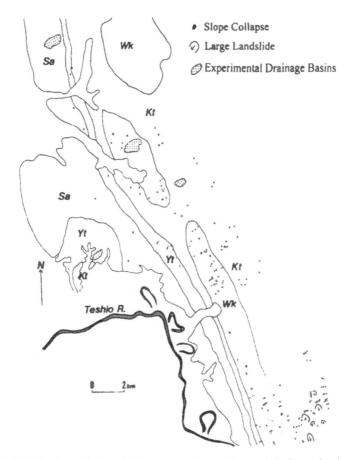

Figure 4. Distribution of slope failures around experimental drainage basins.

Slope failure frequency of **Wk-** and **Kt-** hills is obviously larger than that of **Sa-** and **Yt-** hills(Fig.4).

Climate and Vegetation

Climatic data obtained from AMeDAS(Automated Meteorological Data Acquisition System) station (located at Toyotomi, see Fig.1) of Japanese Meteorological Agency, show that mean annual temperature of this area is 5.6℃ with maximum temperature of 30℃ and minimum temperature of -23.9℃. Mean annual precipitation is 1,100 mm. Hillyland of this area is almost covered with forest vegetation, which contains both coniferous and birch forests partly with Sasa(bamboo) grass.

Setting of Experimental Drainage Basins

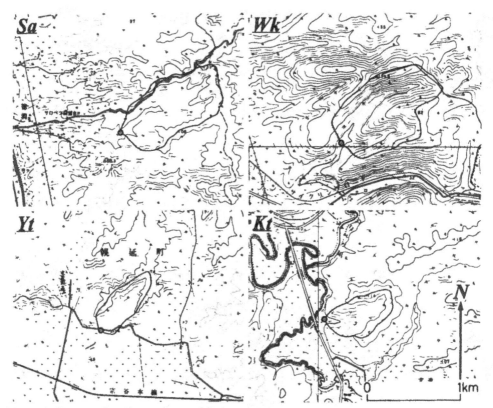

Figure 5. Topography of experimental drainage basins.
Locations are shown in Fig.1. Base maps are 1:25,000 topographic maps with contour intervals of 10 m. The areas surrounded by solid lines show the experimental drainage basins. Solid black circles mean the locations of hydrological observation sites.

Four experimental drainage basins were chosen so that every basin represents the hydrological characteristics of each four rock types(Fig.5). The drainage basin areas of **Wk-, Kt-, Yt-** and **Sa**-basin are 0.690km², 0.242km², 0.167km² and 0.553km², respectively. These experimental drainage basins are located near each other within the distance of less than 10 km. Therefore, the climatic conditions and vegetation of these drainage basins are almost same.

METHODS

Hydrological Observation

Water discharge was measured continuously using a water level gauges and Parshall Flumes. The observation sites are located at the mouth of each experimental drainage basins(Fig.5). Experimental periods were from Jun.12, 1994 to Nov.9, 1994 and from May 2, 1995 to Oct. 27, 1995. No snow covers was found during these experimental periods.

Data Source of Precipitation

Hourly precipitation data of this area were obtained from AMeDAS station(Toyotomi, see Fig.1) for above experimental periods.

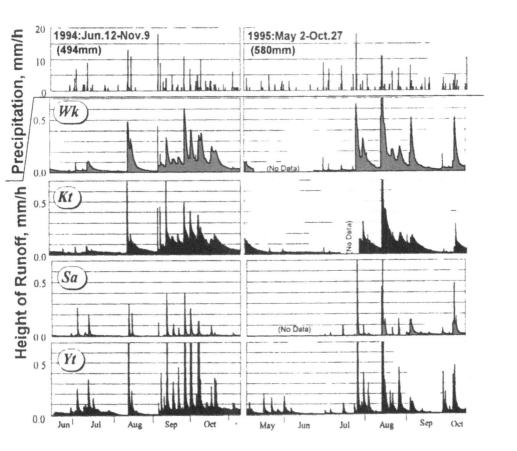

Figure 6. Hyetograph and hydrograph during observation period of 1994 and 1995.

RUNOFF CHARACTERISTICS

Recession Rate of Discharge

Hyetograph and hydrograph during experimental periods are shown in Fig. 6. This hydrograph shows that flood runoff occurs associating with every rainfall events for each drainage basins, and recession rate of discharge of **Wk** and **Kt** is smaller than that of **Sa** and **Yt**.

Difference in Runoff Component Characteristics

Figure 7 shows the results of filter separation and AR method(Hino and Hasebe, 1981, 1984), which divides the daily height of runoff into three components, for four experimental drainage basins. These three components are supposed to correspond to quickflow, interflow and baseflow. This analysis was performed only for the data obtained in 1994, because of many lack of data for 1995.

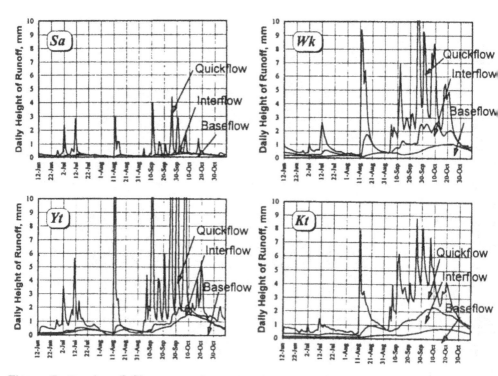

Figure 7. Results of filter separation and AR method for data obtained from four experimental drainage basins in 1994.

Differences in runoff component ratio between four rock types drainage basins are shown in Fig.8. This result clearly shows the runoff characteristics of four rock types drainage basins are divided into two groups; the first is **Wk · Kt** group, whose interflow component is relatively dominant, and the second **Sa · Yt** group, whose qucikflow component is relatively dominant .

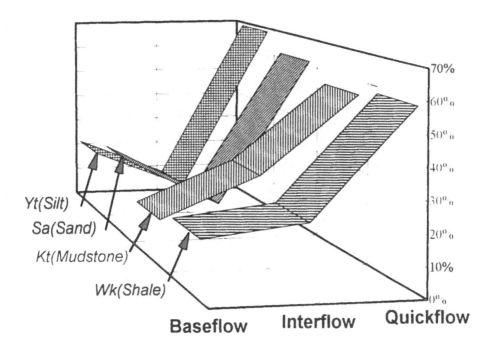

Figure 8. Difference in runoff component characteristics between four rock types drainage basins.

DISCUSSION

The previous studies indicate that valley is formed by the action of water, i.e., surface flow (e.g., Horton,1945; Schumm,1956; Tanaka, 1957; Dunne and Dietrich, 1980), seepage erosion (e.g., Dunne,1980; Laity and Malin,1985) and landslide (e.g., Iida and Okunishi, 1983; Dietrich *et al.*, 1986).

Onda (1994) discussed about the relation between runoff characteristic and slope processes in forested drainage basins in the central Japan, i.e., in the case of the granite subsurface flow generates landslide, resulting many slope failures, whereas in the case of Paleozoic sedimentary rock surface flow is dominant and slope failure is rare.

Based on these previous works, differences in erosional processes associated with runoff characteristics can be described as follows:

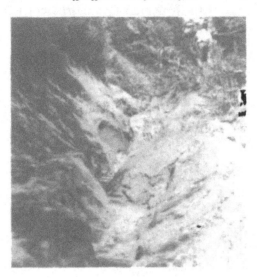

Figure 9. Gully developed on the trail of Sa experimental drainage basin.

In **Sa**- and **Yt**-basins, surface flow would cause severe gully erosion at anywhere, when optimal conditions for gully erosion, such as destruction of vegetation, would be provided (Fig.9). However, under the present vegetation covered condition, slope failure is little in **Sa**- and **Yt**-basins (Fig.4). This means erosional

Figure 10. Slope collapse on the valley side slope of **Wk** experimental drainage basin

actions are inactive in **Sa**- and **Yt**-basins at present. Therefore, the valleys in **Sa**- and **Yt**-basins are thought to be formed not at present but at the past time when vegetation cover was poor in the whole of Soya hills.

On the contrary to the case of **Sa**- and **Yt**-basins, even under no vegetation condition, very little gully erosion is found in **Wk**- and **Kt**-basins. However, in **Wk**- and **Kt**-basins, subsurface flow would cause saturation of regolith or ground water erosion, which generate many slope collapse under the present vegetation covered conditions(Figs. 4, 10). This indicates that in **Wk**- and **Kt**-basins, the erosional actions are active at the present. It can be deduced that in **Wk**- and **Kt**-basins, optimal environments for the generation of erosion are completely different from the case of **Sa**- and **Yt**-basins, associating with the difference in runoff characteristics.

Consequently, the environmental changes are thought to induce the differences in the phase of erosional stage between **Wk**- and **Kt**-basins and in **Sa**- and **Yt**-basins. In other words, the differences in relaxation time of geomorphic responses to climatic changes (Oguchi,1996) between **Wk**- and **Kt**-basins and   **Sa**- and **Yt**-basins would be strongly influenced by the differences in runoff characteristics. This would provide the difference in drainage density.

CONCLUSIONS

The results of hydrological analysis revealed that **Wk**- and **Kt**-basins are characterized by dominant subsurface flow which generates slope collapse, contrary to this, surface flow, which generates gully erosion, is dominant   in **Sa**- and **Yt**-basins. The optimal environment for erosional process of **Wk**- and **Kt**-basins is different from the case of **Sa**- and **Yt**-basins. Therefore, the drainage density difference between **Wk · Kt**-basins and **Sa · Yt**-basins is thought to be generated by the combination of the differences in runoff characteristics and environmental changes.

Acknowledgments

The authors would like to give a special thanks to Prof. Kaichiro Sasa and Prof. Fuyuki Satoh of the Hokkaido University Forests, Faculty of Agriculture, Hokkaido University for their help throughout this study. Thanks also are due to Dr. Yoshimasa Kurashige of Graduate School of Environmental Earth Science, Hokkaido University for his fruitful discussion to setting for experimental drainage basins in the field. The authors thank Dr. Yuichi Onda of School of

Agricultural Sciences, Nagoya University for his helpful advice to hydrological observation.

REFERENCES

1. Dietrich, W.E., C.J. Wilson and S.L. Reneau (1986): Hollows, colluvium and landslides in soil-mantled landscapes: *In* Abrahams, A.D. ed., *Hillslope Processes,* Allen and Unwin, London, 31-53.
2. Dunne, T. (1980): Formation and controls of channel networks. *Prg. Phys. Geogr.,* **4**, 211-239.
3. Dunne, T. and W.E. Dietrich (1980): Experimental study of Horton overlandflow on tropical hillslopes: 1.Soil conditions, infiltration and frequency of runoff. *Z. Geomorph., Suppl. Bd.,* **35**, 40-80.
4. Freeze, R.A. (1972): Role of subsurface flow in generating surface runoff 2. Upstream source areas. *Water Resources Research,* **8**, 1272-1283.
5. Hata, M. and Y. Ueda (1969): *Geological map of "Teshio"* (1:200,000). Geological Survey of Japan.
6. Hino, M. and M. Hasebe (1981): Analysis of hydrologic characteristics from runoff data - a hydrologic inverse problem. *Journal of Hydrology,* **49**, 287-313.
7. Hino, M. and M. Hasebe (1984): Identification and prediction of nonlinear hydrologic systems by the filter-separation autoregressive (AR) method: extension to hourly hydrologic data. In: G.E. Stout and G.H. Davis (eds.), Global Water: Science and Engineering- The Van Te Chow Memorial Volume. *Journal of Hydrology,* **68**, 181-210.
8. Hirose, T., Y. Onda and Y. Matsukura (1994): Runoff and solute characteristics in four small catchments with different bedrocks in the Abukuma Mountains, Japan. *Trans. Japan. Geomorph. Union,* **15A**, 31-48.
9. Horton, R.E. (1945): Erosional development of streams and their drainage basins: hydrophysical approach to quantitative geomorphology. *Geological Society of America Bulletin,* **56**, 275-370.
10. Howard, A.D. (1967): Drainage analysis in geologic interpretation: a summation. *Bull. Amer. Assoc. Petroleum Geologist,* **51**, 2246-2259.
11. Iida, T. and K. Okunishi (1983): Development of hillslopes due to landslides. *Z. Geomorph. N.F. Suppl.-Bd.,* **46**, 67-77.
12. Laity, J.E. and M.C. Malin (1985): Sapping processes and the development of theater-headed valley networks on the Colorado Plateau. *Geological Society of America Bulletin,* **96**, 203-217.
13. Nagao, S. (1960): *Geological map and its explanatory text of "Toyotomi"*(1:50,000). Geological Survey of Hokkaido.

14. Nakagawa, H. (1960): On the cuesta topography of the Boso Peninsula, Chiba Prefecture, Japan. *Sci. Rep. Tohoku Univ., 2nd Ser.(Geology) Special Issue*, **4**, 385-391.

15. Oguchi, T. (1996): Relaxation time of geomorphic responses to Pleistcene-Holocene climatic change. *Trans. Japan. Geomorph. Union*, **17**, 309-322.

16. Onda, Y. (1992): Influence of water storage capacity in the regolith zone on hydrological characteristics, slope process, and slope form. *Z. Geomorph. N.F.*, **36**, 165-178.

17. Onda, Y. (1994): Contrasting hydrological characteristics, slope processes and topography underlain by Paleozoic sedimentary rocks and granite. *Trans. Japan. Geomorph. Union*, **15A**, 49-65.

18. Schumm, S.A. (1956): The evolution of drainage systems and slopes in badlands at Perth Amboy, New Jersey. *Geological Society of America Bulletin*, **67**, 597-646.

19. Sudarmadji, T., Araya, T. and Higashi, S. (1990): A hydrological study on streamflow characteristics of small forested basins in different geological conditions. *Research Bulletin Hokkaido University Forests*, **47**, 321-351.

20. Suzuki, T., E. Tokunaga, H. Noda and H. Arakawa (1985): Effects of rock strength and permeability on hill morphology. *Trans. Japan. Geomorph. Union*, **6**, 101-130.

21. Tanaka, S. (1957): The drainage density and rocks (granite and Paleozoic) in Setouchi sea coast region, western Japan. *Geogr. Rev. Japan*, **30**, 564-578.(*in Japanese with English abstract*)

22. Tanaka, Y. (1990): The relationships between the evolution of valley side slope and rock properties in Soya hill and Shiranuka hill. *Geogr. Rev. Japan*, **63** (Ser.A), 836-847.(*in Japanese with English abstract*)

23. Tanaka, Y., Y. Agata and K. Sasa (1996): Relationships between rock properties of basement rocks and runoff characteristics in Soya hill, Hokkaido, Northern Japan. *Res. Bull. Hokkaido Univ. Forests*, **53**, 269-287.(*in Japanese with English abstract*)

Proc.30*th* Int'l. Geol. Congr., vol. 24, pp. 86-105
Yuan Daoxian (Ed)
© VSP 1997

Karst Cave Development from the Bedding-plane Point of View (Škocjanske jame Caves, Slovenia)

MARTIN KNEZ

Institute for Karst Research ZRC SAZU, SI-6230 Postojna, SLOVENIA

Abstract

Škocjanske jame Caves is one of the most important karst objects on karst, on Classical Karst or Kras and in the rocks that are the most typical and widespread on Kras.

Since the beginning of speleological science the researchers focused the concern on the relation between geological properties (rock, structure) and passage development. The researchers of the Slovene karst underground for a long time register tectonic elements (faults and fissures) on cave surveys; however, lithopetrology and stratigraphy within the studies of cave passage development were underused. Single parameters were partially anticipated, mostly they are guessed. These questions achieved another importance when some researcers stressed the importance of bedding-planes in speleogenesis.

Cave passages, their fragments and other traces of the underground karstification in one of the colapse dolines (Velika Dolina, Škocjanske jame Caves) does not occur at random in the walls but they are controlled by small number of bedding planes.

Thus research concentrated on two basic questions:
- is the concentration of initial channels within limited number of bedding-planes virtual or just apparent;
- is this "enrichment" - if it exists - maybe associated with the properties of rock/bedding-plane.

With other words, I tried to answer the question: "is selective karstification controlled by a rock?"

Keywords: karstology, geology, speleogenesis, bedding-plane, interbedded wrench-fault, phreatic channel, Kras, Škocjanske jame Caves, Slovenia

INTRODUCTION

The main question to which I tried to find an answer arises from a completely lay personal cognition which was born in the collapse doline Velika dolina, part of the Škocjanske jame caves system (Fig. 1). An attentive visitor can easily see that the cave passages, their fragments and other traces of underground karstification do not occur arbitrarily scattered on the walls of Velika dolina - they are collected along a small number of bedding planes. As similar occurence of features may be observed also in some caves of the Notranjska region, it was obvious to put further questions: Is such similarity a widely accepted principle? and Where are the roots of this similarity?

My work was focussed on two basic questions:

Figure 1. Location of Škocjanske jame Caves.

- Is concentration of initial channels within an area with a small
  number of bedding planes real and not only apparent?
- Is such "enrichment" - if it really exsists - probably a
  consequence of the properties of rock/bedding planes and not a
  consequence of tectonics?

My research was initiated on the basis of a probable personal supposition that the answer to the first question would be positive. Detailed mapping confirmed and sharpened my supposition. The second question is much more provoking than was first intended.

The action of water upon a rock is a clearly irreversible process oriented to destruction; the process cannot be repeated within the same area and the same types of rock. The aim of my research was to evaluate the role of bedding planes and changes in rock in their immediate vicinity from the view of sedimentology, and indirectly from the view of inception.

The bedding planes, along which the channels were formed, essentially affected the development of the entrance part to the Škocjanske jame caves, i.e. the Velika dolina collapse doline (Fig. 2). I call them *formative bedding planes*. The remaining much more numerous bedding planes are less important, i.e. they are of no local and speleogenetic significance.

The basic working method was sampling and microscoping of those parts of the beds which border on the bedding planes. In places I treated entire beds if they seemed important for channel formation and the subsequent formation of passages. The beds bordering on the

Figure 2. View of Velika dolina collapse doline at low water level

formative bedding planes were sampled and microscoped interconnectingly, with the total thickness of 100 to 200 cm.

With the aid of sedimentological and other methods I was trying to find differences or similarities in the rock directly below and above the bedding planes; with these methods I was searching for differences between individual beds and those between the bedding planes, in which karstic objects (underground channels) were being formed.

In order to ascertain a possibility of finding relations between the rock and the formation of initial cave channels I decided to find all the underground channels in the entrance part of Škocjanske jame, i.e. in Velika dolina, and to classify them into a stratigraphical profile. I separated those which were formed along the bedding planes from those formed along other geological elements (e.g. fissures) or in their intersections. In the case of Velika dolina it turned out that a major role in the formation of initial cave channels was played by the bedding planes (Fig. 3). For that reason I studied the cave passages (particularly those which were formed along the bedding planes) and the surrounding rock in detail. In a similar way than that used by the scientists of tectonics who ascertained a connection between the final stages of passage formation and tectonic structures, I tried to find out whether there is any connection between the formation of passages and bedding planes, or connection between the formation of passages and the surrounding rock. In other words, I tried to answer the question whether selective karstification is conditioned by rock.

The collapse doline Velika Dolina has vertical, from 140 to 160 m high walls. In its northern wall opens the entrance to Tominčeva Jama, 200 m in length, that have breakdown floor at 315 m and rocky bottom at 308 m a.s.l. In the western wall of the collapse doline there is the access to Schmidlova Dvorana; rocky bottom at 300 m a.s.l. The swallow-hole of the Notranjska Reka river lies at 270 m a.s.l. In all the cited ranges of height there is in the walls a number of larger or smaller phreatic channels.

Authours dealing with the development of caves focussed their attention particularly on geomorphological research of the Kras, and recently also on geological and other researches [3, 1, 2, 11, 12, 18, 19, 22, 23, 27, 41, 42, 45, 46, 47, 48, 52, 53, 54, 55, 56, 35, 36, 37, 38, 39, 25, 49, 64, 40, 61, 58, 29, 30, 31, 32, 33, 34]. With the collected material, the above mentioned authors have presented general review of the karst also within the broader area of the Škocjanske jame caves.

REVIEW OF THE PREVIOUS LITERATURE

The main topic of this paper is completely new in the world literature - there is no direct link between the paper and the literature. Selection of literature from the broader area dealing with the subject is relatively scarce, too. From a theoretical point of view, [43, 44] definitely approached this topic.

In the field of detailed studies of bedding planes as a medium for inicial channels of inception, recently only five volumes have been written [6, 8, 57, 9]. Other researchers as [10, 7, 65, 51, 66, 44] extensively quote the above mentioned authors, perform some precedingly expected conslusions and present new basic karstological starting points, but only few try to opt for this topic.

Figure 3. Beddıng-planes ın Veliḱa dolına collapse dolıne

It has to be pointed out that bedding (being an element of discontinuities, or an area of "weakness points" and lower resistance of rock to corrosional effects of water in the early development phase) particularly in the Anglo-American literature, is regarded as one of important factors.

PROBLEMS OF BEDDING-PLANE RESEARCH

From the very beginning of the speleological science, researchers'view has been orientated to relations between geological control (rock, structure) and the formation of passages. Researchers of the Slovenian karst underground have been putting tectonic elements (faults and fissures) on cave maps for quite a long time now, whereas little has been done from the view of lithopetrology and stratigraphy during the study of passage formation in the karst underground. Individual parameters have only partly been indicated, the majority of them have mostly been guessed. It would be necessary to ascertain actual influence of rock on the development of underground caverns in the phreatic zone.

Today the knowledge of relations between the formation of cave passages and surface karst features, and relations between the formation of cave passages and tectonic phenomena, is considerably better [e.g. 4, 62, 13, 14, 15].

Various authors [5, 16, 17] particularly emphasize the significance of fault zones of the Dinaric and transversally-Dinaric direction (NE-SW and NW-SE), along which numerous water channels were formed, such as those in the caves Planinska jama and Postojnska jama. In this case lithopetrological properties of rock are pushed into the background.

The above put authors often mention water channels which are oriented along the bedding or are directed along the dip. Researchers have mostly interpreted such similarity only from the view of tectonics [24].

In the Dinaric range, limestones are generally far purer than those elsewhere in the world [65]. They contain 1-2%, mostly even below 1% of insoluble residue [12, 28, 32]. Some foreign authors claim that completely pure limestones are most perfectly karstified [65], but on the other hand some others claim that for karstification most suitable are the limestones with 70% $CaCO_3$ [10]. Limestones of the Dinaric karst have very different percentages of $CaCO_3$ (also) due to their age, which is reflection of the conditions during sedimentation on the platform. The Lower Triassic limestones contain from 80 to 95% $CaCO_3$, the Lower Cretaceous (95-98% $CaCO_3$) and the Upper Cretaceous limestones 98 to 100% $CaCO_3$ [26].

According to [20], the accessible passages of the Škocjanske jame system are located in Turonian and Senonian, prevailingly thick-bedded limestone as well as in thin-bedded limestone of the Maastrichtian and Danian. [20] is of the opinion that the mentioned lithostratigraphical difference of the limestones is reflected in the morphology of the cave passages. He [21] emphasizes that lithologic-petrological composition of mainly Cretaceous beds is interesting from the aspect of karstification. Similar characteristics of the Dinaric Cretaceous beds are described also by Croatian geologists [16].

Beside the genetic connection cave passage-fault zone, lately also the bedding (bedding

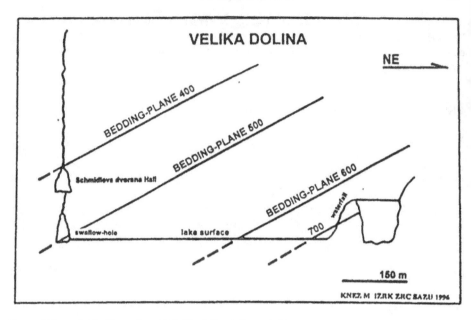

Figure 4. Formative bedding-planes in Velika dolina collapse doline.

planes) [4, 60]has been mentioned in connection with the formation of underground caverns by some Slovenian authors [cf. 30, 33]. Within tectonically less affected rocks, the authors assess a genetic connection between thin-bedded sediments with open bedding planes andthe distribution of cave chambers.

Inception may entirely take place already in the period of early diagenesis, but the formation of cave channels takes place just under appropriate hydrological conditions which do not necessarily follow the inception phase. It seems that in some places the main factors of inception are the contact of fresh and saline groundwater, or the presence of sulphuric compounds [44, 67]. Significance of the influence of fresh- and saline-water mixing for the inception is particularly emphasized in the before mentioned work.

THE STUDIED BEDDING-PLANES

There are three bedding-planes along which the passages developed (Fig. 4). I call them formative bedding-planes. All the others are more numerous (59) and do not bear speleogenetical importance.

The bedding-plane along which the actual swallow-hole of the Reka river developed is labelled by number 500 (Fig. 5). The bedding-plane which caused the former swallow-hole of the river (through Schmidlova dvorana) is labelled with number 400. The breakthrough from Mala to Velika Dolina was influenced by the bedding-plane labelled 600. In addition to these three bedding-planes that are clearly seen on the walls of the collapse doline there outstands the fourth one, labelled 700. This bedding-plane does not show a speleogenetical role in Velika Dolina. Other bedding-planes are labelled by intermediary numbers.

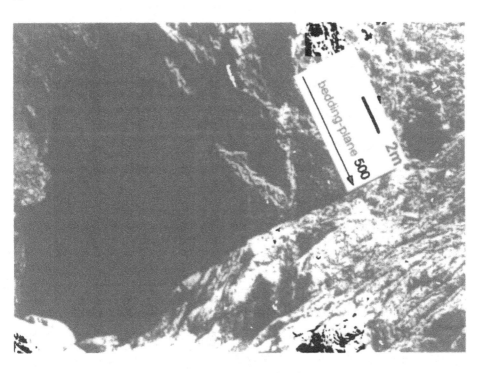

Figure 5. The bedding-plane 500 The Reka swallow-hole in Velika dolina

WORKING METHODS

During my work I tried to use most extensive spectrum of methods or approaches. Although some of them apparently do not fit into the context of research which is limited with the title, they contributed to broader conclusions.

Selection of methods: Regional review of the caves within the broader area of Škocjanske jame, Photography of the terrain, Identification of the karstic objects on the walls of the collapse doline, Identification of the bedding planes in the collapse doline, Rock sampling, Preparation and different forms of the material, Microscoping of the samples, Calcimetry (Scheibler's calcimetre, Complexometry).

The rock was sampled in two ways, which were accorded and supplemented with each other:
- sampling was carried out directly below the bedding plane;
- sampling was carried out directly above the bedding plane;
- sampling was carried out in the lower and upper parts of the beds;
- sampling was carried out in all the parts with differences in sedimentation.

IDENTIFICATION OF THE RESEARCHED KARSTIC OBJECTS, LOCATED

ON THE WALLS OF THE COLLAPSE DOLINE, THEIR CLASSIFICATION AND ASCERTAINING OF POSSIBLE PREFERENTIAL DIRECTIONS OF WATER FLOW

One of the introductory steps for the research of Velika dolina was identification and inventarization of any speleogens in the walls (flanks) of the collapse doline as well as their classification. With this the way was given for ascertaining possible preferential directions of water flow as well as the way for ascertaining the rules and that for ascertaining the system of occurence of speleogen karstic features.

The bedding plane marked 400 is 25 thickness metres above the bedding plane marked 500. Above the bedding plane marked 400 there are about 117 thickness metres of rock.

Below the bedding plane 500 there is the bedding plane marked 600 at a distance of 33 thickness metres, and 14 thickness metres below it there is the bedding plane marked 700. The total thickness of the geological profile below the bedding plane 500 is 52 m.

On the sketch, only three (four) formative bedding planes were identified from the total of 62 for the sake of clarity and simplicity.

Due to a probable genesis I defined four groups of various possible karstic features seen on the walls of the collapse doline:

BP   - passages developed along a bedding-plane
BPF - passages developed along of intersection of a bedding-plane and fissure
BPN - passages developed along a bedding-plane and in corrosion notch within a bedding-plane
N   - corrosion notches developed along a bedding-plane.

GEOLOGICAL COLUMNS OF THE SECTIONS AT A SCALE OF 1:10 AND 1:100

Eight detailed geological columns were completed at a scale of 1:10 and 1:100 and adjusted to the detail of sampling (Fig.6).

At a scale of 1:10, I elaborated geological columns in the immediate vicinity of the formative bedding planes. An average thickness of the interconnectingly taken samples is 140 cm, the largest thickness is 203 cm near the bedding plane marked 600 and the smallest (98 cm) near the bedding plane marked 500L. The total thickness of the interconnectingly sampling rock in the immediate vicinity of the formative bedding planes is 1118 cm.

At a scale of 1:100, I completed geological columns between the formative bedding planes (the bedding planes marked 400, 500, 600 and 700). The thickness of the geological profile

Figure 6. (next page) Geological column including the bedding-plane 500D. Legende: ❶ fenestras, ❷ stilolites, ❸ geopetal, ❹ bioturbation, ❺ gradation, ❿ planctonic foraminefars, ❼ fossil parts, ❽ miliolids, ❾algae, ⑩ ostracods.

AGE	THICK NESS (CM)	LITHOLOGY	BEDD.-PLANE	SAMPLE	TEXT. STRUCT.	FOSSILS	CALCITE VEINS	PORO-SITY
SENONIAN	155	M		78				
	140							
	130	M		76 77 75	❶ ❷	❾	•	I
	120					❿		
	110							
	100	M		71 72 73 74	❸			I
	90	M		68 69 70				
	80	W		67 66				
	70	W		65				I
	60	W W		64 62 63 60 61				
	50	W		59				
	40	W		58				
	30	W		57				
	20	W		55 56	❹			I
	10					..		
	0	W 500D		54 24 25				I
	10	W M		26 27 28 29 30				
	20	W		31 32 33 34				
	30	W		35 36				
	40	W M		37 38				
	50							
	60	W		39 40				
	70	W		41	❺			
	80	W W		42 43 44 45 46				I
	90							
	100							
	110							

below the bedding plane marked 700 is 4.50 m, that between the bedding planes 700 and 600 is 14.20 m, that between the bedding planes marked 600 and 500 is 32.85 m, and that between the bedding planes marked 500 and 400 is 25 m. The total thickness of the (measured up to cm in detail) completely researched geological profile is 7655 cm.

In the geological column I included the following properties of rock: thickness, lithology, marks for the bedding planes, location and serial numbers of the samples, bedding features and structures, fossils, the presence of calcite veins, and porosity. Bedding features, structures and fossils were graphically presented in the column. In some columns, occurence of the specific phenomena is presented by a line.

Analyses of the microscoped thin-sections (10.000 datas gathered from 300 thin-sections) may lead to a conclusion that the environment of limestone sedimentation was quite similar within the entire geological profile. In places, the monotony is broken by frequent planktonic Foraminifera, shell fragments, milliolides and algae. I imagine the primary location of sedimentation to be an area with a calm lagoon and a closed shelf, and only in places to be an area with an open shelf.

In spite of that it should be mentioned that the bedding plane 700, which in the geological profile is quite conspicuous, in reality only morphologically stands out in the bed profile. Within the bedding plane, phreatic channels were not formed in the area of Velika dolina.

Microscopic analyses indicated that it does not have any of the properties which may be noticed in the bedding planes marked 400, 500 and 600.

The determined anomalies do not evidence any direct connection with inception. Before a detailed study of the geological profile it appeared that some sedimentological properties, with which cave-bearing horizons would additionally be evaluated, will become obvious.

Unfortunately, concluding on the basis of the occurence of bioclasts (vertically as well as horizontally) is practically impossible within such a limited area. Palaeoecological conditions in the sedimentatary basin were all the time very uniform during sedimentation of the treated Upper Cretaceous beds. For this reason, the main emphasis of research (while locally evaluating the rocks) should be dedicated to a "non-bioclastic" area.

PERCENTAGE OF CaCO₃

In the entire geological profile, $CaCO_3$ percentages are ranked between 98.18% and 99.82%. The difference between both values is over half a percent (1.64%), so the values are nowhere essentially extreme. Higher concentration of the samples is only in the immediate vicinity of the bedding plane 400, where the contents of all the samples are between 99.60% and 99.80%. Otherwise, the distribution of the samples in all the treated sections of the geological profile is scarce.

By calculating the total carbonate content in the section of the geological profile between the bedding planes marked 600 and 500 it turned out that there is practically no difference between the carbonate content of the upper parts and that of the lower parts within the same beds.

Very clear is the difference (usually the minimum one) between the histograms illustrating the values of the total carbonate content in the immediate vicinity of the formative bedding planes and those presenting sections of the geological profile between the formative bedding planes. In close proximity of the bedding planes, values (with smaller discrepancies) are between 99.40% and 99.99%. The histograms do not illustrate contents below 99.00%.

The histograms of carbonate content between the formative bedding planes are essentially more "elongated", thus individual rectangles of the histogram contain less samples. In all four histograms there occurs at least one value below 99%. The difference between average values between the bedding planes is by 0.33% smaller that that in the immediate vicinity of the bedding planes.

I would like to mention a minimum, but an important difference between an average value of the samples above the formative bedding planes, which is by 0.13% higher than that of the samples below the formative bedding planes.

CYCLICITY OF SEDIMENTATION

Proportionally constant distances between most of the formative bedding planes and a macroscopically discernible reduction in the thickness of beds, when proceeding upwards from the formative bedding plane (400, 500 and 600), give an impression that we deal with cyclic sedimentation. For that reason I in detail examined the contacts of those pairs of the beds, between which formed the caves and those, between which caves were not formed. In the interconnected geological profile I completely inspected several beds along the formative bedding planes.

Although (when being observed with the naked eye) "cycles" start or end with formative bedding planes and although the beds in the lower parts of individual "cycles" are much thicker than those in the upper parts, microscopic analyses did not indicate any striking differences and particularly no characteristic sequences during sedimentation.

It is tempting to connect this "cyclicity" with possible occurances of thin inliers of pelites between the limestones, since they are one of the important factors for inception or the formation of cave channels. These thin layers would definitely be destroyed as first during the formation and growth of a passage, therefore they cannot be observed at the very site any more.

Great uniformity of the bedrock, which is indicated by microscopic observations, evidences that inception here, at the contact of the beds with various lithological properties (trans-bedding contrast), would be impossible.

It has to be mentioned that only the bedding plane 700 optically stands out from the apparent "cyclicity". Macroscopically it is similar to the the formative bedding planes marked 400, 500 and 600, but it is "sterile" (without passages) and obviously did not transmit water.

GEODETICAL MEASUREMENTS OF THE BEDDING PLANES, THOSE OF SOME PRESERVED CHANNELS AND PASSAGES AS WELL AS THE SKETCHES

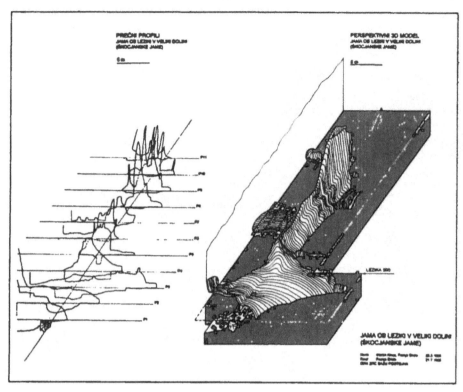

Figure 7. 3D sketch of the *Cave along the bedding-plane* in Velika dolina collapse doline.

OF POSSIBLE DIRECTIONS OF PHREATIC CHANNELS BEFORE THE BREAKDOWN

Within a studied area of Velika Dolina I tried to determine the length, the volume and the direction of channels within a rock before breakdown. The calculations of a direction of channels within the material that was already washed by the Reka through Škocjanske jame Caves towards Adriatic Sea may only be hypothetical.

The transported rock mass among the boulders was probably hollowed out before the breakdown occurred (Fig. 7). An example close to the Reka swallow-hole on its left bank may serve as an illustration. It is formed within 500 bedding-plane. It seems that laminar flow was more or less unimpeded over larger surface of a bedding-plane.

PERMEABILITY

Before karstification, rock permeability is somehow controlled by the process itself. [59] proved that at high secondary porosity, i.e. at high density of fractures, water enters the system in a dispersed way and affects the whole mass of the rock. On the other hand, the flow at low porosity but at discontinuities of larger dimensions is directed only towards some places, which in a way determine the position of karstic channels. The way of karstification is thus inevitably conditioned by the density of rock discontinuities (fissure, bedding plane, fault) which give way to percolating water.

Permeability is influenced also by mechanical relaxation which is a consequence of surface denudation. Rocks elastically expand, closely located bedding planes open wide and transmit substantially larger amounts of water. Deeper located parts of the bedding planes (observed along the dip of the beds and not according to the thickness of the geological profile) within the inclined beds remain close or open up much more slowly.

Transmissivity of bedding planes is increased:

1. mechanically,
2. chemically,
3. both at the same time.

[10] says that rock permeability exponentially decreases with depth. A reason could be a decreased degree of fissure openness and/or a decreasing number of fissures. Consequently, water at some greater depths would penetrate rock discontinuities in a much harder way, although tension strength in that area would be zero [63], and would thus be able to use fewer and fewer primary paths. Water is not capable of travelling downwards for a non-limited time, but if it is, it is much more guided by the paths which were privileged in advance. This exactly leads to the formation and organizing of a tier, as was presumed by [66]. It has to be added that the concept of Milanović was fiercely criticized by [50].

SOLUBILITY

Due to higher $CaCO_3$ contents along the formative bedding planes and due to slightly lower contents further from them, it may be concluded that the rock along the formative bedding planes is more soluble, which was the reason for the formation of the channels. But it would be too impulsive to jump to a conslusion like that. Calcite content is higher in the parts with the largest number of calcite veins, therefore the richness of calcite veins is almost definitely due to vein calcite. Absolutely very small differences between $CaCO_3$ contents speak in favour of this supposition. It is also desputable whether the final effect of calcite richness does not even restrict limestone solubility. The crystals in vein calcite are larger than those in micritic limestone, which means that the former is even less soluble than the latter.

Much more logical seems the explanation that during displacements individual bedding planes partly opened wide. A subsequent flow of water along partly open bedding planes deposited calcite into the fissures. Subsequently cave channels were formed. There follows another question what is the reason that the water, which filled the fissures, did not cement also the bedding planes, which is connected with the fact that the fissures further from the bedding planes are less cemented.

DETAILED REVIEW OF THE OBSERVED SITES

In favour of nonbiassed results, the basis for all the "observations" is measuring and counting. In this way the procedures may be remeasured and repeated.

A.
1. In the broader surroundings of the caves Škocjanske jame, the largest number of caverns (beside "potholes") consists of "inclined caves and multi-level potholes". Channels zigzag in the rock mass and are not oriented along the terraces, as was defined by [56].

2. The largest number of speleological objects is in Cretaceous rocks, among which by far reachest in caves are Turonian limestones.

3. In the recent period, the Škocjanske jame system has been considered to be one of the most important parts of the transmissivity system within the researched area. The major part of Škocjanske jame was formed in Turonian and Senonian limestones.

4. The gained data evidence that an average density of caves within the entire treated area is 1.01 of cave (cave entrance)/km^2.

5. In the area of Velika dolina, karst caverns and their fragments were classified into four groups (BP - passages which were formed along/in the bedding plane, BPF - passages which were formed along/at the point of intersection of the bedding plane and the fissure, BPN - passages which were formed along/in the bedding plane and a notch in the bedding plane, N - notches which were formed along/in the bedding plane), among which group BP (channels which were formed along the bedding planes) is important for this paper. All the channels are located within one tier and are accessible at altitudes between 317 m (the Reka ponor in Velika dolina) and about 350 m.

6. The initial channels are highly concentrated along only three "formative" bedding planes out of 62 observed bedding planes; this similarity definitely cannot be accidental.

B.
1. In the area of Velika dolina, the beds within the bedrock dip at an average of 210/28, which is also the dip of the bedding planes.

2. An average thickness of the beds is between 50 and 100 cm; apparent cyclicity may also be observed.

C.
Fissures filled with vein calcite are much more numerous in the immediate vicinity of the formative bedding planes marked 400, 500 and 600 than in the parts between them.

In the parts further from the formative bedding planes, secondary (postdiagenetic) porosity is much higher than that in their close proximity (up to two metres above or below the bedding plane). Fissures are frequently not filled. In the parts between the bedding planes marked 500 and 400, the voids of secondary porosity have partly been filled with calcite; between the bedding planes 600 and 500 there are no traces of cement.

Further from the formative bedding planes, calcite veins are substantially thinner than those in their immediate vicinity. Microtectonically crushed inliers occur directly (some 10 um) above the formative bedding planes 400, 500 and 600. Directly below the bedding planes, calcite veins are substantially less numerous than in the parts directly above them. Below they mostly occur in the form of a "sheaf", above they are "interwoven", whereas further from the formative bedding planes (between them) calcite veins occur in the form of a "tuft".

Directly below the bedding planes 500 and 600 there is a large number of calcite veins which are parallel to the bedding, and directly above the bedding planes much more numerous are the calcite veins which are perpendicular to the bedding. On the contrary, a higher number of calcite veins which are perpendicular to the bedding appear below the bedding plane 400.

1. Calcite veins are essentially more numerous in the immediate vicinity of the formative bedding planes than in the parts between them.

2. A bedding plane, above which there is an extremely large number of calcite veins in the rock with no samples without calcite veins, is the bedding plane 500, which at present is the "ponor bedding-plane" for the Reka river.

3. Further from the formative bedding planes, calcite veins are much thinner than those in their immediate vicinity.

4. Directly below the formative bedding planes, the rock contains a substantially smaller number of calcite veins than directly above the formative bedding planes.

5. Secondary porosity of the rock is much higher between the formative bedding planes than in their close proximity (up to two metres above or below the bedding planes).

6. Fissures or calcite veins are much less filled with vein calcite further the formative bedding planes than in their immediate vicinity.

7. In most samples, calcite veins (excluding those which are diagonally oriented) are perpendicular to the bedding. Almost by a half less calcite veins occur parallelly to the plane of the bedding.

8. In the parts directly below the formative bedding plane there prevail calcite veins which are distributed in the form of a "sheaf", and above the formative bedding plane the calcite veins are "interwoven".

9. Between the formative bedding planes, calcite veins most frequently occur in the form of a "tuft".

10. Microtectonics is more conspicuous in the samples directly above the formative bedding planes marked 400, 500 and 600.

11. In comparison to the fissures, stylolites in the most samples are parallel to the bedding. There are over three times less samples with stylolites perpendicular to the bed.

12. Much more frequent are the samples where two tectoglyphs are perpendicular to each other than those where the tectoglyphs are parallel to each other; they are the consequence of the same stress.

13. Stylolites occur along the formative bedding planes as well as further from them. In our case, they are probably of no specific significance to speleogenesis.

D.
1. With regard to bioclasts, the differences between the microscoped thin-sections are minimum.

2. Most of the variables of the biostratigraphical description of the microscoped thin-sections practically do not change. For that part of the geological profile, which is treated within Velika dolina, the microscopic analyses did not indicate any major changes of this kind in the rock.

3. The ascertained biostratigraphical differences (e.g. a substantial reduction in fossil remains directly above the formative bedding plane 500, difference in the selection of fossil remains on both sides of the bedding planes 700, 600 and 500), which in the geological profile are positively correlated with the position of the formative bedding planes, do not show the moments which would decisively affect the inception.

4. Palaeoecological conditions in the sedimentary basin were all the time very uniform during sedimentation of the treated Upper Cretaceous beds.

E.
1. $CaCO_3$ contents directly below the formative bedding planes are mostly between 99.40% and 99.99%. Contents below 90.00% do not appear here.

2. An average calcium carbonate content in the samples above the formative bedding planes is by 0.13% higher than that of the samples below the formative bedding planes.

3. Further from the formative bedding planes, an average content is at least once below 99%. An average content in the samples between the formative bedding planes is by 0.33% lower than that of the samples in the immediate vicinity of the formative bedding planes.

CONCLUSIONS

1. Primary phreatic channels are concentrated along only three "formative" bedding-planes among 62 observed; less than along 5% of all bedding-planes. This concordance cannot possibly be only apparent.

2. *Formative bedding-planes* differ from the others at least by the following properties:

a. calcite veins are essentially more numerous in immediate vicinity of formative bedding-planes but less numerous below them;

b. calcite veins are essentially thicker close to formative bedding-planes;

c. the rock along formative bedding-planes is less porous (cracks are filled with calcite) than farther on;

d. the rock along these bedding-planes is typically damaged, indicating interbedded wrench-fault.

We did not find the concordance between the inception reasons (for instance "trans-bedding contrast, low grade of organic substances in the rock, reduction environment in the sediment, the influence of strong acids etc.) and the actual state in Velika Dolina collapse doline. But, it was clearly evidenced that the inception, although the rock was very pure, was concentrated on few bedding-planes only.

REFERENCES

1. D'Ambrosi. Sviluppo e caratteristiche geologiche della serie stratigrafica del Carso di Trieste, Boll. della Soc. Adriatica di Scienze Naturali, **51**, 145-164, Trieste (1960a).
2. D'Ambrosi. Sul problema dell'alimentazione idrica delle fonti del Timavo presso Trieste, Tecnica Italiana, **25**, 8, 6-23, Trieste (1960b).
3. Boegan. Il Timavo. Studio sull'Idrografia Carsica Subaerea e Sotterranea, Mem. dell'Istituto Ital. Speleol., Mem. **2**, **16**, 251 p., Stabilimento Tipografico Nazionale, Trieste (1938).+
4. Čar. Geološka zgradba požiralnega obrobja Planinskega polja, Acta carsologica, **10**, 75-105, 2 pril., Ljubljana (1982).
5. Čar and R. Gospodarič. O geologiji krasa med Postojno, Planino in Cerknico.- Acta carsologica, **12**, 91 - 106, Ljubljana (1984).
6. Davies. Origin of Caves in Folded Limestone, Bulletin of National Speleological Society, Vol **22**, Part 1, 5-22, Alexandria, Virginia (1960).
7. Dreybrodt. Processes in Karst Systems. Physics, Chemistry, and Geology, Springer-Verlag, XII+288 p., Berlin, Heidelberg (1988).
8. Ewers. Bedding-plane Anastomoses and Their Relation to Cavern Passages, Bull. Nat. Spel. Soc., **28**, 3, 133-140, Arlington (1966).
9. Ford and R.O. Ewers. The development of limestone cave systems in the dimensions of lenght and depth, Canadian Journal of Earth Sciences, **15**, 11, 1783-1798 (1978).
10. Ford and P.W. Williams. Karst Geomorphology and Hydrology, XV + 601 p., Unwin Hyman, London (1989).
11. Gams. Tiha jama v sistemu Škocjanskih jam, Proteus, **30/6**, 146-150 Ljubljana (1967/68).
12. Gams. Kras (Zgodovinski, naravoslovni in geografski oris), Slovenska matica, 1-359, Ljubljana (1974).
13. Garašić. Neotektonika u speleološkim objektima, 9. jugoslovanski speleološki kongres, 51-58, Beograd (1981a).
14. Garašić. Neotectonics in some of the speleological objects in Yugoslavia, Proceedings, 8th International Congress of speleology, **1**, 148-150, Americus, Georgia (1981b).
15. Garašić. Neotektonske aktivnosti kao jedan od uzroka geneze i morfologije jednog od največih spiljskih sistema u Hrvatskoj, 9. jugoslovanski speleološki kongres, 457-465, Zagreb (1984).
16. Garašić. Dominantan utjecaj geoloških uvjeta na morfološke i hidrogeološke tipove speleoloških objekata u hrvatskom kršu, Naš Krš, **12**, 21, 57-63, Sarajevo (1986).
17. Garašić. New concept of the morfogenesis and hydrogeology of the speleological objects in karst area in Croatia (Yugoslavia), Proceedings, 10th International Congres of Speleology, **1**, 234-236, Budapest (1989).
18. Gospodarič. Škocjanske jame, Guide book of the Congress Excursion through Dinaric Karst, 137-140, Ljubljana (1965).
19. Gospodarič. Škocjanske jame, Ekskurzije, 6.kong. spel. Jug., 20-26, Postojna (1972).
20. Gospodarič. Jamski sedimenti in speleogeneza Škocjanskih jam. Acta carsologica SAZU, **12**, 27-48, Ljubljana (1984).
21. Gospodarič. O geološkem razvoju klasičnega krasa, Acta crasologica, **14/15**, 19-29, Ljubljana (1986).
22. Habe. Katastrofalne poplave pred našimi turističnimi jamami, Naše jame, **8**, 45-54, Ljubljana (1966).
23. Habič. Divaški kras in Škocjanske jame, Ekskurzije, 6. kong. spel. Jug., 26-33, Postojna (1972).

24. Habič. Kraški relief in tektonika, Acta carsologica, **10**, 23-44, Ljubljana (1982).
25. Habič, M. Knez, J. Kogovšek, A. Kranjc A. Mihevc, T. Slabe, S. Šebela and N. Zupan. Škocjanske jame Speleological Revue, Internat. Journ. of Speleology, **18**, 1-2, 1-42, Trieste (1989).
26. Herak. Karst of Yugoslavia. Karst, Important Karst Regions of the Northern Hemisphere. Elsevier Publ. Co., 25-83, Amsterdam (1972).
27. Jenko. Hidrogeologija in vodno gospodarstvo krasa, 237 p., Državna založba Slovenije, Ljubljana (1959).
28. Knez. Paleogenske plasti pri železniški postaji Košana, unpublished grad. thesis, 99 p., Fakulteta za naravoslovje in tehnologijo, VTOZD Montanistika, Odsek za geologijo, Ljubljana (1989).
29. Knez. Sedimentological and Stratigraphical Properties of Limestones from the Škocjanske jame area (Outher Dinarids), The Second International Symposium on the Adriatic Carbonate Platform, 105, Zagreb (1991).
30. Knez. Phreatic Channels in Velika dolina, Škocjanske jame (Škocjanske jame Caves, Slovenia), Acta carsologica, **23**, 63-72, Ljubljana (1994a).
31. Knez. Paleoekološke značilnosti vremskih plasti v okolici Škocjanskih jam, Acta carsologica, **23**, 303-347, Ljubljana (1994b).
32. Knez. Prehod karbonatnih kamnin v klastične pri Košani, Annales, **4**, 173-181, Koper (1994c).
33. Knez. Pomen in vloga lezik pri makroskopskih raziskavah karbonatnih kamnin, v katerih so oblikovani freatični kanali, Annales, **7**, 127-130, Koper (1995).
34. Knez, J. Kogovšek, A. Kranjc, A. Mihevc, S. Šebela and N. Zupan-Hajna. National Report for Slovenia, COST Action 65, Hydrogeological aspects of groundwater protection in karstic areas, Final Report, European Commission, 247-260, Luxemburg (1995).
35. Kogovšek. Vertikalno prenikanje v Škocjanskih jamah in Dimnicah, Acta carsologica, **12**, 49-65, Ljubljana (1984).
36. Kogovšek. Flowstone Deposition in the Slovenian Caves, Acta carsologica, **21**,167-173, Ljubljana (1992).
37. Kogovšek. Impact of Human Activity on Škocjanske jame, Acta carsologica, **23**, 74-80, Ljubljana (1994).
38. Kranjc. Transport rečnih sedimentov skozi kraško podzemlje na primeru Škocjanskih jam, Acta carsologica, **14/15**, 109-116, Ljubljana (1986).
39. Kranjc. Recent Fluvial Cave Sediments, Their Origin and Role in Speleogenesis, Dela SAZU, Razred za naravoslovne vede, Opera 27, 167 p., Ljubljana (1989).
40. Kranjc, J. Kogovšek and S. Šebela. Les concretionnements de la Grotte de Škocjanske (Slovenie) et les changements climatiques, Karst et evolutions climatiques, Presses Universitaires de Bordeaux, 355-361, Bordeaux (1992).
41. Leben. Poročilo o izkopavanjih v Roški špilji leta 1955, Arheol. vest., **7/3**, 242 - 251, Ljubljana (1956).
42. Leben. Dosedanje arheološke najdbe v jamah okoli Divače, Acta carsologica SAZU, **2**, 229 - 249, Ljubljana (1959).
43. Lowe. Historical Review of Concepts of Speleogenesis, Cave Science, **19**, 3, 62-90, London (1992a).
44. Lowe. The origin of limestone caverns: an inception horizon hypotesis, XIX+512 p., unpublished PhD thesis Manchester Polytechnic, Manchester (1992b).
45. Marussi. Il Paleotimavo e l'antica idrografia subaerea del Carso Triestino, Boll. Soc. Adr. di Sc. Nat., vol. **38**, 104-126, Trieste (1941a).
46. Marussi. Ipotesi sullo sviluppo del carsismo, Giornale di Geologia, serie II, vol. **15**, Bologna (1941b).
47. Marussi. Geomorphology, Paleohydrography and Karstification in the Karst of Trieste and Upper Istria, Steir. Beitr. z. Hydrogeologie, **27**, 45-53, Graz (1975).
48. Marussi. Ipotesi sullo sviluppo del carsismo. Osservazioni sul Carso Triestino e sull'Istria, Atti e Memorie della Comm. Grotte "E. Boegan", vol. **22**, 241-247, Trieste (1983).
49. Mihevc. Morfološke značilnost ponornega kontaktnega krasa. Izbrani primeri s slovenskega krasa, unpublished MSc thesis, 206 p., Filozofska fakulteta, Ljubljana (1991).
50. Mixon. Cave Geology is not Phisics, Geo², **17**, 1, 2 - 7 (1990).
51. Palmer. Origin and morphology of limestone caves, Geological Society of American Bulletin, **103**, 1-21 (1991).
52. Pavlovec. Startigrafski razvoj starejšega paleogena v južnozahodni Sloveniji, Razprave IV. razr. SAZU, **7**, 419-556, Ljubljana (1963a).

53. Pavlovec. Stratigrafija produktivnih liburnijskih plasti v luči novih raziskav, Nova proizvodnja, **14**, 3-4, Ljubljana (1963b).
54. Pavlovec. Regionalni obseg liburnijskih plasti, Geologija, **8**, 135-138, Ljubljana (1965).
55. Pleničar. Stratigrafski razvoj krednih plasti na južnem Primorskem in Notranjskem, Geologija, **6**, 22-145, Ljubljana (1961).
56. Radinja. Vremska dolina in Divaški prag. Problematika kraške morfogeneze, Geogr. zbornik SAZU, **10**, 157-256, Ljubljana (1967).
57. Rauch and W.B. White. Lithologic Controls on the Development of Solution Porosity in Carbonate Aquifers, Water Resources Research, **6** (4), 1175-1192, Pennsylvania State University, Pennsylvania (1970).
58. Slabe. Cave Rocky Relief and its Speleogenetical Significance, Zbirka ZRC, **10**, 128 p., Ljubljana (1995).
59. Smith, T.C. Atkinson and D.P. Drew. The hydrology of limestone terrains. In: T.D. Ford and Cullingford. The science of speleology, Academic Press, XIV+593 p., London (1976).
60. Šebela. Določitev geološke zgradbe ozemlja nad Škocjanskimi jamami s pomočjo letalskih posnetkov, Annales, **4**, 183-186, Koper (1994a).
61. Šebela. Vloga tektonskih struktur pri nastajanju jamskih rovov in kraških površinskih oblik, Doktorska disertacija, 129 p., 19 att., Fakulteta za naravoslovje in tehnologijo, VTOZD Montanistika, Odsek za geologijo, Ljubljana (1994b).
62. Šebela and J. Čar. Geološke razmere v podornih dvoranah vzhodnega rova Predjame, Acta carsologica, **20**, 205-222, Ljubljana (1991).
63. Šušteršič. Jama Kloka in začetje, Naše jame, **36**, 9-30, Ljubljana (1994).
64. Zupan. Flowstone Datations in Slovenia, Acta carsologica, **20**, 187-204, Ljubljana (1991).
65. White. Geomorphology and hidrology of karst terrains, 464 p, Oxford University Press.- New York (1988).
66. Worthington. Karst hydrogeology of Canadian Rocky Mountains, unpublished PhD thesis, XVII + 227p., McMaster University Hamilton, Hamilton (1991).
67. Worthington and D.C. Ford. High Sulfate Contentrations in Limestone Springs: An Important Factor in Conduit Initiation, Environmental Geology, **25**, 9-15, Berlin (1995).

Proc.30*th* Int'l.Geol.Congr.,vol.24,pp.106-113
Yuan Daoxian (Ed)
© VSP 1997

An Analysis on Formation of the Karst Fengcong Depressions in China

ZHONGCHENG JIANG

Institute of Karst Geology,CAGS,Guilin,541004, P.R. CHINA

Abstract

The karst fengcong depressions in China, known as the most typical symbole of karst landforms in the world , are distributed in the humid and hot karst areas north to the Qinling Mountain. The geomorphological structure is characterized by very orderly polygon networks on a plane which consist of steep peaks and deep dolines. Hard and horizontal old strata , thick and pure massive limestones, upland and deficient autogenic water are the basic conditions for the development of the fengcong depressions . The monsoon climate results in the evident differential solution processes in different geomorephologic positions-i.e,there is the highest solution rate on the bottom of depression, the second highest in the saddle between peaks,the third highest on the top of peaks and the lowest on the slope of peaks. The evident differential solution can rapidly form the relief of landforms. Therefore, the climate is the key factor responsible for the marvelous landforms.In addition, the paleoenviromental history played an important role in their formation. Generally, the development of the fengcong depressions has gone through three important processes,epikarstification, concentrated solution and differential solution .

Key words: Fengcong Depression, Karst Processes, Solution, China

INTRODUCTION

Because "Fengcong" and "Fenglin" in China are very typical karst landforms in the world and studied by karstologists for about half a centenary, they become professional term instead of "Peak-cluster" and "Peak-forest". So far the fengcong karst features (Chen,1981; Zhu,1988; Sweeting,1990), relationship between fengcong and fenglin(Ren et al,1983;Yang, 1982;Yuan, 1985) , the polygonal processes (Williams, 1972; Song,1986) and the geological history of fengcong in China(Williams,1987;et al) have had lots of good research results. However, for lack of a systematic research of the karst fengcong depressions, one cannot answer why the marvelous landscapes just develop well in China and give a reasonable explanation for the special geomorphologic processes.

DISTRIBUTION AND MORPHOLOGICAL FEATURES OF THE KARST FENGCONG DEPRESSIONS IN CHINA

Distribution

The karst fengcong depressions mainly lie in the hot and humid karst areas to south of the Qinling Mountains, especially concentrate in Central and North Guangxi, West and

South Guizhou, Southeast Yunnan, Northwest Guangdong, Southeast Sichuan and Southwest Hubei. According to Qin Houren's statistics(1993), the total area of the fengcong depressions is 125,000 km^2 . There is a biggest area around Yunnan-Guizhou Plateau. The landforms mainly are in the tropic and subtropic zone, with latitude ranging from 20° - 36° N, a mean annual rainfall of 800 - 3000 mm, and a mean temperature of 13 - 20℃ .They are well developed in Paleozoic hard and thick limestone strata, less in older and younger carbonate rocks. Dolomite and unpure carbonate rocks are also not favorable for their formation.The landforms usually develop in the areas with no surface water networks and deep subterranean water table, such as the divide of basin. Though some fengcong depressions are distributed near river, the river cuts down deeply and does not disturb the fengcong depression water system.

Morphological features
(1) High and steep relief. No matter what environments the fengcong depressions formed in, they have high relief. Some of them look like massif uplift, some formed the steep high mountains standing up on a plain. The top of peak raises in several hundreds of meter above the local erosional base .

(2) There is a polygonal network pattern on plane. The edge of each polygon may be different in number, ranging from 3 to 10 and mainly from 5 - 6. The polygonal depressions in a fengcong depression complex trend to distribute uniformly (Zhu,1985; Jiang,1988).

(3) High stone peaks adjoin low depressions orderly. But the section and the profile of the landforms can be delineated as periodic oscillatory waves(Fig.1). The amplitude and wavelength vary with environment.

(4) The stone peaks are clustered and have a common base. The shapesofthe peaks are varied but most of them look like a cone. The peaks are steep, with a slope of over 50° , and the slope of hill under saddle, however, is usually gentle (Fig.1).

The internal structure of the polygonal depressions is quite complicated. The depressions have different depth, mainly with a shape of deep column. There are dolines or sinkholes on the bottom of the depressions, which forms a overlapping depression landscape.

CAUSES OF THE FENGCONG DEPRESSIONS IN CHINA

Favorable geological setting
The fengcong depressions always selectively develop in Upper Paleozoic thick and hard pure limestone. The pure limestone is good in solubility.As for the rocks with low porosity, solution processes only can occur on the surface and in the fissure, which leads to form rugged epikarst landforms. In addition, because the hard thick limestone can't be eroded easily, after the formation of steep landforms, the stone peaks can stand up to the destruction from erosion of exogenic forces and pressure of ravity of their own, so they can also continue to develop the marvelous shape.

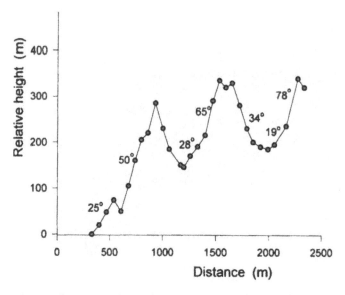

Fig.1,   The slope of the cross section of the fengcong depressions on the east of Yiaji Village

Since Neogene, most of the fengcong depression areas have   intermittently uplifted in different degree.The tectonic movement not only elevated the land of the fengcong areas   but also resulted in a deep cut of river system so that the subterranean water table lowered   greatly   and   the   negative   landforms   cut deeply. Moreover, the existence and development of the fengcong depressions have a closed relation with   the   thick   aeration   zone underneath. The thickness of aeration zone is dependent to a considerable degree on the range of tectonic uplift.

Humid and hot climates

The solution rate of limestone in South China where the rainfall is abundant is much higher than that in the semihumid and semiarid karst areas in North China(Table 1). Dynamically, high solution rate is a foundation for the formation of "pure karst". Field surveys show that,   passing through a fengcong depression system,   the rainwater with low density Ca^{2+} content(less than 5 mg/l) becomes rich in Ca^{2+} up to 70- 110 mg/l only in several hours. Such rapid and strong solution processes not only can promote the formation of microscopic karst landforms such   as karren, lapie, grike, stone   teeth and so on, but also is favorable for the macroscopic landforms to develop in   the   direction of   solutional agents.

The high solution rate and rapid processes are not yet   enough to explain the cause of the marvelous fengcong depressions in China. They   can only infer the polygonal landforms like that in Waitomo area, New Zealand(Gunn, 1981 ;Williams,1985).

The big raining fluctuation events occurring under the high temperature and heavy rain and   storm   have   a   more   important   effect. The   overland flow   out   of seepage, appearing   only in a heavy rain period   can form evident differences   in surface flow quantity of   different geomorpholigical positions , result   in re-formation of   the surface deposits   which cause differences of soil air CO_2 content, and also lead to obvious differential solution. With little surface flow   and no deposits, the top of

Table 1   The   solution rates   of several   karst   areas   with different climates in China

	Guilin of Guangxi	Puding of Guizhou	Central-South Shandong	Beijing
Mean annual rainfall(mm)	1936	1398	650-900	600-650
Mean annual temperature ($°C$)	18.3	15.1	13-15	10-12
Solution rate (m^3 /km^2 .a)	91-200	27.07-93.96	22.23-41.47	9.58-29.7

peak has a   solution   rate is low . The   bottom   of   depression is   the catchment
center of the whole depression where, with rapid confluence of storm,   the storm can't be
drained   quickly and a lot of water is usually accumulated .   Not only   the largest
discharge but also water-filling solution process occur on the bottom of depression.
Moreover,   a large amount of gaseous CO_2 in soil can be produced for the deposits
accumulated and   the thick soil layer on the bottom of depression. Therefore, the
solution rate   there   is much higher than that near the top   of peaks. At the time of
heavy rain,the saddle between two peaks can also receive runoff water from both sides.
Due to the special landform structure the saddle becomes a draught that can receive
directly rather more rainfall. In addition,owing to the air circulation gap, the living beings
in   soil   are quite active   and   can product a great quantity of soil air CO_2 content
(up to 42,000 ppm).As a result, the solution rate in the saddle   is also high(Table 2).

Paleoenvironments
In Neogene   Period, the climate in   China was   hot and   humid. The boundary of
subtropic   zone   was inferred to be in the Yellow River basin (Weng,1992). The
fengcong   depressions   were   widespread and were thought to   extend to Central-
South   Shandong (Jiang,1992). Dating of the residual red clay in   the depression in
Shandong can show the age of the fengcong depressions.

In Quaternary, the mountain glaciers but no continent glaciers were developed in China,
so then there was no the large glacier planation   of the landform surface. The landforms
of the fengcong depressions   formed under the hot and humid climates in Neogene can
remain and develop in an inheritance way.

GEOMORPHOLOGIC PROCESSES OF THE FENGCONG DEPRESSIONS

Though   the   processes   in   different   fengcong depressions may be quite different
owing   to   the   variant   geomorphologic   setting   and environmental history, the
fengcong depressions which formed under the normal solutional condition mainly go
through three correlative stages.

Table 2   Results of the solution   test   in different geomorphologic positions of   the fengcong depressions   unit: mg per tablet

Position	Rock or soil surface		20 cm below soil surface		Mean value
	Number of tablets	Solution rate	Number of tablets	Solution rate	
Peak top	5	76.38	/	/	76.38
Peak slope	3	63.27	4	41.62	52.45
Saddle	6	81.55	3	104.50	93.03
Base slope	3	69.10	3	88.40	78.75
Depression	4	84.10	2	107.05	95.58

Epikarstification processes
Epikarstification processes not only are an   initial stage but also are accompanied with the development of fengcong depressions   all the time.It is the intense solution processes in   vertical direction but not the flowing water   erosion that created   the karst landforms characterized by negative forms. The   rapid   solution   processes in the direction of raining   but not in the direction of surface water flowing direction resulted from the low carbonate   ion content in rainwater, abundant rainfall and high temperature . The porous media formed in   tectonic and sedimentary processes, such as original pores,joints and faults make rainwater possible   to seep deeply into the carbonate rocks and strengthen the solution,which can form lots of surface solutional pores, karrens and solutional pans on the one hand,and lead rock fissures to extend and form grikes or lapies on the other hand. In the course of solution processes rainwater will rapidly be saturated with respect to of Ca and Mg ions,because the atmospheric CO_2 (about 0.03% content) can only provide a low solution potential for rainwater. So grikes of   rocks get closed rapidly with depth, and it is difficult for rainwater to seep and dissolve the rocks below the depth.That is why the strong solution processes always happen in epi-rocks. When the solutional pores, grikes, karrens and lapies etc develop well and become a karst network in the surface rocks, a typical epikarst is formed.

The epikarst phenomena are very popular in South China.But the epikarst scale varys with its environments, for example, the epikarst in Guilin can be about 10 m deep , but only about 2 m deep in Puding, Guizhou. In Zhengan, Shanxi Province where it is in border of the Subtropic Zone,with a about 800 mm annual rainfall, you can not see the zonal epikarst and only find some shallow karrens and lapies.

Concentrated solution processes
Williams (1985) used the focus solution theory to delineate the formation of the polygonal karst. The theory is also available to some processes in the fengcong depressions in China, but the   two factors should be considered.

Firstly,   the embryonic form of the fengcong depressions was formed in the epikarst zone. The universal epikarsts in South China are the base of concentrated solution   processes and also make   the fengcong depressions distribute widely.

Secondly, the differential intermittent uplift of the earth crust within   plate tectonic zone is of important significance for the formation of the fengcong depressions. The uplifts can form upheaval blocks and promote the vertical hydrodynamic processes, which can form the inclined ground conduits and springs outside the fengcong mountains or hills. as a result, the depressions cut down rapidly and form some new doline (or sinkholes) (Fig.2).

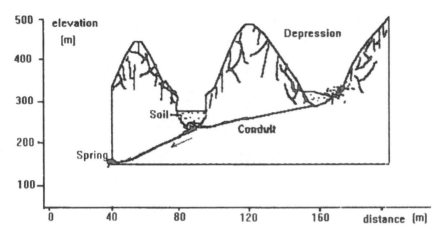

Fig.2, A sketch map of the concentrated solution processes in uplift of the earth crust

Differential solution processes

Along with the concentrated solution processes   with doline as a main feature, the differential solution processes take place in a depression owing to the difference in lithology, fracture, rainfall and runoff in the different geomorphologic positions. Taking a block between two dolines as an example, though it is far away from the center of the water catchment, being in a tectonic zone,it can be corroded more rapidly than the surrounding peaks. Under the high temperature and rich rainfall environment it can be corroded rapidly to become a saddle between two stone peaks.The saddle is a key geomorphologic position of the fengcong depressions in South China and is genetically related with the local special geology and climate.

The more important   differential solution processes follow depressions. That is, there is the highest solution rate on the bottom of the depressions, the second highest in the saddle between two peaks, the third highest on the top of stone peaks and the lowest on the steep slope of the stone peaks.

With the continuing differential solution processes, the bottom of the depression and the saddle lowered rapidly but the stone peaks relatively slowly, so the height difference between the peak and the bottom of the depression is gradually increased, and the stone peaks became tall and steep by degrees, which finally results in the marvelous landscape of the fengcong depression.

In addition, it was reported that there are revived fengcong landforms in some karst areas of Guizhou Plateau because of the large scale uplift of the earth crust. The fengcong depressions there directly come from the uplifted fenglin valleys (Xiong,1992).

CONCLUSIONS

The fengcong depressions in China are distributed in the karst areas to south of Qinling Mountains. In general, they have high and steep peaks, deep depressions, the polygonal network pattern on plane that trends to be distributed uniformly, and the highly orderly geomorphologic structure.

There is a complicated relationship between the development of the fengcong depressions and the environmental factors.Hard,thick and horizontal pure limestone and uplifted massif are the geological foundation. The continuing humid and hot climate is an important exogenic condition for solution dynamics.Besids, the paleogeological and paleogeomorphological histories have also an important effect.

The fengcong depression system is a process-response system that is far away from the equilibrium.The major karst processes are the solution occurring in the unequilibriumal water-limestone-CO_2 (gas) three phase system. The humid and hot climate is one of the key causes that make the system far away from equilibrium,because it can form big fluctuations with the overlap of the high temperature and the abundant rainfall and lead the system to product a strong differential solution potential.

Under normal solution conditions the fengcong depressions go through three major processes, epikarstification,concentrated solution and differential solution. The epikarstification is a base of the development of fengcong depressions, the concentrated solution processes can form the polygonal pattern,and the differential solution processes can resulted in the steep and marvellous landforms.

Acknowledgments
This paper is funded by National Natural Science Foundation of China and the project of the Ministry of Geology and Mineral Resources(8502218) . I am grateful to Mr.Zhang Cheng ,Xie Yunqiu and He Siyi for their much help in the field works.

REFERENCES

1.Chen Zhipeng. Study of features of the karstic depression in South China, *Proc. 8th Int. Speleo.Cong.*,499-500(1981).
2.Zhu Xuwei. Guilin karst (picture album).Shanghai Scientific-Technial Press (1988).
3.M.M.Sweeting. The Guilin karst, *Z.Geomorph.N.F, Suppl.-Bd.*,77,47-65 (1990).
4.Ren Mie et al. Introduction to karstology,Shangwu Publ. House,Beijing (in Chinese) (1983).
5.Yang Mingde. The geomorphological regularities of karst water occurrences in Guizhou Plateau.*Carsologica*,1:2, 81-92(in Chinese) (1982).
6.Yuan Daoxian. On fenglin landforms,*Guangxi Geology*, 1,79-85 (in Chinese) (1984).
7.P.W.Williams. Morphometric analysis of polygonal karst in New Guinea, *Bull.Gelo.Soc.Am.*, 83,761-796 (1972).

8. Song Linhua. Mechanism of karst depression evolution and its hydrogeological significance, *Acta Geographica Sinica,* 41(1),50-58 (1986).
9. P.W.Williams. Geomorphic inheritance and the development of tower karst. *Earth surface processes and landforms,* 12(5),453-465 (1987).
10. J.Gunn. Limestone solution rates and processes in the Waitomo district, New Zealand, *Earth Surface Processes and Landforms,* 6,427-445 (1981).
11. Jiang Zhongcheng. A study on the modern karst effect intensity in the central-southern Shandong Province. *Karst and Karst Water in North China,* Guangxi Normal University Press, Guilin, 8-25(in Chinese)(1993).
12. Weng Jintao. Developmental phase and formation environment of karst since Cenozoic Era in the central-south Shandong province,*Karst and Karst Water in North China,*Guangxi Normal University Press,Guilin,10-17 (in Chinese) (1993).
13. P.W. Williams. Subcutaneous hydrology and the development of doline and cockpit karst, *Z.Geomorph.N.F.,*29:4 (1985).
14. Yuan Daoxian et al.,Karst of China, Geologic Publishing House, Beijing (1991).
15. Zhu Dehao. Some ideas of morphology and evolution of fengcong--taking some regions in Guangxi as example,*Karst Geomorphology and Speleology,* Science Press, Beijing,57-64 (in Chinese) (1985).
16. Xiong Kangning. Morphometry and evolution of fenglin karstin the Shuicheng area, Western Guizhou,China.*Z.Geomorph.N.F.,*36(2):227-248(1992).

Proc.30th Int'l.Geol.Congr.,vol.24,pp.114-126
Yuan Daoxian (Ed)
© VSP 1997

The Geochemical Cycle of Carbon Dioxide in a Carbonate Rock Area, Akiyoshi-dai Plateau, Yamaguchi, Southwestern Japan

KAZUHISA YOSHIMURA

Department of Chemistry, Faculty of Science, Kyushu University, Ropponmatsu, Chuo-ku, Fukuoka, 810 Japan

YOUJI INOKURA

Research Institute of Kyushu University Forests, Sasaguri, Fukuoka, 811-24 Japan

Abstract

The geochemical cycle of CO_2 in a carbonate rock area, Akiyoshi-dai Plateau (Yamaguchi Prefecture, Southwestern Japan), one of the biggest karst plateaus in Japan, has been studied. From the daily data of the calcium concentration and the runoff data obtained by continuous measurements of the runoff issuing from Akiyoshi-do Cave, which has the biggest drainage basin in Akiyoshi-dai Plateau, 18.5 km^2, the following results were obtained: the calcium concentration of the baseflow in the area showed seasonal fluctuations, following changes in CO_2 concentration in the soil. The calcium concentration in the groundwater is controlled by the water-limestone dissolution equilibrium, under open system conditions depending on the meadow's soil CO_2 concentration. At the groundwater runoff peak, the calcium concentration increases, because long-residence water is flushed out from the deeper phreatic zone. During 1983 - 1986, a yearly average of 2,100 tons of limestone was dissolved in 2.1 x 10^7 m^3 of groundwater issuing from Akiyoshi-do Cave, whose catchment basin includes 16.5 km^2 of a limestone area: the mean solutional denudation rate is 47 mm/ka, if the average specific gravity of carbonate rocks is 2.7 g/cm^3. The total amount of CO_2 utilized by chemical weathering in carbonate rock areas all over the world, corresponding to the same amounts of chemically weathered carbonate rocks in mol, was estimated by using a limestone denudation rate of 50 mm/ka and found to be 8.9 x 10^{11} kg/y. The role of chemical weathering of carbonate rocks cannot be ignored in the geochemical cycle of CO_2. To estimate more precisely the total global carbon flux volume from limestone areas, further studies requiring international collaboration are necessary.

Keywords: carbon dioxide, geochemical cycle, chemical weathering, carbonate rock area, solutional denudation rate

INTRODUCTION

Fossil-fuel burning releases into the atmosphere about 5 billion tons of carbon as carbon dioxide, but the net atmospheric gain is only about 3 billion tons annually [5]. Although extensive efforts have been made for tracing the missing carbon,

Table 1. Size of sink of carbon (10^9 metric tons) of the world [5, 11]

World vegetation	560
World soils	1,500
Atmosphere	735
Oceans	36,000
Fossil fuel	5,000 to 10,000
Carbonate rocks	61,000,000

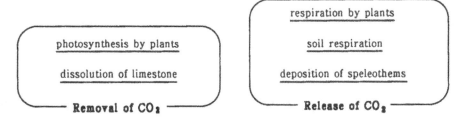

Figure 1. CO_2 flux in karst areas from and into the atmosphere.

the explanation is still unclear. Table 1 lists the world's major carbon reservoirs. The biggest sink of CO_2 consists of carbonate rocks, such as limestone and dolostone, whose combined amount is estimated to be 6.1×10^7 billion tons of C [11]. Carbonate rock areas occupy about 12% of the entire land surface of the earth [2]. In karst areas, the basic principle of landform evolution is the chemical reaction expressed by Eqn (1).*

$$CaCO_3 + H_2O + CO_2 \rightleftarrows Ca^{2+} + 2\,HCO_3^- \tag{1}$$

From the viewpoint of the removal of CO_2 from the atmosphere, the forward reaction, the dissolution of carbonate rocks, is important, because 1 mol of CO_2 is removed when 1 mol of carbonate is dissolved.

Fig. 1 shows the CO_2 flux in karst areas: photosynthesis and the dissolution of carbonate rocks are related to the removal of CO_2 from the atmosphere. On the other hand, soil respiration mainly due to activities of microorganisms and respiration by plants in the soil zone, and the formation of speleothems, are related to the release of CO_2 into the atmosphere. It is very difficult to estimate the CO_2 flux at individual sites and to calculate the total amount. For example, the dissolution rate of limestone tablets in the soil indicates only the CO_2 exhausted in the soil. In the water in which limestone is dissolved, the reprecipitation of $CaCO_3$ may occur. On the other hand, the net amount of CO_2 removed from the atmosphere in a given drainage basin is equivalent to the total amount of limestone dissolved

----

*Here, the discussion is restricted to limestone for simplification.

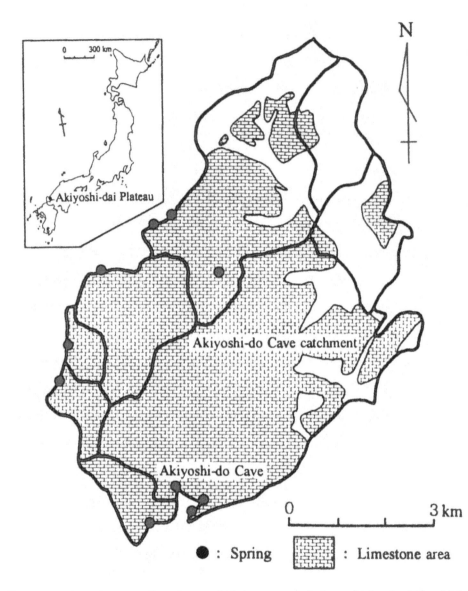

Figure 2. Location map of springs and their assumed drainage basins in Akiyoshi-dai Plateau.

and transported outside of the basin via groundwater. Therefore, the net amount can be estimated by the limestone solubility and the discharge of groundwater.

In this paper, case studies in Akiyoshi-dai Plateau, Southwestern Japan, are reviewed using published [7, 8, 15-18] and unpublished data, from the viewpoint of the geochemical cycle of CO_2. First, the limestone solubility in groundwater during base flow and storm runoff periods, and then the solutional denudation rate and the amounts of C removed from the atmosphere in the Akiyoshi-do Cave drainage basin will be discussed. Finally, we will extend the estimation to the global scale.

STUDIED AREA AND METHODS

Akiyoshi-dai Plateau is located in the southwestern part of the main island, Honshu, in Japan (Fig. 2). Akiyoshi-dai Plateau, most of which is a Quasi-National Park in Yamaguchi Prefecture, is one of the biggest karst plateaus in Japan. Almost all the plateau is covered with grass, which in the past was grown for pasture, but now only because tourists come to watch the grass set ablaze every early spring. At the southern foot of the plateau, the Akiyoshi-do Cave is located. Its total length is about 5 km and a part of the cave (about 1 km) is opened for tourists.

The drainage basins in Akiyoshi-dai Plateau are divided into 8 sections on the basis of topographic divide and the existence of two big underground drainage systems [7]. The Akiyoshi-do Cave catchment basin, the biggest in Akiyoshi-dai Plateau, occupies about half of the plateau, 18.5 km^2, including 16.5 km^2 of a limestone area. In this area, the average annual precipitation is 1,969 mm (1964 - 1988) [8].

The groundwater runoff issuing from Akiyoshi-do Cave was calculated from the observed records of the stream water level. A water gauge (Ogasawara Keiki, WL205) was set at a point of the stream issuing from Akiyoshi-do Cave without altering the natural cross-sectional profile of the stream. The stream water level was measured at hourly intervals from 1983 to 1986. The discharge was calculated from the hourly water level data using the stage-discharge rating equation which was obtained by several measurements of the discharge (cross-sectional area multiplied by flow velocity). The average annual runoff from Akiyoshi-do Cave was 955mm [8].

The groundwater issuing from Akiyoshi-do Cave was filtered through a 0.45 μm membrane filter with a disposable syringe filter unit (Millipore) at the site where the water sample was collected. The Ca^{2+} concentration in groundwater was determined by atomic absorption spectrophotometry with a Nippon Jarrell-Ash apparatus, Model AA-8500. The ionization and chemical interferences were suppressed by addition of caesium chloride and lanthanum chloride to the sample solution to 0.01 and 0.001 mol/dm^3, respectively.

The method for measuring the CO_2 concentration in soil air using a gas detection device was quite similar to that proposed by Hamada and Tanaka [3]. To extract soil air from the soil, a soil air collecting probe (acrylic resin pipe with small holes

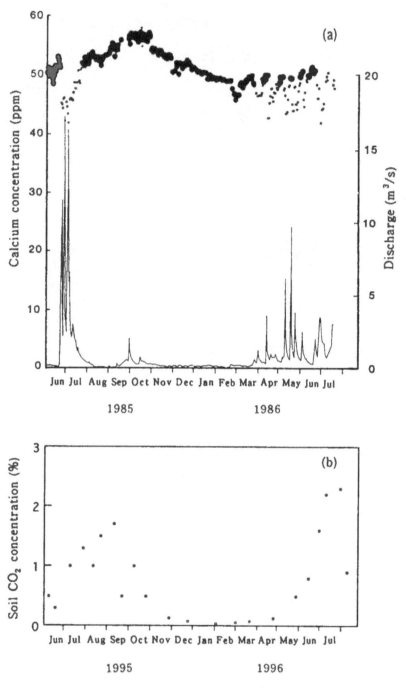

Figure 3. (a) Daily data of the calcium concentration and discharge of the groundwater issuing from Akiyoshi-do Cave (1985 · 1986) [8]. ●: at baseflow; •: at storm flow. (b) Seasonal change of the soil CO_2 concentration in the grassland in Akiyoshi-dai Plateau (1995· 1996; Depth: 40 cm) [9].

for the diffusion of soil air, 3.0 cm i.d.) was installed in the soil and soil air was directly extracted by the gas detection device (Gastec, Model 801) through the collecting probe (Gastec, No. 2LL for 300 · 5000 ppmv CO_2 and No. 2L for 0.25 · 3 %(v/v), and the CO_2 concentration was measured *in situ*.

The chemical equilibria calculation for the $CaCO_3 \cdot H_2O \cdot CO_2$ system was performed using the computer program WATEQF [14].

SOLUBILITY OF LIMESTONE IN GROUNDWATER

Solubility of limestone in groundwater at baseflow
Fig. 3 shows the daily data of the calcium concentration and the runoff data of groundwater issuing from Akiyoshi-do Cave at baseflow. The calcium concentration of the baseflow in the area showed seasonal fluctuations, which followed changes in CO_2 concentration in the soil of Akiyoshi-dai Plateau with a one to two months' time lag, as also shown in Fig. 3. Soil CO_2, measured in the grassy soil which covers most of the area at the depth of 40 cm, varied from a minimum of 0.04 % at a soil temperature of 4.9 ℃, to a maximum of 2.2 % at 22.0℃.

The logarithm of partial pressure of CO_2 vs. soil temperature can be expressed by a linear regression equation. Table 2 shows the intercept and slope values of the equation, Pco_2 / atm = a + b $(T / ℃)$. They are somewhat different depending on many factors such as vegetation, soil moisture, soil horizon and exposure to the sun light. Soil CO_2 concentration in East Yorkshire was reported as a function of soil temperature: a = -0.25 and b = 0.044 in arable land; a = -3.12 and b = 0.042 in grassland [13]. From the calculated Pco_2 of the water, Drake and Wigley [1] have shown that soil CO_2 generally may be expected to have a temperature dependence described by a, which is characteristic of a soil CO_2 concentration associated with a particular land use type, and by b = 0.04. It is necessary to compile more data on soil CO_2 concentration to clarify the discrepancy in b values between Akiyoshi-dai and other areas.

Table 2. Soil CO_2 concentration in Akiyoshi-dai Plateau, Yamaguchi [9]

Locality		Depth (cm)	n	a	b	r	Temperature range (℃)		Year
A Kaerimizu	forest	30	9	·2.68	0.044	0.927	6.1 ·	19.0	1982-1984
B	grassland	30	9	·3.24	0.059	0.891	3.8 ·	20.8	1982-1984
C Museum	grassland	30	50	·3.62	0.072	0.964	4.5	28.3	1993-1996+
D		40	50	·3.65	0.070	0.932	4.7 ·	28.1	1993-1996+
E		50	50	·3.47	0.074	0.948	6.4	26.7	1993-1996+
F		75	50	·3.49	0.067	0.939	7.4 ·	25.8	1993-1996+

log (Pco_2/atm) = a + b T; T: temperature (℃), r: correlation coefficient

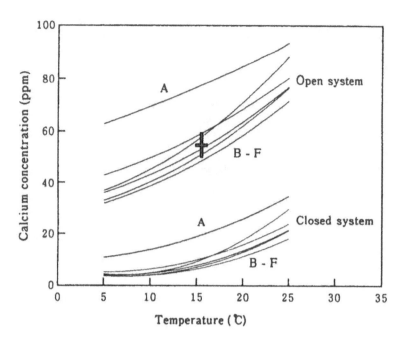

Figure 4. Solubility of calcite equilibrated with the soil CO_2 under different conditions. Forest soil (A) and grassland soil (B · F). The bars indicate the standard deviations of annual changes in water temperature and the calcium concentration of groundwater issuing from the Akiyoshi-do Cave.

In Akiyoshi-dai Plateau, the solubility of limestone was estimated as a function of soil temperature under open and closed system conditions by the chemical equilibrium calculation using equilibrium constants in the WATEQF program [14] and all the equations in Table 2. Fig. 4 shows the solubility range of limestone under different conditions. Under the open system conditions [2,16], the solubility represented by the calcium concentration, $[Ca^{2+}]$, is expressed by

$$[Ca^{2+}] = (K_0 \, K_1 \, K_{SO} \, Pco_2 \, / \, 4 \, K_2)^{1/3} \qquad (2)$$

where K_0, K_1, K_2 and K_{SO} are the equilibrium constants for CO_2 dissolution into water, the first and the second acid dissociation of solvated CO_2, and the solubility product of calcite, respectively, if the activity coefficients of the species concerning these chemical equilibria are unity. This equation depends on the cubic root of soil CO_2 contents under the open system conditions. In this way, the solubility is predicted by using the regression equations. In order to treat more exactly, the ion-pair formation of Ca^{2+} and OH^-, HCO_3^- and CO_3^{2-}, and the activity coefficient terms of all ions concerned, were taken into consideration. In a similar way, the solubility of calcite was calculated as a function of soil temperature under closed system conditions. In this system, the initial concentration of total inorganic carbon is approximately expressed by

$$[CO_2(aq)] + [HCO_3^-] = K_0\, Pco_2 + (K_0\, K_1\, Pco_2)^{1/2} \tag{3}$$

and therefore at equilibrium

$$[Ca^{2+}] + K_0\, Pco_2 + (K_0\, K_1\, Pco_2)^{1/2}$$
$$= [CO_2(aq)] + [HCO_3^-] + [CO_3^{2-}]$$
$$= K_{SO}\,([H^+]^2/K_1K_2 + [H^+]/K_2 + 1)/[Ca^{2+}] \tag{4}$$

Among the solutions of this equation at a given soil temperature, a pair of $[H^+]$ and $[Ca^{2+}]$, which satisfy the electroneutrality ($2\,[Ca^{2+}] + [H^+] = [HCO_3^-] + 2\,[CO_3^{2-}] + [OH^-]$), was chosen by trial and error using the WATEQF program and plotted

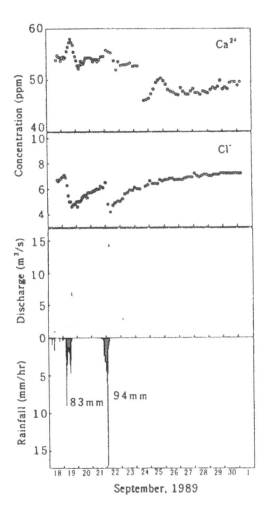

Figure 5. Hydrograph and chemograph of the groundwater issuing from the Akiyoshi-do Cave [17].

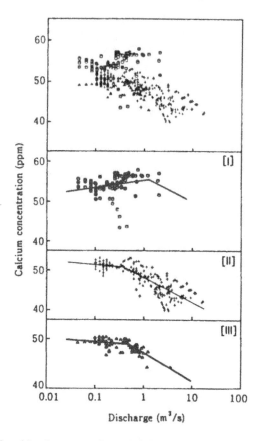

Figure 6. Relationship between the calcium concentration and the discharge of groundwater issuing from Akiyoshi-do Cave (1985 - 1986) [8]. I : high Ca^{2+} concentration period (August - November); II : average Ca^{2+} concentration period (May - July and December); III : low Ca^{2+} concentration period (January - April).

in Fig. 4. The annual mean solubility of limestone and temperature of groundwater issuing from Akiyoshi-do Cave were shown also in this figure, together with their variations as error bars of standard deviation. It can be seen that the calcium concentration in the groundwater is mostly controlled by the water-limestone dissolution equilibrium under open system conditions. Because there is no evidence of geothermal and/or volcanic activities in this area, the CO$_2$ used for the dissolution of limestone must originate from the atmosphere. In general, the CO$_2$ for the dissolution of limestone is supplied from the soil. This is supported by the ^{13}C/^{12}C isotope ratio [4].

Solubility of Limestone in Groundwater at the Stormflow
For the underground river issuing from Akiyoshi-do Cave, the changes in discharge and in the contents of Ca^{2+} and Cl$^-$ after a storm are shown in Fig. 5. The runoff response is similar to that for a surface stream, except for the longer time lag

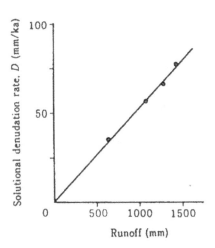

Figure 7. Relationship between the solutional denudation rate and the runoff of the Akiyoshi-do Cave catchment (1983- 1986) [8].

between the runoff peak and the storm peak. Of the major components, Ca^{2+} and HCO_3^- are supplied by the dissolution of Akiyoshi limestone in the presence of dissolved CO_2, whereas half of the Cl^- is supplied by the wet and dry deposits, mainly in the form of sea salt, and the other half by cattle breeding on the plateau [19]. The increase in Ca^{2+} concentration and the decrease in Cl^- concentration at the start of the flood are due to the flushing out of water with a long residence time in the deeper phreatic zone and the subcutaneous zone beneath the soil, as already described by Ford and Williams [2].

Estimation of Solutional Denudation Rate of Akiyoshi-dai Plateau
Fig. 6 shows the relationship between the Ca^{2+} concentration and the discharge of groundwater issuing from Akiyoshi-do Cave. Taking the seasonal variation of limestone solubility into consideration, we divided a year into 3 periods on the basis of the Ca^{2+} concentration in the groundwater. The period of high Ca^{2+} concentration is from August to November, the period of low Ca^{2+} concentration from January to April, while the concentration for the other months is average. For each period, the Ca^{2+} content at a given discharge is predicted by using the regression equations.

The solutional denudation rate D (mm/ka) can be calculated by using Eqn. (5).

$$D = 2.5 \ \Sigma \ Ca(Q) \cdot Q(t) \ / \ (G \cdot A) \tag{5}$$

The discharge of the Akiyoshi-do Cave, $Q(t)$ (m^3/hr), was measured every one hour from June 1985 to July 1986, and its Ca^{2+} concentration, $Ca(Q)$ (g/m^3), was estimated from $Q(t)$ mentioned above. G is specific gravity of limestone: 2.7 g/cm^3

[12]) and A (km^2) the catchment area of limestone. Fig. 7 shows the relationship between the solutional denudation rate and the runoff of the Akiyoshi-do Cave catchment, indicating that D is almost controlled by runoff. As the average runoff was 955 mm, that is, 21 million tons of groundwater issuing from Akiyoshi-do Cave, a yearly average of 2,100 tons of limestone was dissolved: the mean solutional denudation rate is 47 mm/ka.

Table 3 shows the mass balance for the groundwater issuing from Akiyoshi-do Cave. The input in the Akiyoshi-do Cave basin was evaluated by measuring the amounts of chemical components of dry and wet deposition in this area. And the concentration by evaporation was taken into consideration. The values in the bottom line in Table 3 are the average dissolved contents of major chemical components in the groundwater. Almost all the Ca^{2+} and HCO_3^- are supplied from Akiyoshi-dai Plateau. However, the possibility that the acidic deposition may

Table 3. Factors affecting the evolution of groundwater chemistry of Akiyoshi-do Cave (concentrations in ppm)

	Na$^+$	K$^+$	Mg^{2+}	Ca^{2+}	Cl$^-$	HCO$_3^-$	NO$_3^-$	SO$_4^{2-}$	SiO$_2$
Wet deposition*	0.59	0.07	0.08	0.23	1.00	0	0.81	1.92	0
Dry deposition*	0.04	0.37	0.03	0.09	0.21	0	-	0	0
Evaporation**	0.63	0.07	0.09	0.35	1.08	0	0.85	2.05	0
Total	1.26	0.51	0.20	0.57	2.29	0	1.67	3.97	0
Akiyoshi-do Cave***	4.1	0.51	1.21	54.7	7.1	157	4.4	5.2	5.9

*average value in 1989-1993 [19]. **averaged annual precipitation: 1,969 mm; averaged runoff: 955 mm [8]. ***[16]

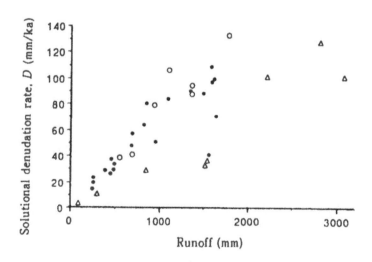

Figure 8. Relationship between solutional denudation rate of carbonate rocks and runoff [10]. △: arctic or alpine (little or no soil); ●: mid-latitude (soil covered); ○: tropical (soil covered).

dissolve limestone should be taken into consideration. According to another experiment, the effect of acidic deposition on the dissolution of limestone is 5 % of the total amount [15].

NET CO_2 FLUX REMOVED FROM THE ATMOSPHERE ON A GLOBAL SCALE

Fig. 8 shows the worldwide data compiled by Jennings [10], together with the present result, showing the relationship of the solutional denudation rate against runoff in the world. The amounts of chemically weathered carbonate rocks all over the world were estimated under the following assumptions:

(i) the average solutional denudation rate is 50 mm/ka.
(ii) the carbonate rock area displaying groundwater circulation is 10 % of the land, i.e., $1.5 \times 10^7 \, km^2$ [2].
(iii) the average specific gravity of carbonate rocks is 2.7 g/cm^3 [12].

At present, it is very difficult to estimate the average solutional denudation rate in carbonate areas all over the world, an attempt adopting 50 mm/ka was made as the first approximation. The amount obtained was 2.0 billion tons of $CaCO_3$/y, corresponding to the CO_2 in mol removed from the atmosphere, and was found to be 0.89 billion tons of CO_2/y, i.e. 0.24 billion tons of C/y. These values correspond to about 1/20th of the release of CO_2 by fossil-fuel burning and 12 % of the missing CO_2. Of course, if the solutional denudation rate is 100 mm/ka, this value should be doubled.

Ichikuni [6] and Kitano [11] estimated the amounts of CO_2 removed from the atmosphere by chemical weathering, using the averaged chemical composition of river water and the total discharge all over the world (Table 4). The values for carbonate rock areas are reasonably in agreement with those obtained in this study. In case of non-carbonate rock areas, although the chemical weathering rate is not so high because of the low reactivity of silicate minerals with CO_2-containing water, the area is about 5 times wider than that of non-carbonate rock areas.

In conclusion, the role of chemical weathering of carbonate rocks cannot be ignored in the geochemical cycle of CO_2. To estimate more precisely the total global carbon flux volume in carbonate rock areas and also in silicate rock areas, more field

Table 4. CO_2 flux removed from the atmosphere in the global scale (10^9 metric tons)

carbonate rock areas	silicate rock areas	Ref.
0.35	0.97	[6]
0.70	0.68	[11]
0.89	-	present study

measurements of soil CO_2 concentration and solutional denudation rate, both of which require more intense international collaboration, are necessary.

REFERENCES

1. J.J. Drake and T.M.L. Wigley. The effect of climate on the chemistry of carbonate groundwaters. *Water Resources Research* 11, 958-962 (1975).
2. D. Ford and P. Williams. *Karst Geomorphology and Hydrology.* Unwin Hyman, London (1989).
3. Y. Hamada and T. Tanaka. A simple method for measuring CO_2 concentration in soil air using a gas-detection device. *Hydrology (J. Jpn. Ass. Hydrol. Sci.)* 25, 123-130 (1995).
4. M.G. Gross. Variations in the O18/O16 and C13/C12 ratios of diagenetically altered limestones in the Bermuda Islands. *J. Geol.* 72, 170-194 (1964).
5. R.A. Houghton and G.M. Woodwell. Global climate change. *Sci. Amer.* 260, 18-26 (1989).
6. M. Ichikuni. Role of water in geochemical systems. in Chem. Soc. Jpn. (Ed) *Ions and Solvents.* Tokyo University Publishers, Tokyo (1976).
7. Y. Inokura, K. Yoshimura, A. Sugimura, and T. Haikawa. Drainage basins of springs in Akiyoshi-dai Plateau evaluated by their discharge and chemical compositions. *J. Speleol. Soc. Jpn.* 10, 14-24 (1985).
8. Y. Inokura, K. Yoshimura, A. Sugimura, and T. Haikawa. Run-off and mass flux from the Akiyoshi-dai Plateau(I) Evaluation of solutional denudation rate based on run-off and calcium flux from Akiyoshi-do Cave. *J. Speleol. Soc. Jpn.* 14, 51-61 (1989).
9. Y. Inokura, K. Yoshimura, and A. Sugimura, unpublished data.
10. J.N. Jennings. *Karst Geomorphology.* Basil Blackwell, Oxford (1985).
11. Y. Kitano. *Environmental Chemistry of the Earth.* Shokabo, Tokyo (1984).
12. B. Mason. *Principles of Geochemistry, third edition.* John Wiley, New York (1966).
13. J.L. Pitman. Carbonate chemistry of groundwater from chalk, Givendale, East Yorkshire. *Geochim. Cosmochim. Acta* 42, 1885-1897 (1978).
14. L.N. Plummer, B.F. Jones, and A.H. Truesdell. WATEQF · a FORTRAN IV version of WATEQ, a computer program for calculating chemical equilibrium of natural waters. *Water-Resources Investigations* 76-13, 1-63 (1978).
15. K. Yoshimura and H. Nakamura. Acid rain in the Akiyoshi-dai (Plateau) limestone area. *Bull. Akiyoshi-dai Mus. Nat. Hist.* 26, 9-20 (1991).
16. K. Yoshimura and Y. Inokura. Evolution of groundwater chemistry in the hydrological cycle of the Akiyoshi-dai Plateau limestone area. *J. Groundwater Hydrol.* 34, 183-194 (1992).
17. K. Yoshimura, S. Matsuoka, Y. Inokura, and U. Hase. Flow analysis of copper by ion-exchanger absorptiometry with 4,7-diphenyl-2,9-dimethyl-1,10-phenanthrolinedisulfonate and its application to the study of karst groundwater storm flow. *Anal. Chim. Acta,* 268, 225-233 (1992).
18. K. Yoshimura and Y. Inokura. The role of chemical weathering of carbonate rocks in the geochemical cycle of CO_2. *Chikyukagaku* 27, 21-28 (1993).
19. K. Yoshimura, Y. Inokura, and H. Nakamura. unpublished data.

Proc.30th Int'l.Geol.Congr.,vol.24,pp.127-133
Yuan Daoxian (Ed)
© VSP 1997

The CO_2 Regime of Soil Profile and Its Reflection to Dissolution of Carbonate Rock

SHENGYOU XU   SHIYI HE

The Institute of Karst Geology, 40 Qixing Road,Guilin, 541004, China

Abstract

The studies on the CO_2 regime on a soil profile underlain by carbonate rock, the fluctuation of pH and HCO_3^-, and the dissolution of the standard limestone tablets provide us with some new knowledge   as follows: (1) CO_2 in the soil profile changes with the seasons; (2) Generally, two CO_2 concentration gradients, which exist soil-atmosphere and soil-carbonate rock, are formed. However, in winter, only the soil-atmosphere   gradient   is formed; (3)The dissolution intensity of the limestone tablets relates closely to the CO_2 concentration gradient rather than the absolute concentration of CO_2; (4) There is larger difference in time between the regime of the pH and HCO3- of karst underground water and the changes of the absolute concentration of CO_2 on the overlying soil profile, but the former nearly coincides with the regime of CO_2 concentration gradient. The studies demonstrates that the gradient of CO_2 concentration, rather than the absolute concentration of CO_2 can reflect karst process in the karst dynamic system.

Key words: Carbonate rock, CO_2 gradient, Dissolution, China

INTRODUCTION

Carbonate rocks have a close relationship with CO_2 of the atmosphere. On the one hand, they influence changes in CO_2 concentrations in the atmosphere through karst processes(carbonate rocks dissolutes or tufa deposits). On the other hand,the fluctuation of atmospheric CO_2 exerts a great influence on the karst process in intensity, model and direction in the carbonate rocks area (Daoxian Y 1993, 1994). Therefore, studying   the behavior and regime of CO_2 in the carbonate rock-atmosphere system and its response to karst process could not only enhance our knowledge about the gaseous CO_2 dynamic mechanism of   karst   process, but also provide   some data   and   methodology   for approaching the sink-source relationship between the "greenhouse gas" CO_2 of   the atmosphere and karst process. In the past, the studies on the regime of CO_2 were mainly centralized in the non-carbonate rocks area (Amundson and Davidson 1990; Suchet and others 1993; Striegl and others1990). Very little attention was paid to the behavior and regime of CO_2 profile on the interface between carbonate rock and the atmosphere. It is difficult to observe the regime of CO_2 profile on the interface between carbonate rock and atmosphere. Because the carbonate rock-atmosphere   system is completely open and is controlled by various environmental factors, specifically climate and human activities. For this reason, this study uses the soil profile as the medium for CO_2 production, storage and transport. One natural soil profile overlying carbonate rock was selected to study its CO_2 distribution and fluctuation, the dissolving intensity of carbonate rock by means of the standard limestone tablets buried at its different depths, and   the regime   of   some hydrochemical parameters of   karst underground water.

Combined with the data of other regions, this study discusses the relationship between the CO_2 regime on the soil profile and the dissolution of the underlying carbonate rock.

LOCATION AND METHODS

The soil profile in this study is the unsaturated zone, near the boundary between the peak cluster depression and peak forest plain within Yaji Karst Hydrogeological Experimental Field, 9km southeast to Guilin City, Guangxi, China(Fig.1). It is about 3m thick, underlain by the pure limestone of upper Devonian and covered by sparse herbvegetation. The soil is classified to brown limestone soil of which some chemical parameters are showed on Table 1.

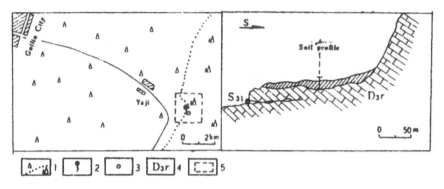

Figure 1. Sketch map of the position of the soil profile. 1. Peak forest plain/peak cluster depression; 2. Karst Spring(S_{31}); 3. Soil profile; 4. limestone of upper Devonian; 5. karst Hydrogeological Experimental Field.

To understand the distribution and regime of CO_2 on the soil profile, 6 plastic tubes were installed in the same profile at different depths such as -20cm, -50cm, -120cm, -180cm, -250cm, -290cm. The tubes were designed to measure the monthly CO_2

Table 1. Some chemical parameters of the soil profile

depth (cm)	pH value	Org.C(%)	Ca^{2+} (%)	Mg^{2+} (%)	K^+ (%)
-10	5.62	5.10	0.073	0.435	0.86
-50	5.45	2.42	0.062	0.432	1.01
-120	5.80	0.70	0.078	0.433	1.09
-180	5.91	0.65	0.094	0.444	1.08
-250	5.45	0.70	0.052	0.462	1.06

concentration with a GASTEC-CO_2 detector (from June of 1994 to June of 1995) and hence generate the distribution curve on the profile every month. Moreover, the short-term fluctuation of CO_2 concentration on the profile was continuously observed from July 21 to 22, 1994 by use of the same methods mentioned above.

RESULTS AND DISCUSSION

Based on observations(Fig.2), the properties of the CO_2 regime on the profile can be summarized as follows:

(1) The concentration and distribution of CO_2 the profile is closely tied to the climate. Not leached by rain, the larger difference in CO_2 concentration on the profile occurs in the dry season. For instance,the results observed show that the curvatures of distribution of CO_2

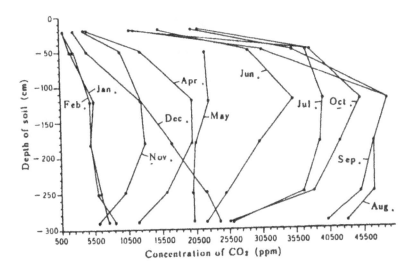

Figure 2. The curves of CO_2 concentration on the profile from June, 1994 to May, 1995

from June to September of 1994 are bigger than that from January to May (Fig.2). On the contrary, there is a less pronounced difference in CO_2 concentration from top to bottom in the rainy season (e.g. in May, 1995).   It is indicated that the leaching process can change the law of distribution and diffusion of CO_2 on the soil profile and make the gas trend to distribute more evenly from top to bottom. Moreover, the concentration of CO_2 on the profile changes with the seasons. The data illustrates that the higher content of CO_2 appeared July to October of 1994 and the top value(40166 ppm)occurs in August of the year,and the lower value in concentration from November,1994 to February, 1995 with the minimum of 4583 ppm.

(2) It is not evident that CO_2 changes in concentration and distribution in a short time. Fig.3 indicates that the curves of CO_2 concentration are comparatively close to each other. However, the CO_2 concentration of the profile measured at 6:45 of July 22 decreased relatively to other four data measured   in the period at the depths of   -20 cm and   -50cm, and increased   below the depth of   -120cm. The reason why the CO_2 concentration decreases in the upper position   is probably the activity of microbes' decreasing with the soil   temperature   down in the morning.   It is not clear why the CO_2 concentration in the lower position is   different   between   day (curve1,2 and 5) night (curve 3 and 4).

(3)Generally,   the   concentration ofCO_2 increases   with   depth   to a maximum value at the depth of -100cm to -120cm, and then decreases with depth. Accordingly, two concentration gradients, i.e. soil to atmosphere direction and soil to carbonate rock direction, are formed. Probably, the formation of two gradients of CO_2 concentration relates to following aspects.   First, the upper   portion soil (from 0 cm to -20 cm) is dry, where   microbes is not   favorable to   active.   Secondly, the CO_2 in upper portion soil

130   S.Xu   S.He

diffuses continuously into the
atmosphere. Two reasons outline
results in the soil to atmosphere
gradient of CO_2 concentration. It
leads to the formation of soil to
carbonate rock gradient of CO_2
concentration that underling
carbonate rock extracts intensely
the CO_2 from the soil in the process
of dissolution.

During the period when the temperature is low from December of
1994 to February of 1995, the concentration of CO_2 on the profile
increases from top to bottom
forming a soil-atmosphere gradient.
It's probably because that the upper
portion of the profile is drier and
has lower temperature than the

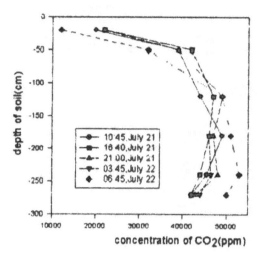

Figure 3. The curves of CO_2 concentration in
short time during July 21-22, 1994

lower portion. The microbes are downwards, therefore, more CO_2 concentrated in the
lower portion. Simultaneously, the underlying carbonate rock less extracts the CO_2 from
the soil due to the process of dissolution weakens by the drop in the temperature.

*The relationship between the dissolution of the standard limestone tablets and the CO_2 of
the soil profile*
To understand the dissolution intensity of carbonate rock at different depths, the standard
limestone tablets were installed at different places such as 100cm (above the ground),
0cm (on the ground), -30cm, -60cm, -150cm, -220cm and -270cm (on the profile). They
were weighed after one year to get their loss.Using the dissolution rates of the tablets as
the dissolution intensities of carbonate rock at different depths, and comparing them with

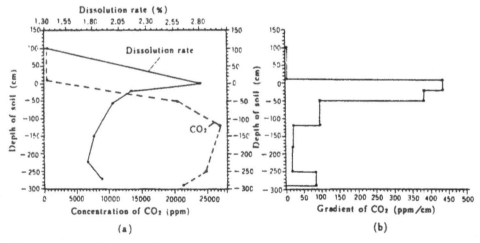

Figure 4a,b. a The dissolution rate and the average content of CO_2 b The gradient of CO_2
concentration on the soil profile

the average   concentration of CO_2 and the concentration gradients of CO_2 (Fig.4 a,b and Table 2), some new knowledge are available follows:

(1)   The tablets in the air (100 cm above the ground) have the least dissolution rate(1.35%), the ones on the ground (0 cm)have the largest dissolution rate (2.78%). The dissolution rates of the tablets in the soil change from 1.71% to 2.11% and decrease with depth. The rate at -220cm is just higher than that in the air, and then the rate increases with depth from -220 cm to the bottom of the profile.

(2)   The dissolution rate of the tablet at certain depth has no relationship with the absolute concentration of CO_2.   For example,the tablets on the ground has the largest dissolution rate, but the   absolute CO_2 concentration at this   place is only 500ppm. On the contrary,   the dissolution rate at -220 cm is only 1.71%, but the CO_2 concentration

Table 2. The relation between the dissolution intensity of limestone tablets and the regime of CO_2 on the soil profile

dissolution rate of tablets (%)	1.35 (100cm)	2.78 (0cm)	2.11 (-30cm)	1.94 (-60cm)
annul mean content of CO_2 (ppm)	330 (100cm)	500 (0cm)	9150 (-20cm)	20500 (-50cm)
annul mean gradient of CO_2 (ppm/cm)	1.70 (100~ 0 cm)	432.50 (0~ -20cm)	378.30 (-20~ -50cm)	95.40 (-50~-120cm)
dissolution rate of tablets (%)	1.77 (-150cm)	1.71 (-220cm)	1.84 (-270cm)	-
annul mean content of CO_2 (ppm)	27000 (-120cm)	25900 (-180cm)	24727 (-250cm)	21317 (-290cm)
annul mean gradient of CO_2 (ppm/cm)	21.20 (-120~ -180 cm)	16.90 (-180~-250cm)	85.20 (-250~-290cm)	-

there is about 50.6 times as much as the one on the ground, and 2.7 times as much as the one at -20 cm.

(3) The change of dissolution intensity of the limestone tablets nearly coincides with the gradient of CO_2 concentration(Fig.4).

The relationship between CO_2 of the soil and the regime of the hydrochemical parameters of karst underground water

The pH value and HCO_3^- content of one karst spring (S_{31}) near the soil profile were measured during the same period. Though only 7 months' data was obtained due to the destruction of some equipment, some important information was obtained from the research and   illustrated   in Table 3 and Fig.5. The absolute concentration of CO_2 in the soil reaches a maximum value in August, while the maximum HCO_3^- content was observed in October. There is a difference of   two months between both sides. It can not be   explained   that the difference in time is resulted by the sensitivity of epikarst zone.

Table 3. The regime of CO_2 and HCO_3^- content changes of S_{31}

month	average CO_2 (ppm)	gradient of CO_2 (ppm/cm)	HCO_3^- (mmol/l)
6	25916.6	276.5	4.28
7	34833.3	264.4	4.60
8	40166.7	325.5	4.81
9	37166.7	428.4	5.02
10	33156.6	454.4	5.56
11	9167.7	123.2	5.45
12	13750.0	167.6	5.35

Comparing HCO_3^- content of the spring water to the gradient of CO_2 concentration, both of them nearly synchronously change and come to the maximum in October. The fact is supported by another case in Zhenan County of Shaanxi Province. In the period from 1993 to 1994, the pH value, HCO_3^- content of Yudong Underground Stream and the CO_2 regime of soil profile nearby were monthly measured. The data shows that the pH and HCO_3^- does not change completely along with the absolute concentration of CO_2, there is a difference of two months(Fig.6). However, contrasting them with the gradient of CO_2 concentration, it can be seen that the pH value and HCO_3^- content fluctuate synchron-ously with the gradient of CO_2 concentration. The result indicates only absolute concentration of CO_2 can not state clearly the relation between the regime of CO_2 and dissolution, but gradient more available.

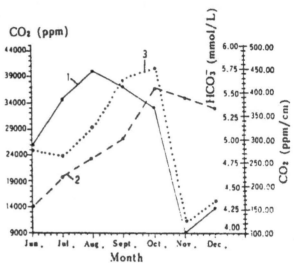

Figure 5. The Curves of absolute concentration of CO_2, and gradient of CO_2 and HCO_3^- of Spring S_{31}
1-CO_2 content　　2-HCO_3^-　　3-CO_2 gradient

CONCLUSIONS

(1) The distribution and regime of CO_2 on the soil profile in carbonate rock area is closely related to the climate and its seasonal changes.

(2) In general, there are two gradients of CO_2 concentration, i.e. one from soil to atmosphere, and another from soil to carbonate rock. In winter however, only soil to atmosphere exists.

(3) The dissolution intensity of limestone tablets in soil coincides closely with gradient of CO_2 rather than the absolute concentration of CO_2. Furthermore, the pH value and HCO_3^- content of karst underground water change synchronously with the gradient of

CO_2 concentration. The fact indicates that karst process in karst dynamic system not only relates to the concentration of CO_2 but also relates to the activity of CO_2.

Acknowledgments

This is supported by Prof.Yuan Daoxian, an academician of the Chinese Academy of Sciences, and funded by the key project of Ministry of Geology and Mineral Resources of China(No.8502218)and the project of National Natural Science Foundation of China(No.49070155). Also Cao Jianhua of the Institute of Karst Geology do some work in the field. This paper is the contribution for IGCP 379 Project "Karst processes and carbon cycle".

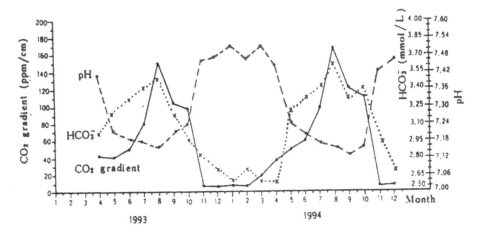

Figure 6. The relation between pH, HCO_3^- and the absolute concentration of CO_2 in the soil near Yudong Underground Stream, Zhenan, Shaanxi,(1993-1994)

REFERENCES

1. Y. Daoxian. Carbon cycle and global karst, *Quaternary Sciences*, **1**:1-6, (1993).
2. R. G. Amundson and E. A. Davidson. Carbon dioxide and nitrogenous Gases in the soil atmosphere, *J.Geochem. Explor*. **38**: 13-41 (1990) .
3. P.A. Suchet et al. Modeling of atmospheric CO_2 consumption by chemical weathering of rocks: Application to the Geronne, Congo and Amazon Basins, *Chemical Geology*, **107**:205-210, (1993).
4. R. G. Striegl and D. E. Armstrong. Carbon dioxide retention and carbon exchange on unsaturated Quaternary sediments, *Geochim. Cosmochim. Acta*, **54**:2277-2283, (1990).

Proc.30th Int'l.Geol.Congr.,vol.24,pp.134-142
Yuan Daoxian (Ed)
© VSP 1997

Studies On Oxygen Isotope Thermometry Of Cave Sediment And Paleoclimatic Record

JIAMING   QIN

Institute of Karst Geology, 40 Qixing Rd, Guilin,541004   P.R.CHINA

Abstract

A model has been presented for the climate changes in the past   36,000 years in Guilin area by means of the karst sedimentation study, AMS-14C,U series and stable isotope experiments based on a big stalagmite from Panlong cave in Guilin. During the past 36,000 to 32,000 years,the lowest annual mean temperature on the surface was from 8℃ to 9℃,while the highest was 13℃. During the past 32,000 to 12,000 years, an interruption of sedimentation occurred and the lowest annual mean temperature reached 6℃ to 8℃ in the late Dali sub-ice age. 11,000 years ago, the Younger Dryas in this area showed clearly and the lowest annual mean temperature came to 9℃.After the Younger Dryas, the temperature increased rapidly and till the past 10,700 years,it reached the nowadays level(19℃). From the past 9,000 years to now, three big cycles could be divided, each included a pair of warm and cold, and lasted about 3,000 years. The highest temperature was 3~4 ℃ higher than today's,while the lowest one was 3℃ lower than today's. The records of oxygen isotope of the stalagmite are basically consistent with the climate information provided by the carbon isotope and the growth rate of stalagmite. They can also be compared with the results of the study of phenology in our country and the section of loess in northern China.

Key words: Stalagmite, Isotope, Paleoclimate, Guilin

INTRODUCTION

To forecast future climatic changes has to understand past climatic change history and study the change regularities on various scales.   In view of the climatic change on large scale, the climatic changes in China show no difference from the global climatic changes. But local or regional climatic changes (especially on small scale) may not be identical. The difference includes both phases and changing range.   Accordingly, to study the global or the national climatic changes must study the regional change trend.
There have been a lot of researches on the ice cores, the lacustrine deposit on the Qinghai-Tibet Plateau, and the loess profiles in the north China, and some important progresses have been made.   In the karst areas of the south China, for lack of similar climatic information sources, it is necessary to extract climatic record from karst formation or speleothem, but there have not been systematic researches so far.
  Guilin is a famous karst area in the world, the karst speleothem and formation are well developed.   Accordingly, a big stalagmite at Panlong Cave in Guilin, with favorable cave environmental conditions, was selected for the research.   On the basis of monitoring the present carbonate deposit, the drip water and the environmental conditions near the sedimentation, and the detailed sedimentological study on its cut sections a synthetic study including stable isotopes, AMS-C14 dating and U-series dating was applied for extracting the climatic change information.

GENERAL SITUATION OF THE ENVIRONMENT OF PANLONG CAVE AND THE SEDIMENTOLOGICAL CHARACTERISTICS OF THE STALAGMITE

Panlong Cave is located on the western side of Guilin-Yangsuo highway about 37 km to the south of Guilin City.   The elevation of the cave's entrance is about 190m in a. s. l. . . The country rock is the Upper Devonian limestone.   The cave is about 251 m   long, 6-12 m high (only 2 m at the northern entrance), and 11-15 m wide (only 5 m at the southern blocked end).   The thickest part of the overlying roof is up to 120 m.   The cave's floor is flat.   Stalagmites, stalactites, columns and so on are widely distributed. No. 1 stalagmite selected for the research is located at the part about 191 m away from its entrance, and 122 cm high, with a diameter of 45 cm at its bottom.   Based on one-hydrologic year's monitoring, the lowest air temperature is 18. 9 'C , the highest 19. 9-20. 0 'C , averagely 19. 5'C.   The temperature of the drip water is more stable, the lowest temperature 19. 0 'C and the highest 19. 8 'C, averagely 19.5 'C, that is close to the ground annual mean air temperature (19. 1. 'C) over 10 a of Guilin area.   The relative humidity is 100%.   Accordingly, the conditions above could be favorable for the reconstruction of paleoenvironment.

The inside growth ring of the stalagmite is clear.   The rings consist of the calcite with various grain size and tones.   According to the tone rhythm (from light to dark) and the sedimentation hiatus characteristics, it can be layered from bottom to top as follows:

I. Yellow-white and fine grained, with cerebrum-like microlaminae and uniform color and texture.   The layer ranges from 103. 2 cm to 122 cm from its top.   On the top of this layer there is a brown weathering crust that was caused by a long period of the interaction of solid -gas (and biologic) phase under the interruption of drip water.   During this period, the carbonate was leached out and Fe, Mn, Cu, Zn, etc. were relatively enriched.

II. White and fine grained, 5. 1 cm thick and with 4 intercalation layers of dark laminae. On the layer top , the boundary of the dark lamina is obvious, which shows a long period of sedimentation hiatus.

III. White and a little coarse grained, with wavy laminae.   The layer ranges from 75.8 cm to 101. 5 cm from its top.

IV. Pink, grey-white hybridized, with 4 secondary tone rhythms, sitting at 36. 2 -75.8 cm from its top.

V . Grey-white, a little coarse grained, with not-clear laminae, sitting at 30.6-36.2cm from its top.

VI. Pink, with obvious laminae, sitting at 15. 6-30. 6 cm from its top.

VII. Grey-white, with many dark laminae, sitting at 0-15.6 cm from its top.

There are two hiatus surfaces with a long period of duration at the lower part of the sec-tion (on the top of layer I and II). There are some hiatus at the middle and the upper part, but the duration is short. Accordingly, these parts may be thought as continuous deposition.

CHECK OF THE ISOTOPE DEPOSITION EQUILIBRIUM

When calcium carbonate (calcite and aragonite) precipitates from water, the exchange equation of oxygen isotope is as follows:

$1/3\ CaC^{16}O_3 + H_2^{18}O \Leftrightarrow 1/3\ CaC^{18}O_3 + H_2^{16}O$

where the equilibrium constant of the reaction is $\alpha_{cw} = Rc/Rw$.. Rc is the $^{18}O/^{16}O$ ratio of carbonate, Rw is the $^{18}O/^{16}O$ ratio Of water.

As the isotopes in two phases are in equilibrium, the equilibrium constant only depends on the temperature during precipitation, and isn't related to other environmental factors, Accordingly, the temperature during precipitation may be inferred from the isotopic composition. The precipitation temperature can be determined by O'neil's oxygen isotope temperature equation, that is, $1000\ln\alpha_{cw} = 2.78 \times 10^6\ T^{-2} - 2.89$

In order to check if the isotopes of sediment are in equilibrium with those of drip water, two clear laminae, sitting at 41. 5 cm and 76. 9 cm from the top of the stalagmite respectively, were sampled in a certain distance from the section axis to the two sides for carbon and oxygen isotope analysis. The results are shown in Fig. 1.

Fig. 1 shows that the change of $\delta^{18}O$ is small, the standard deviation of lamina 24-1 is 0. 15‰, that of lamina 35-2 is 0. 3‰. If the samples could be restricted in a lamina, the variation range may be much smaller. Since the $\delta^{18}O$ values in a growth layer are well identical and haven't simple correlation with $\delta^{13}C$, it can be thought that there wasn't dynamic isotopic fractionation in the process of sedimentation.

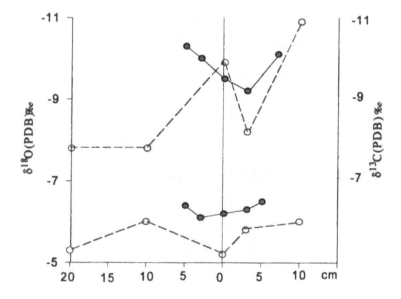

Fig. 1   Variation of $\delta^{18}O(\delta^{13}C)$ in the same growth lamina of No. 1 stalagmite, Panlong Cave
Solid line: lamina 24-1; Dotted line: lamina 35-2

Based on monitoring at different time and sites in the cave, the average carbon and oxygen isotope values of 9 present carbonate samples are: $\delta^{13}C(PDB)$= -8. 5 9‰ , $\delta^{18}O$ (PDB)= - 5. 89‰, and those of the relevant present drip water are: $\delta^{18}O(SMOW)$= - 5. 66‰. Based on O'neil equation and replaced by $\delta^{18}O$ values of the drip water and the carbonate, the calculated sedimentation temperature is 19. 8 'C , that is completely identical with measured mean temperature.

On the basis of the environmental conditions of the cave, and the equilibrium check above, No. 1 stalagmite is thought to be formed in isotopic equilibrium with the drip water.

RADIOACTIVE DATING AND GROWTH RATE OF THE STALAGMITE

Usually, speleothem is pure $CaCO_3$ precipitate. For several hundred thousand years old samples, the ideal dating methods are ^{14}C and U-series dating methods. This study mainly applied AMS-C14 dating, and the results were checked by β-counting ^{14}C dating method and α-counting $^{230}Th/^{234}U$ dating method.

The selection of the dating samples was mainly based on the changing points of sedimentation rate, especially, the hiatus surfaces and the characteristic points of climatic changes.

The reliability of the ^{14}C ages of speleothemes mainly depends on if the carbon isotope in water and in carbonate reaches full exchange during the precipitation of the carbonate. If the full exchange doesn't reach, ^{14}C specific radioactivity of the sediment is on the low side, and the age is older. If the full exchange reaches, the age is reliable. In order to prove this, a growing sediment near No. 1 stalagmite was collected in 1993 for the determination of ^{14}C specific radioactivity, and the ratio to the recent carbon is 1. 22 0. 021 that is close to recent atmospheric ratio. The result shows that the full exchange has reached.

The AMS-^{14}C ages of the stalagmite are in normal order from top to bottom. Meanwhile, the data at the bottom are well identical with the data of U-series dating method. Accordingly, the ages of No. 1 stalagmite are believable.

Fig. 2 shows the relationship between the measured ages and the growth height. The numbers in the figure are the calculated growth rate(mm/100 a).

Fig. 2 shows that the growth rates at different time are different. According to the division from large sedimentation cycle to small sedimentation cycle, the stalagmite can be divided in five cycles (A-E) from bottom to top. Each cycle can be divided into 2-3 stages based on the rates. The large sedimentation rates are in the 3 stages of Cl, Dl, El, reaching up 38-80 mm/ 100 a. The small sedimentation rates range from zero to several mm/100 a. There are two obvious black-brown hiatus at the bottom, that is, stages A2, B2. A2 is between 11080 a. B. P. and 32440 a. B. P. , and the interrupted time is 21000 a. B2 ranges from 6140 a. B. P. to 6930 a. B. P. , and the interrupted time is 790 a. The two hiatus resulted evidently from the interruption of drip water under dry and cold climate.

The changes of the local environment of caves, e. g. earthquake, roof collapse, blocking-up of drip water passages, displacement of drip water and so on, can also result in the interruption of drip water to form hiatus. In addition, growth rate can reflect regional environmental changes. For example, a larger sedimentation rate may reflect a relatively warm, humid climate and abundant rainfall, and a smaller sedimentation rate may reflect a

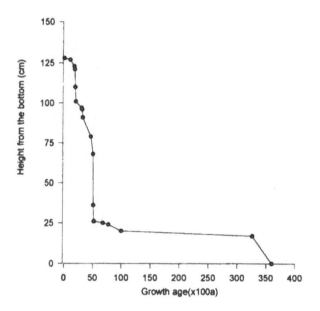

Fig. 2   Relationship between growth ages and growth height of No. 1 stalagmite, Panlong Cave

relatively dry and cold climate.   The   changes in the growth rates of No. 1 stalagmite are
basically identical with the climatic characteristics reflected by oxygen and carbon isotope, which shows that the growth of the stalagmite is mainly controlled by regional environmental factors.

RECORD OF STABLE ISOTOPES

During stable isotopic analysis, the carbonate samples react with 100% phosphoric acid to produce CO_2 that is purified to be determined by mass-spectrograph (MM-903E of England VG Company).   The relative standard of $\delta^{18}O$ and $\delta^{13}C$ is PDB standard.   The systematic error is less than 0.1 ‰.   The oxygen isotope of the water samples is processed by H_2O-CO_2 equilibrium method, and the hydrogen isotope is got by zinc-reduction method.   The decrepitation and zinc-reduction method is used for the process of inclusion hydrogen.   The relative standards of δD, $\delta^{18}O$ are SMOW standards.   The systematic error of δD is less than 1 ‰, that of $\delta^{18}O$ is less than 0.1 ‰ ,and that of δD of inclusion water is less than 2 -5 ‰.

The samples of carbon and oxygen isotopes were taken in a certain distance along the axis of the cut section of the stalagmite. The results show that the $\delta^{18}O$ during Holocene Epoch ranged from -5.5 to -6.8 ‰., the relevant temperature ranged from 5-6 °C. Many low peak areas (with high value) show the existence of some cold periods in the climatic changes.

Most of sedimentation hiatus are related to glacial periods.   The high $\delta^{18}O$ value from 32,000 a. B. P. to 36,000 a. B. P. and the suddenly high $\delta^{18}O$ value round 11,000 a. B. P. show the low temperature at those periods that are analogous to the Ashihe glacial stage and the Fuping glacial stage of the north China respectively.

During Holocene Epoch,   except for the high $\delta^{13}C$ values in several cold periods of the Middle and the Late Holocene Epoch, the $\delta^{13}C$ values in other periods were generally low. The $\delta^{13}C$ generally ranged from -8. 0 to -12. 0   ‰ , which shows that the general climatic characteristics during Holocene Epoch were warm, and humid, the ground vegetation was luxuriant, and the carbon isotope composition of the stalagmite resulted from the mixture of the isotopes of vegetation, atmosphere and source rock.   During the last glacial period, the global climate was dry and cold, the $\delta^{13}C$ of the carbonate was generally between - 6. 0 ‰ and -8. 0 ‰ or more, which shows that vegetation was difficult to grow under dry and cold climate, and the carbon isotopic composition of the stalagmite was formed by the carbon isotope exchange between atmospheric CO_2 and source rock.

THE CHANGING PATTERN OF THE CLIMATE SINCE 36,000 YEARS B.P.   IN GUILIN AREA

Because speleothemes are usually composed of pure calcite and there is not hydrogen-bearing mineral in them, the hydrogen exchange between water and rock can't take place and the original hydrogen composition can be kept.   Accordingly, we directly measured the hydrogen isotopic composition of the inclusion water, and calculated the $\delta^{18}O$ value of the original drip water by applying the meteoric water isotope equation ($\delta D = 8. 39\delta^{18}O + 16.04$) of the local area.   The results are as follows:

During the normal precipitation period of Holocene Epoch, the average δD was - 31. 4 ‰, accordingly the calculated $\delta^{18}O$ was - 5. 65‰, that is equal to the measured average $\delta^{18}O$ value (- 5. 66 ‰) of the present drip water near No. 1 stalagmite.

During the last glacial period (32000-36000 a B. P.), δD was -42. 4 ‰ and the calculated $\delta^{18}O$ was - 6. 97 ‰,   that is 1. 31‰ lower than the $\delta^{18}O$(- 5. 66 ‰) of the normal pre-cipitation periods, which shows that the sea level greatly descended during that period and Guilin area was much farther from the coast-line at that time than it is today.   Because of the influence of continental-oceanic effect, the precipitation enriched light isotopes.

During the periods disturbed by storm effect, the sedimentation rates are larger and the $\delta^{18}O$ of the carbonate usually is less than -7. 0   ‰ that have to be corrected by synchronously sampling analysis.

On the basis of $\delta^{18}O$ of the carbonate and $\delta^{18}O$ of the medium water, as well as O'neil e-quation, the temperature during formation of the sediment can be calculated, and the yearly $\delta^{18}O$ changing curve can be transformed into   yearly temperature changing curve.

Fig. 3 shows the climatic record extracted from No. 1 stalagmite.   The temperature on the ordinate has been corrected by the present mean temperature (19. 1°C) of Guilin. The climatic changes since 36000 a B. P. can be concluded as follows:

The Last Glacial Period
Because of the influence of the global dry and cold climate, the karst processes in this karst area were greatly weakened and the growth of stalagmite was slow.   During some stages, the drip water might stop and the deep brown weathering front was formed.

(1) During 36,000-32,000 a B.P., that corresponds to the coldest period of the middle Dali sub-glacial period,   the lowest ground annual mean temperature was 8-9°C,   the highest

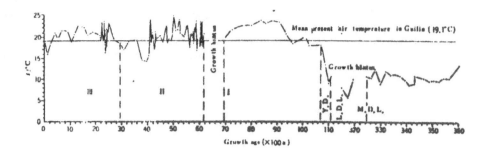

Fig. 3   Curve showing air temperature variation since 36,000 years B. P. in Guilin area
M. D. L: The middle Dali sub-glacial period; L. D. L - The late Dali sub-glacial period;
Y.D. : Younger Dryas; I . II . III : 3 climatic cycles during Holocene Epoch

temperature was 13 °C.

(2) During 32,000 a B. P. to 12,000 a B. P. that corresponds to the late Dali sub-glacial period, the sedimentation was interrupted.   On the basis of the oxygen isotope of the fine carbonate lamina on the hiatus surface, the lowest annual mean air temperature was 6-8 °C that was the lowest air temperature in this area during the studied time interval.

(3) Around 11,000 a B. P. , the expression of Younger Dryas was obvious in this area and the lowest annual mean air temperature was up 9 °C . After that period, the air temperature rapidly went up and was close to present level to 10,700 a B. P. . The air temperature was going up 9°C during 300 a.

Holocene Epoch
The jump-like rise of the air temperature in Younger Dryas resulted in a distinct leap between the last glacial period and the Holocene Epoch.   The period from 9,000 a B. P. to now can be divided into 3 climatic cycles from warm to cold, namely, 9,000 a B. P. to 6000 a B. P. , 6,000 a B. P. to 3,000 a B. P. and 3,000 a B. P. to now.   Each cycle lasts about 3,000 a. The highest annual mean air temperature during warm period was up to 22-23 °C that is about 3-4°C higher than that of today.   The lowest annual mean air temperature during cold period was 16°C that is about 3°C lower than that of today.   The distinguishable cold periods can be compared to the phenology records of China.   For the modern small glacial periods, the stalagmite records are fully identical with the historic records and the conclusion of the tree ring index of the cypress of Qilian Mountain.

The oxygen and carbon isotopic records of the stalagmite are also fully identical with the climatic information from the growth rates itself, and can be compared to the research results of the loess profiles in north China and the isotopic records of the global ocean.

CONCLUSIONS

(1) AMS-^{14}C dating method has the distinctive advantage of high resolution and sensitivity.   Combining with the stable isotopic studies, we can get the records in stalagmites and reveal the changing events of climate.   The setting up of the changing pattern since

36,000 a B. P. in Guilin area provides a reliable way and method for the studies of the land paleoclimatic changes in south China.

(2) The temperature changing records from the oxygen isotope are checked by the information of the carbon isotope and the growth rates. The study results show that the low δ^{18}O and δ^{13}C and the large growth rates usually reflect warm and humid climatic characteristics. On the contrary, the high δ^{18}O and δ^{13}C and the small growth rates often reflect cold and dry climatic characteristics.

(3) Compared with the similar researches at home and abroad, this study applied a large stalagmite and advanced AMS-^{14}C dating method, and the sampling interval can reach up millimeter magnitude. Except for the interrupted periods of sedimentation, the resolution of climatic changes has been up 100-200 a, even during cold period, it can be up 500 a. For a time interval of tens of thousand years, other researches can't reach this resolution. A Franch Duplessy et al. (1987) selected the mono-species foraminiferal fossils from the two, deep - sea borehole cores (SU 8 1 - 1 8 , CH 7 3 - 1 3 9 C) in the North Atlantic Ocean , made AMS -^{14}C dating and δ^{18}O analysis and achieved the outstanding research results of the oxygen isotopic records (in the time interval since 20000 a B. P.). But because the selection of the fossils was difficult, the sampling interval was large (usually 100 mm) and the resolution of climatic changes could only be 500-1000 a.

There have been achieved many results in the studies of paleoclimatic changes by using stalagmites, e. g. in Cold Water Cave, Iowa of USA; in Uamhan Tartair Cave , Scotland (cooperation of England and USA); in Drotsky Cave of Botswana, Africa (made by Americans). These researches have reached up high resolution, but all time intervals are confined to Holocene Epoch because of the small stalagmites (only 16-40 cm high).

(4) The karst areas are widely distributed and the cave resources are abundant in the south of China, which provides a superior position for paleo-climatic studies by using large stalagmites. The study in Panlong Cave shows that speleothem is a good carrier of age and paleoctimatic information. Applying the comprehensive method of AMS-^{14}C dating (TIMS U-series method for the samples older than 40000 a) and stable isotope study, we may get the details of climatic changes with a resolution of 10-100 a, and provide reliable regional data for the studies of the global climatic changes and the forecast of the future climate.

Acknowledgments

This paper is funded by the project of the Ministry of Geology and Mineral Resources(8502218) and the project of National Natural Science Foundation(49070155).

REFERENCES

1. O'neil, J. R..Oxygen isotopefractionation in divalent metal carbonates. *J.Chem.Phys.*,**51**,(1969).

2. Zhang Zigang. Stable isotopic geothermometer. In: *Application of Stable Isotopes in Geological Science.* Shanxi Science and Technology Publishing House, 23-26 (in Chinese with English abstract) (1983).

3. Sun Jianzhong et al. Paleoclimatic environment of loess plateau during last glacial period. *Quaternary of Loess Plateau.* Science Press, Beijing, 154-185 (in Chinese with English abstract) (1991).

4. State Science and Technology Committee of China. Possible climatic changes of future 60 years in China. In: *Chinese Science and Technology Blue Book (No. 5, Climate)*. Scientific and Technological Literature Press, Beijing,128-136 (in Chinese with English abstract) (1990).

5. Duplessy, J. C., Bard, E. , Arnold, M. , Maurice, P.. AMS-^{14}C-chronology of the deglacial warming of the North Atlantic Ocean. *Ibid*, **B29:1-2**,223-227 (1987).

6. Dorale, J. A. *et al.* . A high resolution record of Holocene climate change in speleothem calcite from Cold Water Cave, Northeast Iowa. *Science*, **256**, 1626-1630 (1991).

7. Andy Baker *et al.*.Annual growth banding in a cave stalagmite. *Nature*, **364:5**, 518-520 (1992).

8. L. Bruce Railsback *et al.*. Environment controls on the petrology of a Late Holocene speleothem from Botswana with annual layers of aragonite and calcite. *Journal of Sedimentary Research*, **A64:1**,147-155 (1994)

Proc.30*th* Int'l.Geol.Congr.,vol.24,pp.143-154
Yuan Daoxian (Ed)
© VSP 1997

Development and Harnessing on China's Rivers and Environmental Geology

SI FUAN

Water Resources and Hydroelectric Planning and Design General Institute, MWR, China

SHAO WEIZHONG YANG GUOWEI

Tianjin Investigation Design and Research Institute, MWR, China

Abstract

China is not abundant in water resources. The distribution of precipitation in time and space varies largely, therefore, water logging and drought disaster happen frequently. China's government have invariably insisted to attach great importance to water engineering construction. In the past years, the water projects have brought great social and economic benefits to our country, such as, in flood controlling, water-logging control, drought fighting, disaster reduction and urban water supply, etc. Recently, China's government uplift the water projects at the first level of infrastructural facilities and industry of national economy and social development, and decided to develop and harness rivers and lakes rapidly. This paper will discuss the environment geological factors and its effects on the construction of water project in China, the environment geological problems caused by the construction of water project, especially the reservoir. Finally, some countermeasures are put forward in order to prevent/control these environmental geological problems associated with the development and harnessing of water resources.

Keywords: water resource development, environment geology, reservoir works

INTRODUCTION

China lies in the eastern part of Asia with a land area of about 9.6 million km^2. There are numerous rivers and streams in the land, among them rivers with a catchment area of over 1000 km^2 amount to more than 1500. The development and formation of rivers in China are in a close relation to the geological conditions and the topographic features of China. China is in the east of the Eurasia Plate, which is sutured to the Indian Plate by the rift of the Yaluzangbu River in Tibet and connected in the east to the Pacific Plate through deep oceanic trench on the east of Taiwan. The interaction of the above-mentioned three plates has governed the formations of geological structure systems and geomorphy as well as the development trend and setup of river systems in China since the Mesozoic-Cenozoic era.

The relief in China is characterized by high west and low east. Most of the rivers in China originate from the west and flow east into the Pacific, and the minority turns south into the Indian Ocean. The Qinghai-Xizang Plateau, over 4000 m in elevation, is the original place of the Yangtze River, and Yellow River and the rivers in the southwest China. To the east and north of the Plateau, the relief is down to 2000 - 1000 m in elevations and forms the secondary giant plateaus and scattered basins, where are the sources of the Pearl, Huaihe, Haihe, Liaohe and Songhuajiang rivers, etc.. On the east of the Daxinganling, Taihang, Wu mountains and the Yunnan-Guizhou Plateau, being below 500 m in elevation and crisscross hills and plains, where the large rivers enter in their middle and downstream reaches on flood plains.

China is one of the countries with strong monsoons in the world. The precipitation is very uneven in temporal and spatial distributions, progressively decreasing to an amount of below 200 mm in the northwest inland from 1600 mm in the coastal area in southeast. Also, the runoffs in large rivers have very disparate changes annually and within a year, resulting in a variation coefficient of $C_v=0.2 - 0.6$, and low water or high water periods going on 3-8 years occurred in some rivers. Within a year, extreme runoffs took frequently place in rain season and formed floods so as to cause flood and drought diasters from time to time in China.

Rivers in China have an annual runoff of about 2700 billion m^3, corresponding to 2250 m^3 per capita, which is only 1/4 of that in the world. Therefore, China is not rich in surface water resources.

The large rivers in China have distant sources and long streams, and there are large and concentrated water drops in many rivers. Especially, in the southern part of the country water-power resources are rich due to a plenty of rain. According to the investigation, the hydropower potential in the whole country is 676,000 MW, of which the exploitable is 378,000 MW, and up to now, developed hydropower amounts only to 10% of the total.

China has a population of about 1,200 million persons, which makes up 1/4 of the world population, in which 90 percent reside in the middle and east parts where the middle and downstream reaches of major rivers lie. Therefore, to control flood diasters and develop water projects have been taken as an important task of river harnessing since ancient times. Since, 1950, a great achievement in the construction of water projects has been attained, and on statistics, that are:

- dyke regulation, 0.24 million km;
- reservoir construction, over 80,000 sites;
- irrigation area, increasing to 740 million mu from original 240 million mu;
- installation of hydropower units, 37,800 MW;
- increasing water supply for 140 million persons and 83 million domestic animals;
- controlling soil and water loss, 0.61 million km^2.

In 1994, the benefits from flood control, draining water logged areas and drought resistance provided by water projects were up to 210 billion RMB Yuan.

Nevertheless, river harnessing and water resource development still require arduous efforts, because of the facts that the targets for flood control and disaster resistance have not been achieved to a great extent, tasks for irrigation and interbasin water diversion are still heavy, and development of most hydropower resources and soil and water conservation yet remain to be carried out.

The construction of water projects, with which mankind remakes nature, should be both suited to complex environmental geological conditions and given prevention and control of damages caused by the harmful environmental geological conditions so as to bring their benefits to mankind.

MAJOR ENVIRONMENTAL GEOLOGICAL PROBLEMS WITH DEVELOPMENT AND HARNESSING OF RIVERS IN CHINA

China covers a vast expanse of land with different physicogeographical belts and geological zones, various geological disasters took place from time to time. The development and harnessing larger rivers in China, therefore, have to accept a challenge of various environmental geological problems.

Seismic Disasters
Most of China lies in the contact area of two major seismic zones in the world, the Himalayas and circum-Pacific, where seismic activities occurs frequently. The destructive earthquakes in the records available were as much as over 1000 times, making up 35% of the total in the world, and since 1949, the earthquakes greater than or equal to 7 in magnitude amounted to 52 times. China is one of the countries with frequently-occurring earthquakes. The earthquakes took place mainly in:

- the sea area around Taiwan Province;
- Xinjiang;
- Qinghai-Xizang Plateaus;
- Jinshajiang, Yalongjiang, Daduhe, Minjiang and Bailongjiang rivers in the upstream of the Yangtze River;
- Lanzhou, Zhongwei, Yinchuan Basin, Fen-Wei Valley in the upstream of the Yellow River;
- rift zones of Yan Mt. and Taihang Mt., and Hebei Plain in the Haihe River Basin;
- Tan-Lu Seismic Zone running through the plain in downstream of the Huaihe River, Bohai Sea Bay and the downstream of the Liaohe River.

During the seismic activities, besides the earthquake force itself causes the failures of hydraulic structures, some secondary damages would be induced by earthquake as well.

- Ruptures of buildings due to the revived seismic faults (inclusive of active faults). Such a seismic rupture took place during the Tonghai Earthquake (7.7 in magnitude) in 1970, Luhuo Earthquake (7.9 in magnitude) in 1973, Haicheng Earthquake (7.3 in magnitude) in 1975 and Tangshan Earthquake (7.8 in magnitude) in 1976. During the Luhuo and Tangshan earthquakes, the rupture lengths extended for 90 km and 8 km, horizontal displacements 3.6 m and 1.53 m, and vertical one 0.5 m and 0.7 m, respectively.
- Landslides due to earthquakes. Since this century, the landslides and collapses induced by earthquakes took place in the instances as much as 37 in China, in which, the known Diexi Earthquake (7.5 in magnitude) in the upstream of the Minjiang River in 1938, caused extensive collapse in the epicentre area and the River was blocked off, in result, 3 large lakes formed, which burst after 45 days,

resulting in a death of over 2600 persons owing to the subsequent flood disaster.
- Liquefaction induced seismically. The liquefaction occurred mainly in the sand, sandy loam and light loam foundations, which resulted in the foundation failure. Such a liquefaction took place during the Haicheng Earthquake in 1975 and Tangshan Earthquake in 1976. The investigation after the earthquakes showed that the liquefaction just could be induced in the area with intensity of VI, the damage ratios of buildings by the earthquake were up to 90% and 100% in the area of intensity VII and VIII separately.

Landslide, Avalanche and Mud Rock Flow
The landslide, avalanche and mud rock flow are common geological disasters in the mountain region. In China, the statistics indicated that the large-scale landslides with volumes of over 10 million m^3, avalanche with volume greater than 1.0 million m^3 and mud rock flows of 1.0 million m^3 above, which are observable, were 140-odd, 51 and 149 sites, respectively. They cover an area of over 1.73 million km^2, making up 18.1% of total land area of the whole country, and distribute mainly in:

- the area of the east edge of the Qinghai-Xizang Plateau;
- Jinshajiang, Yalongjiang, Daduhe rivers, and upstream area of Minjiang and Bailongjiang rivers in the upstream of the Yangtze River;
- mountains in east Sichuan, west Hubei and west Hunan;
- the Three Gorges in the main stream of the Yangtze River and upstream area of the Hanjiang River;
- loess plateau in the upstream of the Yellow River.

Besides the topography, lithology, geological structures, earthquakes and rainfalls, the landslide, avalanche and mud-rock flow are in relation to the engineering activities by mankind. Since 1950, as a result of a great amount of the economic and engineering activities, such as reclamation of slope land, clearing of forest, mining, road construction, the landslide, avalanche and mud-rock flow, which may cause damages have got a trend to intensify.

They are very harmful to the harnessing and development of rivers. For example, in 1983, the Sale Mt. Landslide in Dongxiang, Gansu Province, which has a volume of 31 million m^3 with a debris cover area of 2 km^2 and a slipping length of 1500 m, instantly collapsed 3 small reservoirs. The information from Yunnan Province indicated that between 1950 and 1990, the medium and small reservoirs and hydropower stations damaged by the landslide, avalanche and mud rock flow amounted to 60-odd and 360-odd sites, respectively.

The river blockings caused by these actions took place on the Jinshajiang, Yalongjiang, Daduhe, Minjiang and Bailongjiang rivers, and on some rivers (e.g. on the Diexi in the upstream of the Minjiang River), especially, the flood disaster or the secondary mud-rock flow were induced by a burst of the blocking dam and the damages were caused in the downstream.

The Xintan Landslide with a volume of 2.6 million m^3 on the mainstream of the Yangtze

River in 1985, extended into the River, formed a tongue-like slide mass of 250 m wide and 93 m high above the water surface, and raised the water level resulting in a swell of 54 m in height.

Therefore, in developing the water-power resources in high mountain and valley regions, it is important to pay great attention to resultants from these geological phenomena.

Soil and Water Loss and Associated Sedimentation in Downstream Rivers and Lakes
The soil and water losses in China come into existence to varying degrees of region-dependence. According to the statistics, the soil and water losses cover an area of 1,823,700 km² in whole country, which distributes mainly on the east to the line of the Daxinganling Mt.--Yinshan Mt.--Helanshan Mt.--the east edge of Qinghai-Xizang Plateau, the most serious soil and water losses exist on the middle and upstream reaches of the Yellow River, around the Sichuan Basin in the upstream of the Yangtze River, on the hilly land in the south China and the upstream regions of the Haihe and Liaohe Rivers. The soil and water loss on the loess plateau amounted to an area of 430,000 km² with an annual mean erosion modules of 8000 t/km².a, and the value in the north Shaanxi Province and the east Gansu Province is up to 10,000 - 30,000 t/km².a, and the sediment in the Yellow River comes mainly from these regions. The observations by the Sanmenxia Hydrological Station showed an annual mean sediment amount to 1.6 billion t and an annual mean sediment concentration in runoff reaches 37.6kg/m³.

Most of the sediment was accumulated in the downstream river channels and lakes and their flood flowing, flood storage and navigable capacities are affected. For example, the downstream reaches of the Yellow River have been changed into"elevated river" along with the raised river beds year by year due to the accumulation of a great amount of sediment, and the flood water flows in the channels formed by dykes on its both banks, as a result of the safety of which the Huang-Huai-Hai Plain during the flood season is threatened.

The Dongting Lake is a key regulating lake in the middle reaches of the Yangtze River, which receives the flows from the Xiangjiang, Zishui, Yuanjiang, Lishui rivers and can partially regulates the flood of the Yangtze River. However, since the observed sedimentation of 3.75 billion m³ between 1951 and 1988 and excessive reclamation around the lake, the lake area has been decreased to 2691 km² at present from 4530 km² in 1949. The similar conditions are also observed with the Poyang, Chao Lakes and other lakes in the middle and downstream reaches of the Yangtze River.

The soil and water loss and resultant sedimentation in the rivers and lakes downstream has been, therefore, one of important factors intensifying the flood inundation in China in recent 10 - 20 years.

Leakage Associated with Karst
The karst is an unique landscape in the regions where soluble rocks distribute. In China, the carbonate rock covers an outcrop area of about 910,000 km², and distributes mainly in Yunnan, Guizhou, Sichuan, Guangxi, Guangdong, Hunan, Hubei provinces/regions in the middle and upstream areas of the Yangtze and Pearl Rivers, and in the partial regions

of the Yellow, Haihe and Liaohe Rivers. These regions are also more rich in water-power resources in China. However, the leakage associated with karst, collapse and water bursting during tunnelling, which exist in the carbonate rock regions, added to the difficulties in the development and utilization of water resources in these regions.

Since 1950, in the karst regions, many large-scale reservoirs and hydropower stations, such as Wujiangdu, Lubuge, Dahua, Geheyan, etc. have been constructed successfully and a wealth of experience accumulated for the further development of water resources in the karst regions. In addition, there is abundant groundwater resources in the karst region, developing and utilizing the groundwater for industrial and agricultural productions is also one important task in the construction of water projects.

Loess Collapse

In China, the loess zones have an area of about 640,000 km². They are distributed mainly in the most of Henan, Shanxi, Shaanxi and Gansu provinces and partial Qinghai Province, Ningxia and Inner Mongolia Autonomous Regions in the middle and upstream of the Yellow River. The accumulated thickness of the loess changes greatly, from max. over 200 m in the loess plateau of Shaanxi-Gansu-Ningxia to only 20 - 30 m on the valley terraces and hillside slope lands. The Wucheng loess (Q_1) in the lower Pleistocene and the Lishi loess (Q_2) in the middle Pleistocene have largely degenerate macro porous structure with light or no collapse; but the Malan loess of the upper Pleistocene (Q_3) is of large thickness, loose nature with large pores, developed vertical joints and more serious collapse. These characteristics of the loess have a great influence upon the stability of water projects, such as dam foundation, channel, diversion tunnel and reservoir bank slope.

ENVIRONMENTAL GEOLOGY PROBLEMS WITH RESERVOIR WORKS

Large and medium reservoirs are key projects in the harnessing of rivers. A reservoir plays an important role and has large benefits in flood control, irrigation, water supply, power generation and improvement of environment, however, it also changes the natural conditions in part of a river. The cyclic water level fluctuation and continuous water volume change in reservoir would result in the changes of the natural geological conditions in and around the reservoir area, and induced a series of the geological and environmental geological problems that exert unfavourable influences on environment.

Reservoir-induced Earthquake

The reservoir-induced earthquake is a seismic phenomenon caused by the reservoir filling. Since the first induced earthquake was taken place in the Xinfengjian Reservoir at the end of 1959, the reservoir-induced earthquakes occurred in 24 reservoirs scattered over 14 provinces/regions of the country. These earthquakes have the following features:

- Earthquake strength: 1.7 - 6.2 in magnitude, max. intensity up to VIII;
- Depth of focus: generally 3 - 6 km, max.11km, min. below 1 km;
- Seismic characters: having foreshock--mainshock--aftershock, the value b being higher than that of the natural earthquake in the locality, i.e. $b \geq 1$;
- Epicentral location: mostly in the tails of reservoirs and close to reservoir banks,

- only a few in the locations close to the dams (e.g. the Xinfengjiang reservoir);
- Water depth in the mainshock zone: mostly 50 m, max. about 80 m (the Xinfengjiang reservoir);
- Time of the earthquake occurrence: in 1 - 6 years after the impoundment, mostly during the impoundment, but few during the discharging of reservoirs (e.g. the Shengjiaxia Reservoir).

60% of the above-mentioned reservoirs are located in the regions with soluble rocks, 40% in the insoluble rock regions, which have the following one or several geological conditions:

- Soluble rock regions, nearly horizontal strata, very developed karst, having old karst or deep karst system;
- Transitional belt between the mountain and plain regions, obvious topographic changes in the vicinity of the reservoir;
- Active fault passing through the reservoir area and being a seismogenic one;
- Major regional fractures, especially tensile and strike-slip, passing through the reservoir area;
- Dislocation basin in the Tertiary or Quaternary;
- Obviously unusual geothermal area, such as warm spring, effusive rock, etc.

When the construction of reservoirs in the regions with above-mentioned geological conditions, it is necessary to give special investigation to the reservoir-induced earthquake. The influence of the reservoir-induced earthquake on environment depends on its magnitude. In China, the intensity of most reservoir-induced earthquakes was less than that of the earthquakes in the locality. However, the Xinfengjiang Reservoir-induced Earthquake with 6.2 of magnitude and the epicentral intensity of VIII caused the cracks in the dam abutments, and the Danjiangkou Reservoir-induced Earthquake, with 4.7 of magnitude damaged the dwelling houses around reservoir. Therefore, in the assessment of the damages due to reservoir-induced earthquake, not only should the dam area but also the surrounding area of the reservoir be paid attention to, and both the strong and weak shocks should be considered, because the ground shock was relatively strong due to shallow hypocentre of most reservoir-induced earthquakes and the rock and soil were loosed for high shock frequency.

Although the reservoir-induced earthquakes could not be avoided in some conditions, they occurred at a very small probability, and the statistical information indicated only a few reservoirs in China induced earthquake. The damages from the reservoir-induced earthquake can be reduced to a minimum extent provided a full investigation is carried out early and proper preventive measures are taken.

Slope Sliding in Reservoir
The earth and rock excavation in construction, reservoir filling, fluctuation in stage and leakage from hydraulic structures would result in, directly or indirectly, destabilization of reservoir slope. The slope sliding in reservoir is a process of reservoir bank reformation caused by the reservoir impoundment and water level fluctuation. The slope sliding in the Zhexi Reservoir on 6, March, 1961, and that in the Huanglongtan Reservoir

in 1974 - 1983 are typical examples.

The slope sliding take place generally in the sections of reservoir bank having the following geological features:

- There are weak structural planes dipping towards the river in slope rock mass, e.g. the Zhexi, Fengtan, Huanglongtan Reservoirs, etc.
- There are very thick loose accumulative formation, strongly weathered rock mass and creeping loosed rock mass on the slope, e.g. most of slope slidings in some reservoir areas in the south of China.
- There are old slope slidings which might be inundated or partially inundated by the reservoir, e.g. the Xintan and Huanglashi sections in the area of the Three Gorges Reservoir under construction.
- There are harmful structures of slope rock mass and mankind activities, e.g. the bank slope composed of weak coal bearing formation and shale overlain by solid thick limestone in the Permian, which is common in the Yangtze River Basin, forming a large-scale unstable rock mass and very thick colluvial deposit.
- Others, e.g. probably collapsed cliffs, as seen on the Dahuangya cliffs in the Wujiangdu reservoir area.

The reservoir slope sliding is in relation to the following factors:

- Water-logged softening or argillization, increased pore water pressure and lowered shear strength in the interbedded layers or structural planes in unstable soil and rock mass due to the raised groundwater tables in the reservoir bank and slope, forming potential slip surfaces.
- Lowered sliding resistance of the front edge on the anti-slid body due to uplift from the raised reservoir water level.
- Decreased volume of the front edge of sliding body due to erosion and local collapses by wave in reservoir.
- Continuous rainfall.
- Sudden drawdown of reservoir.

The high speed and sudden slope sliding into a reservoir often is accompanied with a very high surge, seriously harming the buildings, towns and shipping traffic in the reservoir area and the reservoir downstream area. For example, a surge of up to 21 m high was induced by the Tangyanguan Slope Sliding in the Zhexi Reservoir.

The stability of reservoir bank, therefore, has been paid great attention to by engineers. At the exploration and design stages, an extensive investigation should be carried out to find the unstable reservoir bank sections harmful to the dam and the town nearby, and to give satisfactory treatments to them before impoundment as far as possible. For those the stability of which is difficult to assess, a long-term observation system should be set up before the impoundment so as to monitor their behaviour and select the treatment measures.

Bank Caving

The bank caving is a process of reservoir bank reformation, caused by the wind-wave, and took place mainly in the plain-basin type reservoirs with wide water surfaces and in the loessic regions in the north of China. Occasionally, it occurred in some reservoirs in the south where the banks are formed of loose accumulative formation or strongly weathered rocks.

The width and progressive speed of the bank caving zone is governed dominantly by topography, soil quality, wave height and fluctuation range in stage, etc. The Guanting Reservoir, for example, located in the Huailai Basin, is set up on the sedimentary deposit composed mainly of loessic soils in the Quaternary. In the Reservoir, the widths of bank caving zones are generally 20 - 30 m with max. 50 - 80 m, having a per meter volume of 300 - 700 m³. Such a serious failure was resulted mainly from steep and high bank slopes, long fetch length of wind wave in the Reservoir. Such is the case with the Sanmenxia Reservoir, where critical bank cavings were caused by the bank composed of loessic soil mingled with silt and wide water surface.

In some reservoir areas in the south of China, the ship wave abrasion would also result in the bank caving. For example, in the Xijin Reservoir area, Guangxi Region, the bank caving with a width of 60 m odd occurred where the bank is closed to the navigation course. The bank caving occurs dominantly in the early stage of impoundment. For example, the Guanting Reservoir, it was impounded in the post-flood period of 1953, and up to 1955-1956 the width of bank caving reached to 50 - 80% of the final one, and then the caving decreased gradually and trened towards to be stable in 1960. Moreover, the observation data of the Longyangxia Reservoir showed that the bank caving took place mostly in the first raise of water level in the Reservoir, after which, the reiteration of water levels in this range had no large influence on the bank formation.

The bank caving will bring unfavourable impacts on the resettlement along the bank and on the reservoir life as well. The prediction of the bank caving, therefore, has always been one of major tasks of the engineering geological exploration for reservoir.

Immersion
By the immersion is meant the groundwater table raise due to the changed discharging datum level of groundwater by reservoir filling, which results in the swamping and salinization in the surrounding land, reduces yield in farmland, and causes deformation or failure of housing foundation and frost boiling of road. The earliest problem with the immersion occurred in the Guanting Reservoir, afterwards, immersions with varying degrees took place also in Sanmenxia Reservoir area and some reservoirs in the north and northwest China. The reservoir bank sections subjected to the immersion have generally the following features:

- The traditional belt between the rear edge of wide accumulated terrace of plain or basin type reservoir and the diluvium in the front of mountain;
- Mostly loessic light silty loam and sandy loam;
- Depressions with shallow groundwater table and surface water and groundwater difficult to discharge free.

The existing reservoirs in the south China have a smaller immersion as compared with those in the north, probably because there are differences in the geologic conditions, soil quality of reservoir bank and in the cropping system between these two regions, as shown below.

- In the south, flood plains or accumulated terraces on the both banks of rivers have mostly a double layer hydro-geological structure, i.e. a lower water-bearing bed formed of sand or sand-gravel and an upper relative water-resisting bed of heavy loam or clay.
- In the north, less rainfall and more evaporation bring about the condition for salt accumulation in the surface layer so as to be susceptible to the secondary salinization. There is no similar problem in the wet and rainy south.
- The upland crops require a less water-bearing capacity in soil. In the Guanting Reservoir area, severe immersion occurred in the large area of orchard, where because of deep fibrous root a great amount of roots died by the immersion after water level being raised. In the south, crops composed mainly of rice have a very strong adaptability to the immersion.
- In the north, low temperature would result in the frost-heaving of the foundation soil after the immersion to cause the cracking and failure of housings.
- The failure of houses in the north, sometimes, is in relation to loess collapse.

In addition, the reservoir area immersion in the karst region has some specific characteristics as follows:

- A reservoir constructed in the karst region takes water from both the surface river and subsurface river systems, therefore, besides the immersion caused by the surface river system, the immersion/inundation problem resulted from backwater of the subsurface river system should be investigated. For example, in the Yantan and Etan reservoirs area, there are solution depressions nearby, having a bottom elevations higher than the normal water level of the reservoir, they are difficult to be discharged due to narrow drain at the bottom. Especially, in continuous storm period, the depressions would be inundated temporarily because the surface runoffs in which could not be discharged free due to the reservoir backwater effect.
- For example, in the Dongping Reservoir in Shangdong Province, that is one of the plain type or low mountain/hill type reservoir in which solution residual hills outcropped, reservoir water leaked out to the Guishan 1 km away from the west bank dyke via the karst passage below the Xiao-anshan and the sedimentary deposit in the Quaternary in the reservoir, and resulted in swamping of the land around the Guishan.
- The leakage from the reservoir having the karst passages to the neighbouring valley would cause the immersions in some depressions on the another side of dividing ridge. For example, the Nagu Basin, 2.5 - 3.0 km southwest from the Shuicaozi Reservoir and with surface elevations 90 - 150 m lower than the reservoir water level, was immersed as a result of considerably raised groundwater table in the borders of the Basin in 2 months after reservoir filling in 1958.

The development of irrigation without drainage would also cause the raise of groundwater so as to induce the immersion and salinizatiion.

The essential measures for controlling the immersion and secondary salinization is to set up an efficient drainage system, so as to limit the groundwater table to below a critical depth at which the immersion and salinization will occur.

Siltation in the Backwater Area
After a river enters to the reservoir, the sediment carried by the river deposits mostly in the reservoir tail or the river entrances to the reservoir as the flow gradient becomes flat suddenly, forming a new delta. Along with continuously extended delta, the bed in reservoir tail and groundwater tables in both banks would be raised gradually, an upwarping tail of reservoir be formed so as to expand the immersion range in the upstream of reservoir. In the Sanmenxia reservoir on the Yellow River, impounded in 1960, up to 1962 the river bed upstream (the Tongguan area) was deposited to over 5 m thick, then the siltation extended upstream to the Weihe River, a tributary of the Yellow River, caused the overflow from the River and inundated the land of 300,000 mu. In 1967, when a heavy flood occurred in the Yellow River, sediment in the tributary Weihe could not be discharged due to the backwater effect and the Reservoir had to be reconstructed.

In the mountain reservoirs, diluvial deposit on the entrances of main stream and tributaries, which are formed of mud rock flows due to storms, would be extended upstream by the backwater effect. For example, since 1961, the 20 tributaries flowing into the Zhexi Reservoir, have their river bed deposited by 3 - 6 m thick, which damaged a large amount of farmland.

In recent years, man's activities, such as forest clearance, slope-reclaiming, mining, road construction and building, have been major factors led to the soil and water losses and reservoir deposits. Strengthening the soil and water conservation and comprehensive treatment of basins is therefore an efficient measure in reducing the siltation in a reservoir.

CONCLUSIONS AND SUGGESTIONS

The harnessing of rivers is in close relation to the environmental geological conditions, and to a certain extent, is limited by them. Construction of the water projects would bring on large comprehensive benefits for national economy and environment conditions in the locality but result in a revival of some geological actions and occurrence of geological phenomena creating specific environmental geological problems with the water projects.

The environmental geology is an important task that attracts the attention of various fields of a water project, such as exploration, environmental assessment, design, construction, operation and monitoring. For the continuous development of national economy, it is necessary to put the environmental preservation on a key place and to further study the environmental geology to relieve and remove its harmfulness. For this, the following suggestions are made.

- Enhancing public awareness of the need to protect environment and better doing comprehensive treatment of environment. In the human activities for remarking of nature it is necessary to keep relative dynamic balancing between humanself and environment nature. In constructing the water projects, it is important to place the environmental preservation on a key point and coordinate it with the development of water resources. In this case, a good comprehensive treatment of environment requires participation from all concerned such as industry, mining, agriculture and forestry, etc. in addition to the water resource sector.
- Better carrying out exploration of water projects. Most of the environmental geological phenomena above-mentioned could be prevented or controlled to minimal harm extent by some measures, though manpower could not stop their occurrence. Good exploration and prediction are of critical.
- Persisting in simultaneous planning and implementation of water project construction and environmental engineering. The major environmental geological problems, possibly resulted in disasters and exerted great influences on the national economy, should be treated prior to the operation of reservoirs and hydraulic structures, because a pretreatment is much more easy than that after the problems taking place.
- Conscientiously planning and construction of reservoir resettlement area. Many environmental geological problems have direct influences on the living environment of relocatees, therefore, careful investigation and planning of the resettlement areas are the most important for both the resident relocation and the continuous development of water projects.
- Strengthening monitoring during operation period. The data obtained from the monitoring would show the complex effects applied by various natural and artificial factors and the dynamic changes of the environmental geology. Through monitoring during the operation period, specially in the early days of reservoir operation, it is possible to exactly predict the trends of their development, find unfavourable signs in time and take the remedial measures.

REFERENCES

1. DUAN Yonghou, LUO Yuanhua and LU Yuan etc. The geological hazards in China, China Architecture & Building Press, Beijing (1993).
2. Institute of Environmental Geology, MGMR, Series of Environmental Geology Map of China, China Cartographic Publishing House, Beijing (1992).

Proc.30th Int'l.Geol.Congr.,vol.24,pp.155-170
Yuan Daoxian (Ed)
© VSP 1997

Experience of Studies of Geoenvironmental Consequences of Hydropower Development on Volga River

PARABOUTCHEV I.A.

Designing, Surveying, Research and Production Share-Holding Company "Institute Hydroproject", Moscow, Russia.

LARIONOV A.D.

Share-Holding Company "Samarahydroproject", Samara, Russia.

Abstract

The paper deals with analysis of long-term observations over nature and scale of negative geoenvironmental consequences of construction and operations of a chain of storage reservoirs (impounded by 8 hydro dams) on the Volga River, the largest stream of the European part of Russia. The main objective of these observations and analysis of such consequences is groundwater rise in the area adjoining the reservoir, which is widely spread in the plain terrain and affecting rather negatively the environment. Analysis is made of the data resulting from analysis of the causes ground water rise peculiarities, as well as differences in the magnitude and forms of their manifestations on the specific projects depending on the extent of impact of technogenic and natural factors in each study case.

Performance of the structures built for control of ground water rise is evaluated and low effectiveness cases are analysed.

Basing on the generalised expertise of the studies the paper outlines recommendations for effective measures ensuring reliable control of ground water rise in the areas adjoining the reservoirs, the most frequent consequence of hydro power development on large plain water sources.

The Volga river is one of not-numerous large lowland rivers in the world with regulated runoff where a cascade consisting of eight hydropower plants with the head of water between 10 and 20 m has been constructed. Creation of reservoirs with the water-surface area of many thousands of hectares (Fig.I) resulted in their large-scale effects on the environment and negative ecological consequences manifested the themselves during many years.

One of the most important ecological problems associated with hydropower development on the Volga river is underflooding of reservoir banks resulted from affluent of ground water drained by the river valley [I] As a basic guideline of distinguishing of territories with higher ground water level the so-called standard of unwatering (minimum permissible depth of occurrence of ground water level) is used, which until recently was taken as 3 m for towns, 2 m for rural settlements, 0,5-1 m for arable lands, I m for recreation and afforestation zones.

In the zone of Volga reservoirs the territories effected by the rise of the ground water level are large [2]. Their area amounts thousands and first tens of thousands hectares, i.e. as a rule 4-10% of the reservoir water surface area (Uglich and Ivankovo reservoirs - 40-77%) - Table 1. Some settlements, arable lands and forests are effected by the rise of the ground water level Damages resulted from the rise of the ground water consist in decreased strength of structure foundations, appearance of dampness in living spaces and production areas, flooding and failure of basements and cellars, deterioration of underground utility systems, deterioration of quality of potable water in wells, decrease in growth, destruction of bushes and trees.

Arising of the problem of ground water level rise on near-shore territories is explained by underestimation of this problem at creation of reservoirs in 1930-1960 on large lowland rivers of the Soviet Union and on the Volga and Dnieper rivers in

Fig. 1 Scheme of largest reserviors of the Volga river

Table I

Information on areas effected by underflooding on banks of Volga reservoirs

Sl.Nos	Reservoir	Years of construction of hydroelectric scheme	NWL elevation	Area of flooder lands, thou.ha	Area of lands effected by underflooding, thou.ha	Ratio of area effected by underflooding to reservoir water-surface area, %
1.	Ivankovo	1933-1937	124	33	25.4	77
2.	Uglich	1935-1941	113	25	10	40
3.	Rybinsk	1935-1950	102	455	36.7	8
4.	Gorky	1948-1957	84	157	26.2 11.3 5.7	17 7 4
5.	Cheboksary	1968-1980	68 (Design) 63 (1994)	215		
6.	Kujbyshev	1949-1957	53	645	28.7 18.3	4 3
7.	Saratov	1955-1969	28	183	19.3	10
8.	Volgograd	1951-1961	15	312	26.9 18.9	9

particular. The main problem raised at that time was to generate cheap electri
power for revival of the national economy. The effect of constructed projects on th
environment, even though it was taken into account in the design, did not receive
fair amount of attention. As a result many methodical problems were not properl
developed or appeared imperfect: development of underflooding criteria, method
of distinguishing of underflooding boundaries and assessment of flooding volum
and size; regularities of the dynamics of this process; natural and technogeneou
factors of underflooding; methods of its prediction (Table 2).

Among numerous factors of underflooding of banks of Volga reservoirs natura
(including geomorphological, lithological, hydrogeological) and technogeneou
(irrigation of lands, backwater effect of the reservoir, town-planning) factors stan
out [3]. Besides a separate group of factors associated with flaws in constructio
and operation of engineering protection works may be distinguished.

However, on Volga reservoirs out of 21 studied objects of underflooding 18 object
were associated mainly with organization-methodical reasons [4]. Errors i
investigations and designing, flaws in construction and operation may b
distinguished among them. the most of protected territories are characterized b
several factors of the rise of the ground water level but the most frequent are flaw
in building and operation of structures. It has been just this reason that the weake
coupling agent in the main common link is investigations - construction
operation. As a typical example operating conditions of drainage systems in th
town of Kineshma on Gorky reservoir may be mentioned [5]. The lower part of th
town 1,5 km in length and 300-400 m in width is protected against the reservoi
effect by the protecting dike, near-dike horizontal drainage and the pumping plan
(I) - Fig.2 and 3. In the first years of reservoir operation (1957-1960) the protectiv
structures ensured the required lowering of the ground water level and allowed th
optimum sanitary situation to be maintained on protected territories. In the cours
of time the drainage system was silted and clogged with domestic garbage and th
protected territory accommodating about 300 houses appeared underflooded an
water-logged.

Unfortunately investigators performing substantiation of designs rarily take part i
construction and seldom, if ever, participate in operation of structures, Therefor
they cannot estimate correctness of their recommendations and prediction
Designers it their turn not always perform designer's supervision of constructio
efficiently. The process of operation as a rule is outside their field of vision, whic
gives no way of checking reliability of adopted design solutions and correcting ther
in proper time.

Thus the possibility of active and skilled supervision of development of unfavourabl
technogeneous and natural processes and elimination of their effects durin
opetation of reservoirs is lost. Therefore the most efficient method of solution of th
given problem is organization of the engineering-geological monitoring system wit
functions of observations, predictions and control of natural and technogeneou
processes in the zone of interaction of the structure with the environment [6].

The main components of such monitoring were elaborated and tested on th
Cheboksary reservoir (Fig. 4, 5).

Monitoring during construction period consisted of formation of data bank
supervision of hydrogcololgical situation and correction of predictions on the bas
of field observations, supervision of quality of construction, engineering-geologic
active support of field designing.

Owing to the formed data bank it was possible to select properly some drainag
structures on the base of standardization of geologicalstructure and use c
analogues.

Table 2

Chief drawbacks of present methods, substantiation of prediction
of ground water level on reservoir banks, construction-period monitoring and
post-construction period monitoring of protection works

	Stages	Chief drawbacks of methods
I.	Study of hydrogeological conditions and assessssment of natural rise of ground water level	1. Absence of average-scale topographic-geodetic control 2. Absence of systematic field observation of hydrogeological situation (pre-construction period monitoring) 3. Absence of regional hydrogeological zoning of reservoir banks by conditions of underflooding 4. Absence of assessment of underflooding on territories under natural conditions 5. Insufficient assessssement of problems related to underflooding in instructive-methodological literature
II.	Engineering-geological substantion of prediction of ground water level rise	1. Innacuracy of regional hydrogeological zoning by conditions of predicted ground water level rise due to absence of proper geodetic control. assessssement of natural ground water level rise, monitoring of the dynamics of hydrogeological situation 2. Insufficient consideration of technogeneus factors at prediction 3. Absence of regulatory framework of assessement and prediction of rising of ground water level in forests 4. Innacuracy of criteria of underflood in Codes of Practice (SNiP 2.01.15-90) 5. Absence of scientifically grounded conception of prediction
III.	Engineering-geological substantion of design solutions and protection measures	1. Innacuracy of prediction of ground water level rise on banks in the absence of reliable initial data 2. Insufficient substantion of compensation for damages resulted from a reservoir due to absence of evaluation of underflooding in natural conditions 3. Inssufficient consideration of analogues at selection of drainage due to absence of data bank on ground water level on territories and protection measures on banks of varios reservoir 4. Insufficient consideration of ecological aspects and social factors
IV.	Construction-period monitoring of engenneering protection work	1. Absence of detail as-built documentation on quality of construction of protection works 2. Absence of monitoring of hydrogeological situation during construction period 3. Absence of analysis of methodological problems of drainage and pumping plants in geotechnical supervision manuals
V.	Post-construction period monitoring	1. Absence of detail monitoring documentation of operation of drinage system and pumping stations 2. Absence of monitoring of hydrogeological situation in the process of operation] 3. Insufficient analysis of problems related to methods of assessement of ground water level rise on the territories in operating rules 4. Absence of data bank and unified methods of supervision of engeneering protection

Paraboutchev I.A. et al

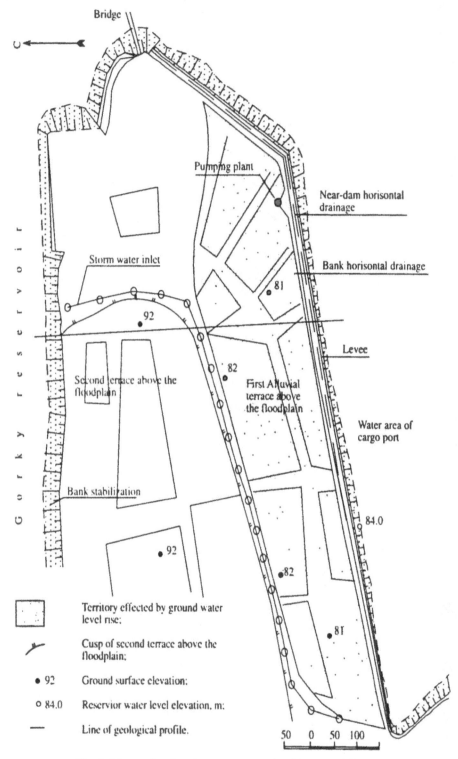

Fig. 2 Scheme of engineering protection works of the town of Kineshma.

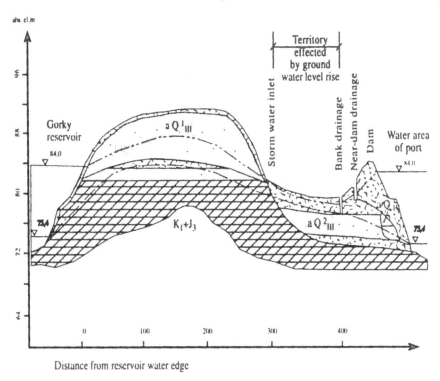

Distance from reservoir water edge

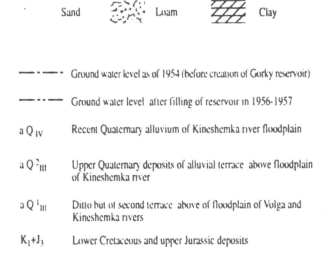

Sand   Loam   Clay

— · — · — Ground water level as of 1954 (before creation of Gorky reservoir)

— · · — Ground water level after filling of reservoir in 1956-1957

aQ_{IV} Recent Quaternary alluvium of Kineshemka river floodplain

aQ^2_{III} Upper Quaternary deposits of alluvial terrace above floodplain of Kineshemka river

aQ^1_{III} Ditto but of second terrace above of floodplain of Volga and Kineshemka rivers

K_1+J_3 Lower Cretaceous and upper Jurassic deposits

Fig. 3 Schematic geological profile through the town of Kineshma

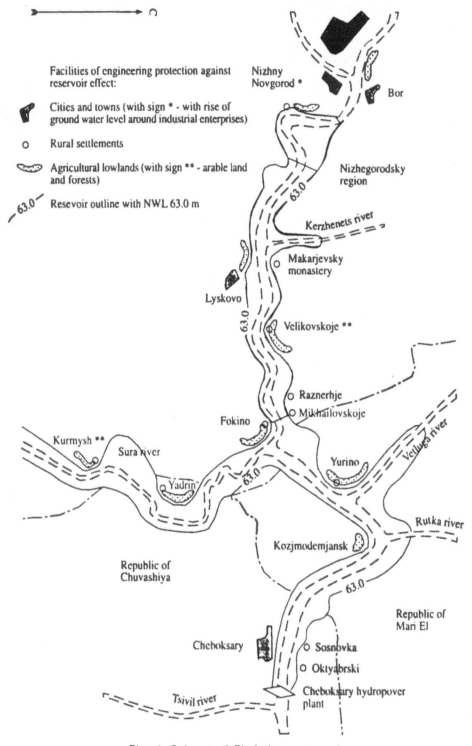

Facilities of engineering protection against reservoir effect:

Cities and towns (with sign * - with rise of ground water level around industrial enterprises)

o   Rural settlements

Agricultural lowlands (with sign ** - arable land and forests)

63.0   Resevoir outline with NWL 63.0 m

Nizhny Novgorod *

Bor

Nizhegorodsky region

Kerzhenets river

Makarjevsky monastery

Lyskovo

Velikovskoje **

Raznerhje

Mikhailovskoje

Fokino

Kurmysh **

Sura river

Yadrin

Yurino

Vetluga river

Rutka river

Kozjmodemjansk

Republic of Chuvashiya

Republic of Mari El

Cheboksary

Sosnovka

Oktyabrski

Cheboksary hydropover plant

Tsivil river

Fig. 4 Scheme of Cheboksary reservoir

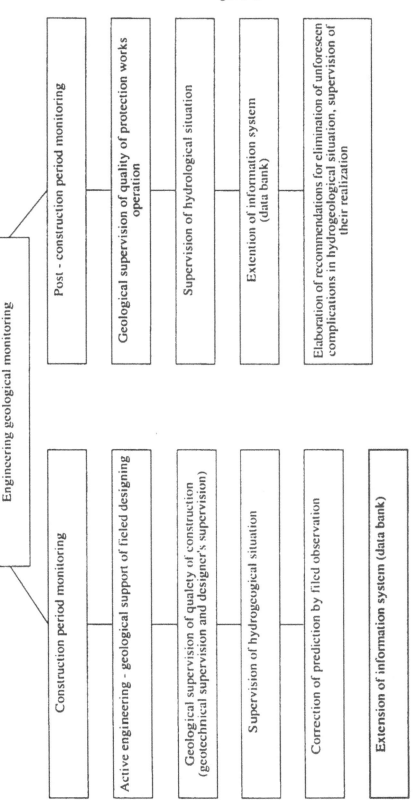

Fig. 5 Main stages and composition of engineering - geological monitoring of territories protected against ground water level rise in zone of Cheboksary reservoir

Supervision of hydrogeololgical situation during construction period was performed by two methods: with the use of stationary observations of ground water regime and by systematic field inspections of territories with appeared manifestation of underflooding. Hydrogelolgical observations were used for assessment of efficiency of constructed drainage systems (Fig. 6, 7, 8) and for check up and refinement of underflooding predictions. These observations made it possible to reveal deviations from some prediction assessments immediately after bringing into service of first drainage structures and to introduce necessary corrections into designs of engineering protection.

Engineering-geological monitoring in the process of operation of protective structures included supervision of operation of structures and supervision of hydrogeololgical situation of protected territories, elaboration of recommendations for elimination of unforeseen complications and supervision of their realization. In this case control forms and records were executed in which three types of parameters were fixed for each type of the structure: design, actual at introduction into service and actual by the moment of field observation (Fig. 9).

The result of performed studies of underflooded territories on banks of Volga territories was elaboration of proposals for methods of engineering geological substantiation of protection of near-shore territories against the rise of the ground water level. The following are five main methodical lines forming their basis:

1. Design investigations (assessment of geological-hydrogeological conditions without consideration for reservoir effects);
2. Substantiation of underflooding prediction;
3. Engineering-geological substantiation of the design of protective structures;
4. Monitoring during construction period;
5. Engineering-geological monitoring during operation. These methodical lines are main stages of engineering-geological studies of underflooding (Fig. 10).

The considered methods and composition of engineering-geological investigations of the rise of the ground water level on banks of Volga reservoirs are a part of the general program assessing the effects of technogeneous processes on the environment and may be used as experience at designing of reservoirs on other large lowland rivers.

PREFERENCES

1. Abramov S.K., Nedriga V.R., Romanov A.V., et al. "Protection of territories from flooding and underflooding" (in Russian)., Gosstroyizdot, M., 1961, 424 pp.

2. Yemelyanov A.G., "Underflooding as physical geographical process". From book "Environmental impact of Ivankovo reservoir" (in Russian), Kalinin, 1975 pp. 5-16.

3. Yemelyanov A.G., "Principles and methods of prediction of reservoir banks underflooding". VNIIG, Tr.coord sovesh po ghidrotekhnike, vyp .107, L., 1976, - pp. 161-169.

4. Zolotaryev G.S. "Objectives of comprehensive investigations for studying the underflooding and bank transformation and other processes in designing and operation of reservoirs". Sb. mauch. tr.Ghidroproyekta, vyp.103, M., - pp. 61-69.

5. Larionov A.D. "Engineering-geological grounds for bank protection of large plain reservoirs from under flooding". (on example of the Volga cascade) (in Russian). Diss. na soisk. uch. st. cand. techn. nank Samara, 1994 - p. 163.

6. Paraboutchev I.A. "Monitoring of interaction of hydraulic structures and geological environs" (in Russian). Inzhenernaya geologiya, 1992, No.2, izd-vo Nauka,-pp.3-16.

Main type of lithology of soils and draininge systems	Drainage analogues for given lithology	Drainage systems adopted in the design of Cheboksary	Names of projects at witch drainage measures are effected	Effeciency of drainage systems
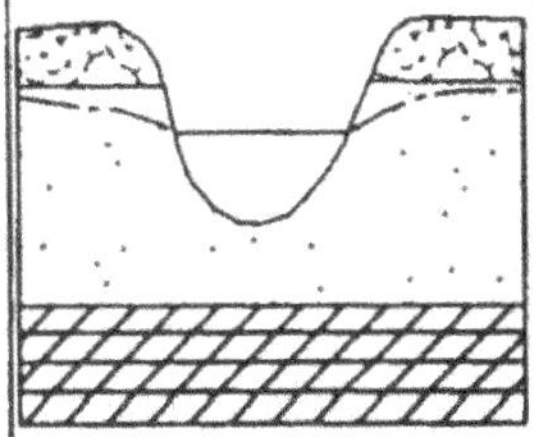	Deep canals constructed by hydraulic excavation (Kiev and Kanev recervoirs on the Dniper river)	Deep canals constructed by hydraulic excavation	Lyskovskaya and Kurmyskaya lowlands, Yurino settelment	High (radii of effect - 0.5 - 3.0 km)
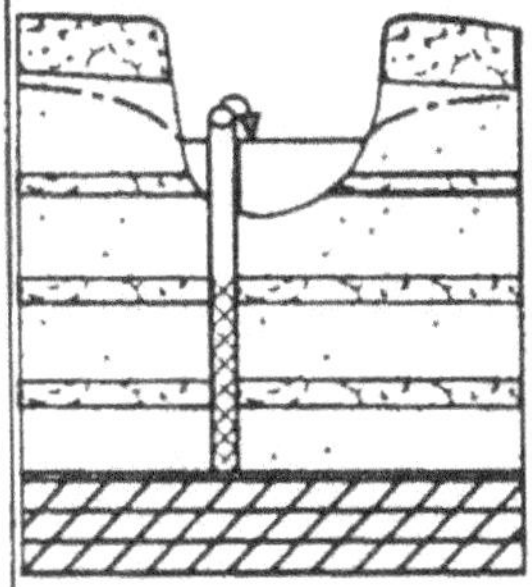	Ditto	Deep canals with flowing wells	Fokinskaya and Kurmyskaya (partially) lowlands)	High (canals operate without wells)
		Blind horizontal pipe drainages with flowing wels	Villadg: Sosnovka, Mikhailovskoye, Raznezhie, Sukhodol, Velikovskaya lowland	Insufficiently high (radii of effect - tens of meters)
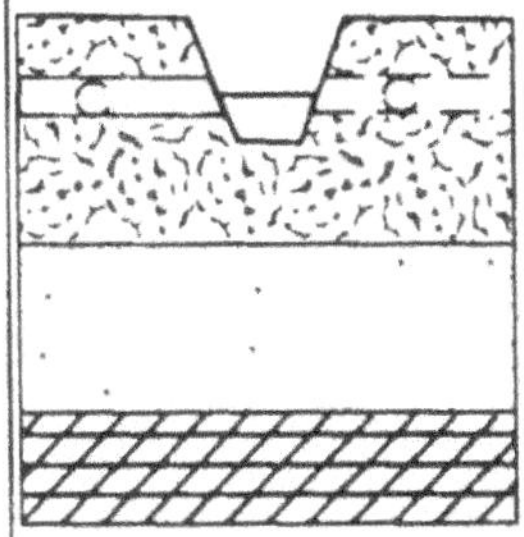	Combination of canals with system of reclamation blind and open drainages ((KOstroma lowland of Gorky reservoir on the Volga river)	Vertical drainage with filtres installed in sand	Fokinskaya lowland (partly)	Very low which requires addi- tional dranage systems to by
		Horizontal open drains (shallow canals)	Jadrinskaya lowland	Insufficienly high

Conventional signs of soils:

Sand 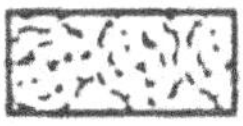   Loam 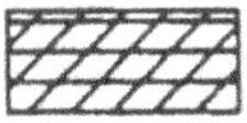   Clay

Fig. 6 Selection of drainage measures at protected projects of Cheboksary resevoir on the basis of standardization of geological structure and utilisation of analogues.

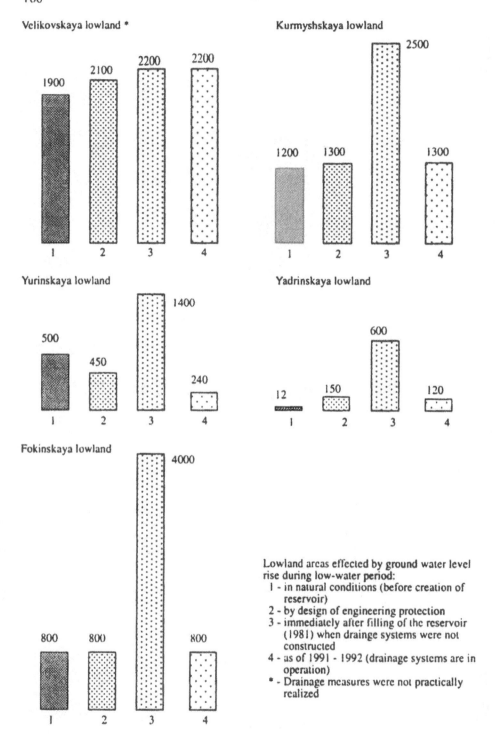

Fig. 7 Dynamics diagrams of ground water level rise on agricultural lowlands in zone of Cheboksary reservoir

a)

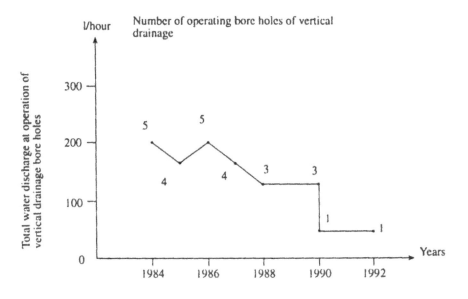

b)

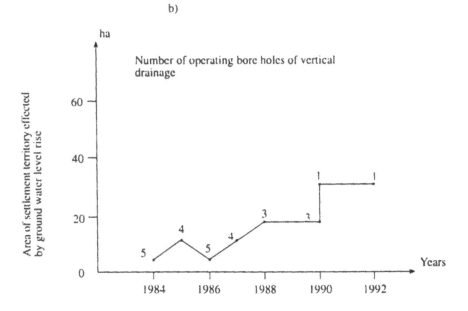

Fig. 8 Curves of variation of water discharge of vertical drainage (a) and areas effected by ground water level rise in Oktyabrsky settlement (b) in 1984 - 1992.

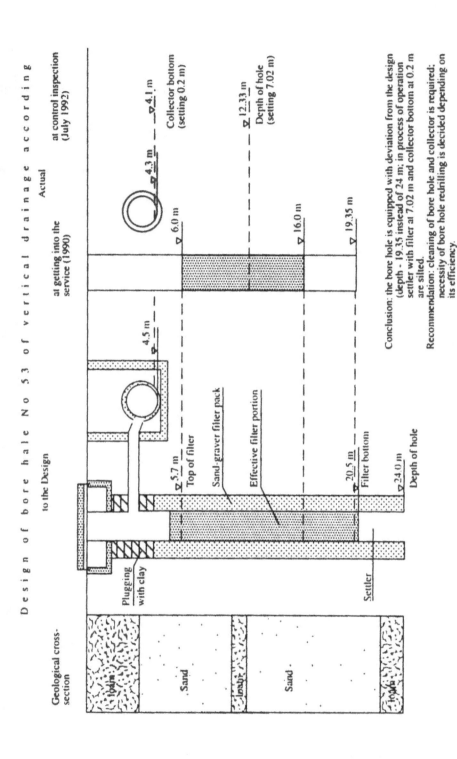

Fig. 9 Sample of control documentation during operation of vertical drainage on Fokinskaya lowland (Cheboksary reservoir)

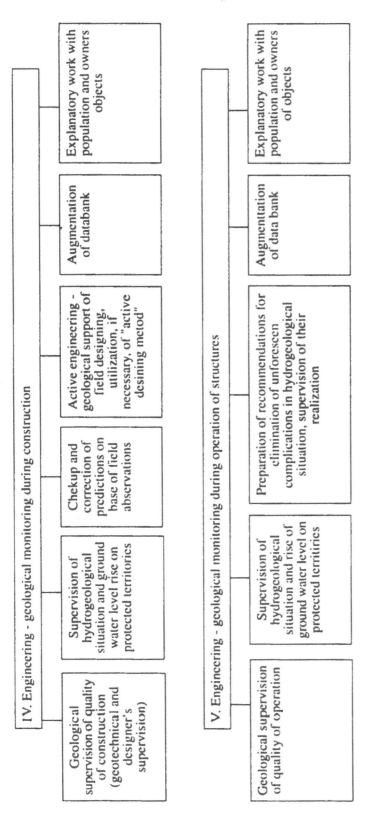

Fig. 10 Main metthodological lines and composition of engineering - geological investigations for substantiation of protection of reservoir banks against ground water level rise (scheme)

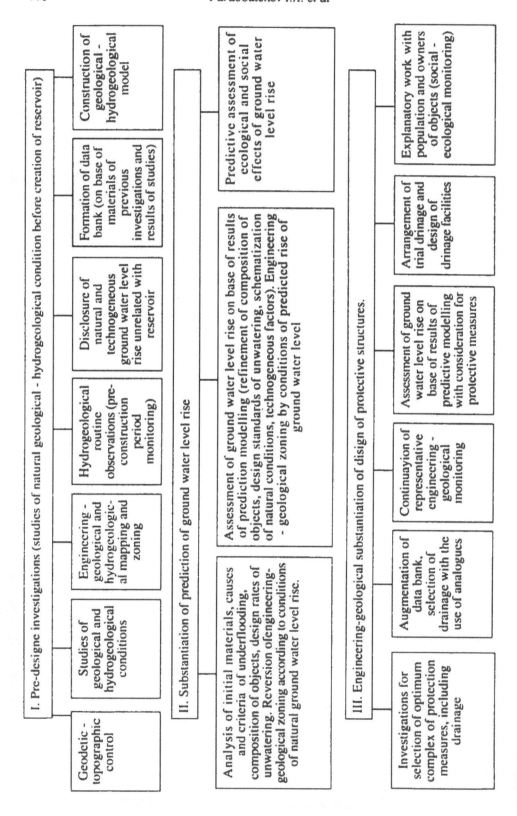

I. Pre-designe investigations (studies of natural geological - hydrogeological condition before creation of reservoir)

- Geodetic - topographic control
- Studies of geological and hydrogeological conditions
- Engineering - geological and hydrogeologic- al mapping and zoning
- Hydrogeological routine observations (pre-construction period monitoring)
- Disclosure of natural and technogeneous ground water level rise unrelated with reservoir
- Formation of data bank (on base materials of previous investigations and results of studies)
- Construction of geological - hydrogeological model

II. Substantiation of prediction of ground water level rise

- Analysis of initial materials, causes and criteria of underflooding, composition of objects, design rates of unwatering. Reversion of engineering-geological zoning according to conditions of natural ground water level rise.
- Assessment of ground water level rise on base of results of prediction modelling (refinement of composition of objects, design standards of unwatering, schematization of natural conditions, technogeneous factors). Engineering - geological zoning by conditions of predicted rise of ground water level
- Predictive assessment of ecological and social effects of ground water level rise

III. Engineering-geological substantiation of disign of protective structures.

- Investigations for selection of optimum complex of protection measures, including drainage
- Augmentation of data bank, selection of drainage with the use of analogues
- Continuayion of representative engineering - geological monitoring
- Assessment of ground water level rise on base of results of predictive modelling with consideration for protective measures
- Arrangement of trial drainage and design of drinage facilities
- Explanatory work with population and owners of objects (social - ecological monitoring)

Proc. 30th Int'l. Geol. Congr., vol. 24, pp. 171-184
Yuan Daoxian (Ed)
© VSP 1997

Environmental Geological Problems In the Development and Harnessing of the Yangtze River

Chen Deji, Yan Peixuan
Bureau of Investigation & survey, CWRC, China

Abstract

The Yangtze River, the longest river in China, is well known throughout the world. Since the People's Republic of China was founded, great progress has been made in the development and harness of the river. However, the present situations of the river can not meet the requirements of economic growth. Nowadays, various natural disasters in the river valley occur frequently, therefore, the steps of development and harness of the river valley must be quickened.

The environmental geologic problems of the Yangtze River being dependent on the original geologic environment and subjected to the impact by engineering and economic activities of human beings, have the distinctive features of regional geographic distribution Reasonable zoning is needed so as to master the genetic and developing rules of the problems and take effective counter-measures

The two great trans-century projects, the Three Gorges Project and S-N water Transferring Project, are under construction and planning respectively. Their environmental geologic problems should be paid enough attention to in the course of demonstration and construction with corresponding protective measures taken

The population is increasing rapidly while the economics is growing, and the whole Yangtze River Economic Zone is coming into being. Comprehensive regulation as well as environmental protection and monitoring should be conducted for the areas and reaches with severe environmental geologic problems in terms of the principle of
" Development, Regulation, Protection "

THE STATUS QUO AND A LONG RANGE DEVELOPMENT OF THE YANGTZE RIVER

The Yangtze River is the largest river in China and also famous in the world. It is 6300 plus km long, with a drainage area of 1.8 million km^2 which makes up about 18.8 % of the total area of China, yielding a mean annual runoff of 979 billion m^3, and water power potentials of 268600 MW.

Since the founding of New China, great achievements have been made in the development and harnessing of the Yangtze river. 3500km of main leeves and 30000 km of small dikes have been remedied, strengthened or heightened along the middle and lower reaches of the river, basically forming a complete flood — control leeve system. 48000 plus reservoirs of large, medium and small size with a total volume of 122.2 billion m^3, have been built, of which large size reservoirs number 105, totalling 73.3 billion m^3 in volume. Within the Valley the ever built hydropower stations with installed unit capacity over 0.5 MW each add up to 13490 MW. The total installed capacities of hydropower stations presently under construction reach up to 21680 MW (including the Three Ggorges Project which commenced to construction officially in Dec.

1994). The everbuilt and under construction hydropower stations total 35000 MW in capacities in the whole valley.

The development level of the Yangtze River, however, lags far behind and can not meet the demands of social and economic development in China. Up to 1985, the total installed capacities of ever built hydropower stations within the Yangtze River basin made up only 8.6 % of the total exploitable hydropower. Untill now the Yangtze River is still a rainy and floody river with frequent natural disasters such as flood, waterlogging, drought, etc. More than 84 percent of the Yangtze River basin is located in highland, mountainous and hilly area, as a result, the environmental geological problems are rather complex such as landslides, rockfalls, debris flow, soil erosion and the like.

In order to catch up with China's reform and openning to the outside world and the national economic development, the development and harnessing of the Yangtze River is required to speed up. In the report entitled " A brief planning report in the Yangtze Valley comprehensive development " , approved by the state on Sept. 21,1990, a detailed plan and development goal was proposed on flood control, power production and navigation, etc[1].

In flood control, the stem stream of the Yangtze River is required to control 100 — year frequency flood in the Jiangjian stretch, while to control the flood as in 1954 in other stretches. The main tributaries are required to control 20 to 50-year frequency flood,a few of them to control 50 to 100-year flood. In hydropower development, it is anticipated that by the year 2000, 33130 MW of hydropowers are to be developed within the Yangtze valley, mainly concentrating in the stem stream and its tributaries east to the Yibin city; By the year 2030,1117000 MW are to be developed, mainly concentrating in the upper reaches, including the stem steam JingshaJiang River, he tributaries Dadouhe River and YalongJiang River, etc.

In navigation, it is planned that the navigation conditions of stem stream of the Yangtze River and its tributaries will be improved by means of river regulation, dredge, (river) regime stablization, navigable depth increase and water way canalization by building hydroproject, etc. In terms of long range planning, by year 2030, together with the construction of water resources projects on the XianJiang River, GenJiang River, harnessing and development of the Zhu Jiang (Pearl) River as well as the openning of the arterial water conveyancy canal heading from the Hanjiang River to the North (i.e. the South-North water transporting Project), a complete waterway navigation system will be formed taking the Yangtze River as an artery, characterized with free passage between stem stream and its tributaries, and between rivers and sea, connecting the east with the west, and linking up the south and the north.

THE PRINCIPAL ENVIRONMENTAL GEOLOGICAL PROBLEMS AND ENVIROMNENTAL GEOLOGICAL ZONING OF THE YANGTZE VALLEY

The basic features of geological environments within the Yangtze Valley[2].
The environmental geological problems within the Yangtze valley result from following conditions: one is that they were generated and progressed under the natural

geological environments and exclusively controlled by natural factors; The other is that they are formed under certain geological environments but their occurrence and progress are affected in different degrees by human activities. Both of the two conditions are related to the geological environments where they occurred.

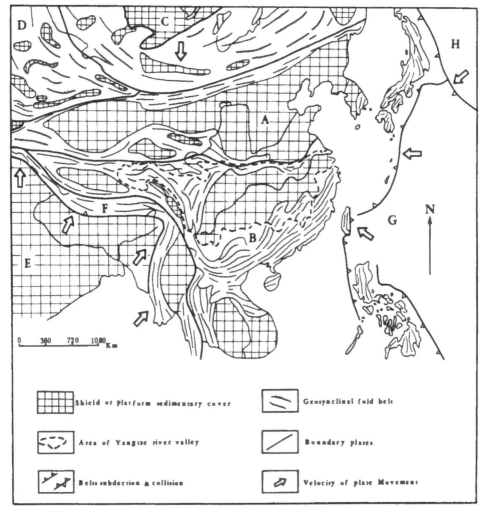

Figure 1 . Sketch Map showing location area of Yangtze River Valley in the Asian Plate

Simbols : A, Tarim - Sina - Korean plate, B, South China - Southeast Asian plate.

C, Siberian plate, D, Kazakhstan plate, E, Indian plate.

F, Mid - Iran-Gangdise plate, G, Philippine plate, H, Pacific plate.

The Yangtze River originates from the southern west side of the Geladandong snow-capped Mountain, the main peaks of the Tangula Mountain range of the Qinhai-Tibet plateau. From its origin to the mouth the Yangtze River is characterized geographically by the entire topography lowering from the west to east, forming three gigantic steps.

The stratigraphic formations developed within the Yangtze valley are rather complete, exposed from the late Archean era to the recent Quaternary period.
The Yangtze valley is situated on the northern tip of south China and southeast Asia plate, and connected closely with the Talimu, Sino-korea plate on the north, and is adjacent to the India plate on the west and southwest.Since the Cenozoic Era, the India plate has been moving northward successively and compressing, which directly controls the geological settings of the upper reach region, the Jinshajiang reach in particular, of the Yangtze Valley.To the east, and southeast side of the Yangtze valley is the Pacific and Philippine sea plate which has some influence mainly on the east of the valley and the coastal region (Fig. 1)[3].
Since the Pliocene Epoch, including the entire period of Quaternary, the neotectonic movement in the Yangtze valley is characterized generally with the strong uplifting in the west while weakening gradually eastwards, In view of seismicity, 90 percent of the strong earthquakes with magnitudes over six ever happened are located in the west region.

The principal environmental geological problems within the Yangtze Valley.
The principal environmental geological problems within the Yangtze valley are of many types, and can be classified into three major types as follows according to their geneses.Natural geneses,including earthquakes,active faults, wet collapse of loess, freeze-thaw of frozen ground, debris flow, dangerous passages and silting of river channel, bank scouring and caving, piping and subsurface erosion of leeve, etc. Human factors,including reservoir — induced seismicity, rock blasting, surface subsidence, riverbed deformation downstream of dam, reconstruction of reservoir bank,water pollution,and so on. Combination of natural and human factors, or led solely by one of them,including landslides, rockfalls; deformation of unstable rockmass, soil erosion, ground crack, ground collapse of karst, soil swelling and the like.

With the continuous spreading of human activities, the environmental geological problems, caused by human factors or combination of natural and human factors,are becoming more and more prominent and serious.

Zoning of environmental geology[4]
Each type of environmental geological problems has its own genetic envionments, including geological structure, stratigraphy and lithology, topography and geomorphology, hydrology and meteorology, neotectonic movement, endo and exo-dynamic geological process and so on. Therefore their distributions show obvious regional and geographic zonig features, based on which the Yangtze valley can be divided into three principal environmental geological regions, including nine sub regions. Their basic features are described in table 1 and shown on the attached map (Fig.2).

PRINCIPAL ENVIRONMENTAL GEOLOGICAL PROBLEMS IN THE DEVELOPMENT AND HARNESSING OF THE YANGTZE RIVER.

The environmental geological problems in the upper reach region of the Yangtze River[5].

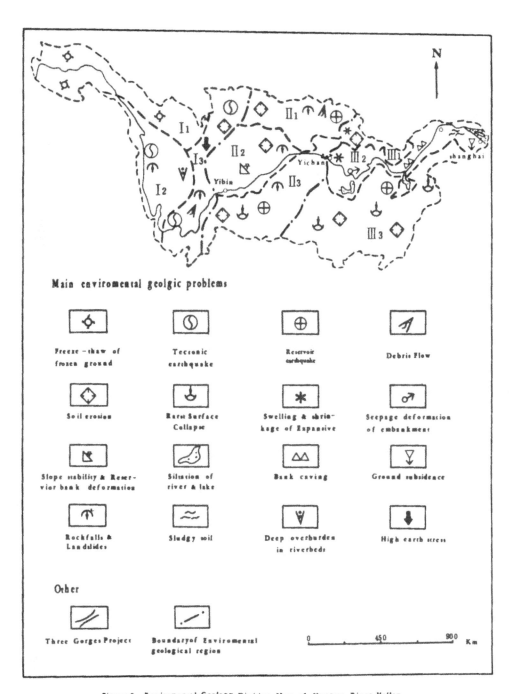

Figure 2. Enviromental Geology Division Map of Yangtze River Valley

Table 1. Basic Features of Environmental Geological Regions

E.G Sub-region	Brief account of geological environments	Principal environmental Geological problems
South Qinhai, west Sichuan plateau E.G sub-region I₁	Hilly plateau with EL. 3500-5000m and consisting of deep faults and compound folds trending NW;with active Quaternary Faults ; In lithology mainly consisting of sandy slate overlied with thick frozen soil	Frozen and thawed soil, thermal thawed mudflow; sliding; active faults problems with engineerings; and geothermal and rock burst in deep seated tunnels
Hengdan-shan mountainous E.G sub-region I₂	high mountain-deep valley region with E.L4000-5000m and consisting of compound mountain- folds zones and deep faults; with very active faults and frequent earthquakes; lithologically consisting of sandy plate inter- layered with phyllite, occassionally containing magmatite, and carbonate rocks	Landslides, rockfalls, debris flows active fault belts , high seismic intensity region , Karst water bursting
South Gansou Sichuan, Yunnan mountainous E.G sub-region I₃	High to medium mountain-deep valley region with EL.1000-3500m and composing the south tip of south-north trending tectonic zone;with active Quaternary faults and strong earthquake zones; lithologically consisting of metamorphoic rocks	Lansides, rockfalls, debris flows active faults , high earthquake intensity , locally Karstic leakage water and soil loss
Qinling Dabashan sub-region II₁	Medium to low mountain-deep valleyed region with EL.1000-2000m; located in the Qinling-Dabashan folded zone; with medium level seismic activities,locally occurred magnitude 6 earthquakes; lithologically consisting of metamorphoic rocks , carbonate rocks and magmatite	Landslides,rockfalls, debris flow, karstic leakage, reservoir-induced earthquakes, etc.
Sichuan basin E.G sub-region II₂	Low mountainous and hilly region with EL. 250-700m , situated in the Sichuan platform depression with weak neotectonic activities ; and lithologically consisting of red clastic rocks except for the core of anticlines where outcrop carbonate rocks	Landslides; rockfalls, water and soil loss; locally karstic collapse

E.G Sub-region	Brief account of geological environments	Principal environmental Geological problems
Hunan,Hubei Guizhou mountainous E G sub-region II3.	Medium to low mountains , deep gorged c plateau with EL.500-2000m;and consisting of the Huangling anyicline and NE trending compound anticlines with weak neotonic activities; and lithologically consisting of carbonate and clastic rocks except for the Huangling area where outcrops magmatite and the TGP is located	Landslides, rockfalls, karsti leakage and collapse, water and soil loss, reservoir-induced earth quakes in some areas
Huaiyang mountainous and hilly E.G sub-region III1	Low mountain and hill region with EL. 200-800m ; situated in the Nanyang depression and Huaiyang uplift zone ;The neotectonic movement is strong in the west while weak in the east; Distributed active faults with NWW and NE trending; occassionally occurred earthquakes over 6 degrees in magnitude; lithologically consisting of metamorphic rocks and magmatite	soil swelling and shrinking; water and soil loss; small scale landslides
middle & lower reaches plain of the Yangtze low mountainous and hilly E.G sub-region III2	Vast alluvial and lacustrine plain , with EL below 200m;situated in the Yangtze peneplatform depression belt; earthquakes are characterized by obvious difference in regions medium strong earthquakes mainly occurred in the locally relatively active region; lithologically consisting of Quaternary loose soil	soil swelling and shrinking ;sand liquefaction; piping soft soil deformation, swampiness ;ground subsidence; bankcaving , river siltation, and water pollution ,etc
South of the Yangtze low-mountainous and hilly E.G sub-region III3	Low mountainous and hilly region with EL 200-600m and a few mountain peaks of EL. 1000-1500m;situated in the south China folded zone and the Yangtze peneplatform uplift and depression zone; Lithologically consisting of metamorphic rocks and granite in the mountainous and hilly region, redbeds in the basin area and carbonate rocks in the central area of Hunan province	water &soil loss; karstic collapse small scale landslides; soil swelling and shrinking; reservoir-induced earthquakes.

The upper reach region refers to the upstream region above Yichang city, including the stem stream and its tributaries, covering an area of 1 million km². Among the 12 construction bases of hydropower development in China, 5 are located in this region, in addition to the developmentable hydropower resources of the Jialingjiang River, the Hanjiang River. The total capacities of the developmentable hydro-power stations will reach up to 130,000 MW, making up 65 percent of the developmentable powers of the entire valley.

The principal environmental geological problems encountered in the development can be summarized as follows:

(1) Regional stability and anti-earthquake provisions. The proposed hydro projects are mostly located in the region with seismic intensities of VII to VIII degrees, some even in the region of IX to X degrees, which will bring about difficulties in anti-seismic provisions and project costs. (2) Natural disasters such as landslides, rockfalls and mudflows, which will affect the stability of reservoir banks and safety of township relocations. (3) Deep overburden in riverbeds, generally over 30-40m in depth, some even exceeding 100m, which will give rise to technical difficulties and financial problems in foundation treatment. (4) Frozen earth. In the highland region in the south of Qinghai Province and in the west of Sichuan province, the permanent frozen soil layer ranges from 20 to 80m in thickness. The frozen and thaw caused collapsing and sliding and thermokarst settlement of frozen soil will harm or affect the canal and underground engineerings of the west route alternative of South-North water transporting project. (5) High earth stress. As experienced by some projects in the upper reach region of the Yangtze River, high initial stress may cause load-rebound and stress-release related deformation and failures, such as rock deformation and blasting of caverns and slopes which will bring about harmfull effects on the stability and construction of structures. (6) karst and related environmental geological problems, such as karst leakage, water bursting, karst collapse and reservoir-induced earthquakes, to which a special attention still has to be paid in construction of various engineerings.

The principal environmental geological problems in the middle and lower reaches' regions of the Yangtze River include:

(1) Dangerous passages in river channel and lake siltation. According to uncompleted statistics, within the strech from Daju (near to the HutiaoXia or Tiger-leap Gorge) to the river mouth about 3806km long, there exist 942 dangerous passages and bars of different types, averaging one in each 4 km length. Among them, there are 794 passages where navigation is seriously hampered, including 759 passages on the upstream above Yichang city and 35 on the downstream below Yichang city.

The formation of dangerous passages in the upper reach is related to natural disasters such as rockfalls, landslides and debris — flows, while in the middle and lower reaches it is related to meandering and braiding of the river and these passages are mostly located in the tectonic depressions.

Due to large-scale reclamation and siltation lake area decreases with each passing day. The area of the DongTing lake has been decreasing from 4530 km² in 1949 to 2740 km² at present time. The Poyang lake is being silted with a weight of 10 million tons and a silting thickness of 2-3 cm each year on average. The water area of the lake has decreased by 1185km² since 1954. The

Cao lake in history used to have a maximum area of 2000 km^2 but now decreases to 820 km^2. In Hubei province there were 1060 lakes with a total water area of 8300 km^2 in the beginning of new China, but yet exist only 326 lakes with a total water area of 2300 km^2 at present time.

(2) Bank caving mainly happens in the plain stretch of middle and lower reaches of the Yangtze river. According to statistics, on the stem stretch from Chenlinji to the estuary there are 12 to 35 percent of bank slopes where bankcaving frequently happens. Bankcaving mainly depends on such factors as the properties and structures of rock and earth that compose the banks,the dynamic conditions of water flows and river configuration. With the intensification of human engineering and production activities along the river, various artificial structures altered the state of flow in local stretches, thus frequently leading to the change of bankcaving in form, location and scale.

(3) The main environmental geological problems with the flood control levees
The flood control levees are the most important structures to control flood in the plain region of middle and lower reaches of the Yangtze River. According to statistics, the flood control leeves on both sides of stem stream, with their total length of 3569km, beginning from LingZhongsi of Songzi County of Hubei Province upstream and ending at Island ZongMingdao of Shanghai city downstream, protect farmlands of 38.73 million mu nd populations of 33.74 million. The most critical portion is the JingJiang levee where environmental geological problems are also prominent. Due to the complexity of foundation soil, environmental geological problems frequently took place, such as dispersing erosion, piping and boiling, levee sliding, bank caving and outburst, etc.

(4) Swelling — shrinking of soil
In the middle and lower reaches region of the Yangtze River Swelling soil is mainly distributed in the Jianghan Plain, Nanyang Basin and Feidong plain. Swelling and shrinking of soil can cause such environmental geological problems as foundation deformation and slope sliding, and the like.

(5) Ground subsidence, mainly resulted from extensive exploitation of ground water from loose layers, is one of the vital environmental geological problems in the lower reach region of the Yangtze River and its delta plains, such as Shanghai city, Changzhou city, Wuxi city and Nantong city.

Studies on the major environmental geological problems for the Three Gorges Project.[6]
At the NPL 175m the Three Gorges Project has a maximum dam height of 175m, total installed capacities of 18200MW, total reservoir volume of 39.3 billion cu.m, and backwater length of around 600Km as well as reservoir area of 1084Km^2.

(1) Studies on reservoir--induced earthquakes[7]
A great attention has all the time been paid to the studies on the reservoir--induced earthquakes for the Three Gorges project. On the basis of analysis of 106 cases of reservoir--induced earthquakes all over the world, and studies on the hydrogeological conditions and activities of fault at the damsite and in the forebay region of reservoir, earth stress measurement in deep boreholes, intensified seismic observations at the faulted regions, and studies on the regional

stress field and its possible changes after filling of reservoir by means of three--dimensional finite-element numerical simulation, and evaluation on the possibilities of reservoir-induced earthquakes according to the lithology, structural geology, permeability conditions, stress state in each reservoir stretch in combination with the background of regional seismicities, a prediction has been made on the locations and intensities of possible reservoir--induced earthquakes and their possible impacts on the hydraulic structures and reservoir environments. The study results show that the JiuWanxi Fault which is nearest to the damsite is most likely to produce reservoir--induced earthquakes at the location crossing the reservoir. But their maximum magnitudes would not go beyond 6 degrees and intensities affecting on the damsite would be VI degrees, less than the proposed design intensity of VII for the dam. In the stretch covered by carbonate rocks karst--type reservoir--induced earthquakes would probably happen but their magnitudes would be under 4 degrees.

(2) Studies on the reservoir bank stabilities
At the NPL 175m the Three Gorges reservoir will have a length of 580 to 667km in the stem stretch and 173 tributaries with backwater length over 1 km each, totalling about 1840 km, and submerged land area of 632 km^2. Since 1950s systematic yet multi--measure and multi--level coordinated researches have been carried out on the key problems. Till now the locations, numbers, scales and stabilities of the rockfalls and landslides on both sides of the stem stream and its tributaries have been basically identified, and their possible influences have been analysized and evaluated item by item. Study results indicate that the total volume of all the rockfalls and landslides and the volume of the earth and rock mass potentially entering the reservoir make up only 4.4 percent and 0.86 percent, respectively, of the total reservoir volume, thus they would have no substantial influence on the life-span of the reservoir. There exist no large scale rockfalls and landslides within the range of 26 km upstream from the damsite. As for those farther than 26km above the damsite their possibly produced surges would attenuate to a small height in front of the dam, as indicated by surge calculation and model tests according to various assumed conditions, which would not endanger project construction and operation of hydraulic structures. After impoundment of the Three Gorges reservoir the river channel of 600 plus km in length will be deepened to dozens even to more than one hundred meters and broadened to 200 to 800 m, and enlarged by 5 to 10 times in wet cross-section. Then the serious situation that the navigation would be likely obstructed by landslides entering the river under natural conditions will be basically improved after impoundment of the Three Gorges reservoir.

(3) Studies on the environmental geological problems of new sites for relocated towns in the reservoir area.
At the NPL 175m there will be 13 counties and /or cities and 140 towns in the Three Gorges reservoir area that need to be relocated entirely or partly, which is one of the vital problems for the Three Gorges project. The existing towns in the Three Gorges reservoir area are basically situated along the Yangtze River-the "golden waterway", on the slopes generally below elevation 150 to 300 m. The Three Gorges reservoir will still be of a typical river gorge type. The majority of the towns is proposed to be relocated directly backwards and still be on the both bank slopes of stem stream and its tributaries. Therefore the major environmental geological problems for relocated towns are still of site stabilities except for some towns and sites which are located on terrace II with the immersion problems .Through two phases of comprehensive engineering geological investigations, including the preliminary investigation and detailed

investigation for the new sites and their surrounding area. each of the new sites has been zoned in terms of suitability based on their basical geological conditions. In addition, emphasis has been paid on the further analysis and check of the environmental geological problems in the new sites and predictions are made to the various problems which will arise possibly during the site construction and operation. Finally a prediction map is compiled of environmental geology for each of the new sites. In recent years, engineering geologic investigation for new sites have been carried out. The main environmental geologic problems encountered included: inundation and immersion, reservoir bankcaving, stability erosion and water pollution.

(4) Studies on the prediction of variations in hydrodynamical field, hydro-chemical field and temperature field of groundwater in the reservoir region.
Before the reservoir impoundment groundwater in both banks discharges into the river in the reservoir region. Yet no significant change to the general pattern of recharge and drainage of groundwater is expected after impoundment of reservoir because the topography and groundwater level on both banks are much higher than NPL. But some variations are expected in hydrodynamical field, hydrochemical field and temperature field of groundwater in the reservoir shore as a result of temporary influence from the back recharge of reservoir water. According to observation data and computation results, the influence width varies with lithology in different reservoir stretches, generally ranging from several dozens to 500 meters.

(5) Studies on the environmental geological effects in the estuary of the Yangtze River
The estuary stretch of the Yangtze River, starting from Erbizui in the JiangYin city, ending at the No.50 light beacon at the river mouth, has a length of 254.4 km. Due to a long term silt accumulation mouth bars are very developed in form of sand islands, submerged sand bars and shore bars.

After impoundment of the Three-Gorges reservoir one of the major concerns is focused on such environmental geological problems as salt water encroachment at the river mouth and erosion resulting from less silt charge from upstream. Through the regulation of the Three Gorges reservoir some improvement will be achieved in the salt water encroachment at the mouth of the Yangtze River during the dry season, and the maximum salinity at Wusongkou will decrease from 11 % to 7.7 %. According to the experiments and computations an equilibrium will be reached between scour rate and deposition about 80 years since operation of the Three Gorges reservoir, but in the first 10 years the silt discharge will be reduced by 300 million plus tons compared to the pre-impoundment of the reservoir because most silt will be depositing in the reservoir. However, the clear water releasing will not lead to scouring of the harbours and docks at Shanghai city because the load discharge will remain the same as before at the Datong hydrometrical station, some 820 Km downstream from the damsite due to the progressive silt supplement.

Soil erosion in the Yangtze valley and its related geological environmental factors

Soil erosion is also one of the vital environmental geological problems in the Yangtze valley. According to the statistic data the water erosion area in the whole valley reaches to about 560,000 km^2, making up around 31.22 % of the total area of the valley. Soil erosion can be attributed to two types of factors--natural and human factors. Natural factors include topography, structural geology, lithology, earthquakes and such physicogeological processes as rockfalls, landslides and debris flows etc. Human factors include various human economic and engineering activities. In the hilly region of redbeds, for example, widespread assart and farming leads to less forested land but more waste land and strong surface erosion and gully erosion on slopes. Prevention and control of soil erosion is a comprehensive systematic engineering and should be combined with land regulation and comprehensive development of the Yangtze valley and state economic reconstruction, so that twice the result could be got with half the effort. Along with the speeding up of water resources and hydropower development in the upper reach region various measures should be taken in the upper reach region and the grave soil erosion areas for soil and water conservation such as afforestation, returning cultivated land back to pasture-land or forest, construction of small size soil saving works and slope engineerings, Besides, to build large number of reservoirs will be able to effectively intercept a vast amount of sediments.

ENVIRONMENTAL GEOLOGICAL PROBLEMS WITH TRANS-VALLEY WATER TRANSPORTING PROJECT

The major environmental geological problems with the middle route of South-North water transporting Project can be summarized as follows: (1) The arterial canal with a length of 600 plus km has to cross the strong earthquake region with intensities ranging from VII to VIII degrees, therefore the problems of sand liquefaction and anti-seismic provisions will arise. (2) The arterial canal with a length of 160 plus km has to cross swelling soil region, so that the problems of canal slope stability must be considered. (3) Part of the arterial canal will have to cross sand-gravels area and highly karstified rocks, such a series of environmental geological problems will be encountered as canal leakage, seepage deformation, immersion and secondary salinization and so on. (4) The problems of canal stability will appear when the canal route has to cross over the coal deposits and the areas of underground mining caverns.

IMPACTS OF THE ENVIRONMENTAL GEOLOGICAL PROBLEMS IN THE YANGTZE VALLEY AND RELATED COUNTER MEASURES

The Yangtze valley is located in the subtropical monsoon zone of eastern Asia, which has abundance of water, plenty of heat, rich resources of all kinds and superior environments, thus possessing various basic conditions for economic development. In recent more than 30 years, along with the rapid development of industry and agriculture and the exploitation of water resources and hydro powers, as well as mineral resources, various environmental geological problems have appeared due to a sharp increase in population and unreasonable engineering and economic activities, especially lacking of public consciousness in environmental protection, even though the development and utilization of resources is still very limited at present time. In the upper reach region destruction of forest and vegetation is very serious. Taking the Sichuan basin as an example, the forest covering rate in it is only 4 % and soil erosion is very grave. Under the impacts of storms and floods, some environmental geological problems and disasters as landslides, rockfalls, and debris flows are very frequent. The middle and lower reaches regions are affected remarkably by silt transported from the upstream, resulting in siltation of rivers and lakes, dense distribution of dangerous passages, unstability of river regime, and severe erosion and caving of bank slopes in some locations.

Besides, farmland making by pond filing and lake reclamation has weakened the regional water storage and flood control capacities, leading to the increase in frequency of flood and waterlogging disasters. In addition, water pollution belts are expanding in the riverside region, and some harmful substances are often checked out with their content largely exceeding the standards, which is critically endangering the health of the people living along the river. Along with the increase in population, and development of economics the Yangtze River development has yet to be picked up speed. The lessons and mistakes in the past should not be forgotten and the natural environments on which human are living should not be destructed randomly. The principles of development, harness and protection must be followed in the development and utilization of the entire Yangtze valley. In pace with the speeding up of exploitation of water power resources and mineral resources of all kinds in the upper reach region and construction of the Yangtze River economic zones, comprehensive harness of the regions and river stretches where serious environmental geological problems exist must be executed in order to strengthen environmental protection. The detailed counter measures can be described as follows: (1) According to the characteristics of resources distribution within the Yangtze valley, the emphasis of hydroelectric construction in the future will be moved to western regions as the Jinshajiang River, the Yalongjiang River, and the Daduhe River. Having the most frequent earthquakes, and most developed active faults, landslides, collapses and debris flows and featured by high mountains and deep valleys, these regions, however, have very complicated geological conditions. Multi-measure has to be applied in harness of the Yangtze River environment to make a comprehensive study of the above mentioned various environmental geological problems, including their geneses and evolution laws, so that relevant monitoring, prevention and engineering measures can be taken to guarantee the smooth construction of projects.(2) Strengthen the control and harness of water and soil loss in the upper reach region, keep on the construction of upper reach shelter-forest in order to further control soil erosion, water and soil loss and siltation. (3) Along with the construction of the Three-Gorges Project and the Pudong development zone, the economic belt from Chongqing to Shanghai along the banks of the Yangtze River will be further developed and large number of large to medium size cities, factories and enterprises, harbours and docks are in the making. How to guarantee the foundation and slope stability along the riverside region and prevent the water pollution zone from spreading is an important yet to be studied issue in geological environmental protection. We must understand and adapt ourselves to the stability conditions of the geological environments in the planning, designing and construction of large scale relocation of cities and towns, and relevant measures or legislations should be taken as soon as possible to guarantee the smooth construction of the Three Gorges project and the riverside economic belt along the Yangtze River. (4) Strengthen the public consciousness in environment protection. Environmental assessment must be made in the future and relevant environmental protection measures should be taken before constructions of any water conservancy and water power projects, large--size factories and enterprises, medium to large cities, and large-scale roads, etc, otherwise state will not approve them. (5) Strengthen the comprehensive planning of development and utilization of water resources along the river, for this purpose, the suppervision management of legislation and administration of justice should be carried out.

REFERENCES

1.   Shi Mingding,Zhang Xiaozhen. Yangtze Valley. Almanac of China Water Resources (1990)
2.   Yan Peixuan. Environmental Geology Within The Yangtze Valley. Scientific Publishing House (1996)
3.   Li Chunyu,Wang Quanliu, Xuya etal.. Geotectionic Map of Asia and its explanatory notes China
      Cartographic Publishing House(1982)
4.   Edited by Institute of Environment Geology MGMR . Explanation of the Map of Enviromental Geology
      Division of China. China Cartographic Publishing House(1992)
5.   Yan Peixuan . Engineering Geology. China Electronic Technological University Publishing House(1993)
6.   Chen Deji . Investigation on geologic condition for the Three Gorges Project Yangtze River Proceedings
      of selected Papers for Three Gorges Project(1987)
7.   H.K. Gupta and B.K. Rastog Dams and Earthquakes. Elsevier Scientific Publishing Company (1976)

Proc. 30ᵗʰ Int'l. Geol. Congr., vol. 24, pp. 185-195
Yuan Daoxian (Ed)
© VSP 1997

Environmental Geology in Developing and Harnessing the Yellow River

MA GUOYAN XU FUXIN CUI ZHIFANG

Reconnaissance, Planning, Design and Research Institute, Yellow River Conservancy Commission, Ministry of Water Resources, Zhengzhou, China, 450003

Abstract

Through a comprehensive study on the environment geological conditions and factors in the Yellow River basin, it is concluded that the most important aspects of environmental geology resulting from the development and harnessing of the Yellow River is the shortage of water resources and loss of soil and water. The most effective way to preserve, adapt and improve the environmental geology is to construct the South-to-North Water Transfer Project, to implement water and soil conservation measures on a large scale and to legally curb man-made disturbances that lead to erosion.

key words: Yellow River basin, environmental geology, siltation caused by man-made erosion, river bed accretion, loss of soil and water, flooding of the suspending river.

INTRODUCTION

The Yellow River is the second largest river in China, crossing nine provinces and autonomous regions. The total length of the main stream is 5,464km, with a drainage area of 0.752 million km^2 and a population of 0.16 billion. There are altogether 20 million hectares of arable land within the basin and the protection area in the lower reach. In order to meet the demand of the national economic development, seven large or medium sized water conservancy projects, such as Longyangxia, have been completed and four large multipurpose projects, including the Xiaolangdi Multipurpose Dam Project, have been under construction since the end of 1993. This practice has indicated that it is very necessary to objectively study the geological conditions and the environmental geology which has close relationship with the activities of mankind, to probe the adaptability and appropriateness of hydroelectric development to the geological environment, and to assess the advantages and disadvantages of the geological environment for the purpose of guiding the plan for the Yellow River development, and of reasonably making use of the environment. Therefore, the following observations are presented for the reference of design, planing and policy-making departments. The authors are looking forward to criticism from the counterparts.

ENVIRONMENTAL GEOLOGY

Topography and Land Feature

Relief in the Yellow River basin lowers progressively from west to east. Three stages could be classified based on the elevation change. The first stage is in the northeast of the Qinghai-Tibet Plateau with an elevation of above 4,000m in average including the river head section upstream of Duoshixia near Maduo and the upstream section between Maduo and Qingtongxia of the Yellow River. The second stage, limited by the Taihang Mountain in the east, includes the Inner Mongolia Plateau and loess plateau, with an elevation ranging between 1,000m and 2,000m, which is the middle Yellow River section between Qingtongxia and the middle route of the South-to-North Water Transfer Project. The third stage is the area to the east of Taihang Mountain, which includes the lower reaches of the Yellow River from the middle route of South-to-North Water Transfer Project to the Bohai Sea.

Strata and Lithology
The deposits of the Pre-Paleozoic era is highly metamorphic gneiss and migmatite, mainly distributed in the Luliang, Taihang and Taishan mountains. The strata of the early Paleozoic era is the carbonate rock and clastic rock in the above mentioned mountains. In the Yinshan Mountain area is slightly metamorphic clastic rock with basic volcanic rock, siliceous rock and limestone scattered between. In the Qinling and Qilian mountains are slight metamorphic clastic rock and volcanic rock. Clastic rock and carbonate rock of the late Paleozoic era are mainly seen in the middle-lower reaches of the Yellow River. The Triassic system of the Mesozoic Era is the clastic rock intercalated by carbonate rock in the lower section in the Qinling and Kunlun mountains, and Cretaceous system of continental, fluvio-lacustrine mudstone, conglomerate and sandstone, and fragmental deposits in small mountain basins are the only deposits of the Mesozoic Era in the east of Helan Mountain, which is red clastic rock containing coal. The Tertiary system of the Cenozoic era is not well consolidated continental lacustrine basin deposits. The Quaternary system is widely distributed, and its facies are very complicated, of which the predominating deposit is continental clastic rocks. The loess in the middle and lower reaches traced to the middle period of the middle Pleistocene. The Yellow River began to take its shape in the middle-late period of the middle Pleistocene.

Geological Structure
Generally, the Bouger continental gravity anomaly is high in the east and low in the west, and the gravity anomaly is gentle in the east and steep in the west. The buried depth of the crustal base ranges between 34km and 44km in the east, and reaches over 65km in the west, and even deeper than 70km in the Qinghai-Tibet Plateau. From the geotechnical point of view, the Yellow River basin could be divided into Aerduosi graben, North China stable massif, Qilian and Qinlian active folding types.

Main tectonic systems in the Yellow River basin are: 41°-43° north latitude Tianshan-Yinshan mountains and 32°-36° kunlun-Qinling mountains latitudinal tectonic system; longitudinal tectonic system of 106°-107° east longitude Helan-liupan mountains and of the south of the Luliang Mountain; NE20° trending Yinchuan faulting graben and NNE trending island-shaped Hebei-Shanxi-Shaanxi Cathaysian tectonic settlement system; Xiyu and Hexi systems consisting of Qilian Mountain tectonic zone, Yongchang-Linxia faulting zone and Yongdeng-Lanzhou faulting zone; reverse "S" shaped Qinghai-Tibet tectonic system, "ω" shaped Qilian-Luliang-Helan mountain system; and Longxi and Luxi rotating tectonic system.

Climate and Hydrology

The Yellow River basin is mostly located in arid or semi-arid of the mid-humid climate zone with exceptions of the lower close-to-sea area which belongs to monsoon climate, of Taishan Mountain area in the lower reaches, Taihang Mountain and the east of Zhongtiao Mountain in the middle reaches which belong to semi-humid climate. The annual average temperature is -4° to -6°C for the plateau and highlands in the upper reaches; 10°C or so for Shaanxi-Gansu-Ningxia, and 3° to 6°C for Ningxia-Inner Mongolia in the middle reaches; and 10° to 15°C for Henan and Shandong provinces. Annual average precipitation ranges between 300 and 700mm for the plateau and high land; between 200 and 300mm for the bordering zone between Ningxia and Inner Mongolia and only dozens of millimeter locally; between 700 and 800mm in Taihang-Zhongtiao mountains; and between 600 and 700mm for the plain area in the lower reaches. Perennial average precipitation amounts to 478mm and precipitation 330 billion m^3 annually for the whole basin. It can be seen from the precipitation distribution that the most area is desert grassland except for Qinghai-Tibet Plateau whose annual precipitation is less than 300mm, the precipitation ranging between 300mm and 500mm is mostly arid grassland, greater than 500mm is forest and grassland.

The Yellow River takes in 150 odds tributaries with more than $1,000km^2$ catchment area each. In its head area, mountains, lakes and swamps are widely distributed, and Animating, Lengling and Shulenan mountains are covered with snow all the year round. Flowing through bedrock mountains, the upper reaches with a natural gradient of about 1.2%, supplied by the rich rainfall and snow, has a constant discharge. The middle Yellow River, running through the loess plateau, has a dense network of waterways since the major tributaries originate from or run through the arid or semi-arid loess plateau. As the ground water does not have much capability of reservoir action, the river is mainly fed by precipitation, of which the flood water takes a larger proportion. The lower reaches, very gentle in relief, has the gradient of only 0.01%-0.02%. Dykes are built along the banks to restrict the river flowing. Most of the runoff is produced by the river section upstream of Lanzhou. The flow is larger both upstream of Lanzhou and downstream of Sanmenxia, and is smaller in between.

Vegetation

The Yellow River basin crosses 4 vegetable zones from west to east.
Vegetable zone in the Qinghai-Tibet Plateau. The plateau is covered by marshy grassland, bush and grassland of frigid zone, except for Huangshui valley where grassland of temperature zone is distributed.

Desert zone, including the area near Zhuozi Mountain in the west of Yikezhaomeng and south of Helan Mountain.

Grassland zone, mainly distributed in the loess plateau.

Deciduous leaf-wood zone. This zone is located in the east and the south of the basin, inclusive of the middle and lower reaches of Yanhe and Weihe rivers, and the areas to the east and south of them. The virgin forest in the loess plateau has nearly been denuded to nothing.

The renowned Ningxia-Inner Mongolia Hetao plain, Fenwei graben basin and the plain in the lower Yellow River are the major agricultural bases in China for irrigation. Furthermore, there is a large area of non-irrigation agriculture in the forest zone and in the sub-zone of grassland. The subzone of desert grassland and desert zone have no agriculture due to no irrigation. The agriculture in the Qinghai-Tibet Plateau is unexceptionably distributed in the banks of rivers or in the damp valleys. The vegetation is all artificially cultivated.

PRESENT STATUS OF WATER RESOURCES UTILIZATION

The Yellow River basin is one of the areas seriously deficient in water resources, which controls and restricts the economic construction and social development in the basin.

Status of Surface Water Resources Utilization
The natural annual runoff of the Yellow River is 58 billion m^3. The total area of land, supplied water by the Yellow River, is $865,000km^2$. Utilization of the perennial average runoff of the Yellow River already reached 47% by the year of 1980 and the percentage of utilization has being increased year by year since then. The development and utilization of the Yellow River promotes the development of agriculture, mitigates the intense situation of water demand in the municipal industry, and supplies drinking water for people and domestic animals which accounts for 60% of the total required by the arid mountain region. Nevertheless, a conflict between water supply and demand is still prominent due to the high sediment content of the Yellow River water.

Status of the Underground Water Utilization
The different types of ground water resources total 43.025 billion m^3. The permissible exploitation amount of the shallow-buried water is 17.55 billion m^3/a. 78% of water supply for 36 major cities along the banks relies on ground water. 80% of the water supply for coal mines and thermal power stations distributed in the loess plateau comes from karst spring and alluvial pore water. Water consumption has amounted to 5.6 billion m^3 for irrigation and 0.8 billion m^3 for urban industry and the people and livestock in the rural area. Excessive exploitation of water is unbearable in a lot of districts, especially in the major or middle-sized cities along the banks. Therefore there is not much potential for ground water exploitation in most areas of the Yellow River basin.

Existing Problems in Development of the Water Resources
Lack of regulating reservoir capacity in the middle-lower reaches and the tributaries heavily affects industry and agriculture production. The use of water for irrigation is large in scale, but the economic benefits lag behind. The efficiency for use of water is low locally, since the existing state of ground water utilization is not rationally regulated, and ground water equilibrium is neglected, resulting in secondary salinization of earth. Irrigation in the lower reaches by diverting the Yellow River has caused silt accretion both in the river and channel. Repeated utilization of water for urban industry is quite low. Average quota of water consumption in industry for the whole basin is $802m^3/10,000$ Yuan, repeating utilization factor is only 29%, which is rather low compared with 70%, 51% and 64% in Qingdao, Tianjin and Xian respectively. The water resources is seriously polluted. The quality of water in 12,550 km long main and tributaries were

assessed. The assessment, compared with the assessment results obtained at the beginning of 1980, shows that the length of fine quality of water (classes I and II) is reduced by 42.8%, those of passable, poor and extremely poor (classes III, IV and V) is increased by 7.2%, 24.7% and 10.9% respectively, which evidently demonstrates that the development of water quality worsening is striking.

The irrigation by making use of the surface water has caused ground water deterioration because of the pollution of the surface water, and concentration of pollution is increased year by year. Meanwhile, waste water released by industry and domestic sewage also contaminate the surroundings of the water resources. Water resources for water supply in many cities are also contaminated to varying degree.

ENVIRONMENTAL GEOLOGY

The following problems existed in the different zones of the Yellow River basin need to be further studied.

Frozen soil

The West Route of South-to-North Water Transfer Project is located in the island-shaped frozen soil of the river head in both sides of Bayankala Mountain (see attached figure), where the average air temperature is below 2°C throughout the year. The maximum seasonal thaw depth of the perennial frozen soil varies from 0.9m to 4.7m in different elevation and lithology. Because the soil has large frozen-heave force when freezing up, the difference in density, compressibility and strength of the soil from freeze to thaw could be severalties. Frozen-heave feature is shown when frozen, and thaw-settlement is displayed when thawed. Because of construction and operation of the projects, vegetation cover will be destructed, hydrothermal condition be changed, water content increased, freezing and thawing frequently alternated, which will result in a series of destruction, such as bank sliding and slumping of the reservoir and further retrogressive erosion, fracturing and dipping out of the retaining wall both in intakes and outlets of tunnels, thaw-settlement of water passage section and frozen heave failure of non-water passage section of the canals, thaw-settlement of the surface structures such as pumping station, power plant and aqueduct. Ice pingo, ice-heave mound, ice plug and ice wall will directly affect the normal operation of the projects.

Earth Flow and Avalanching

They are mainly distributed in the upper-middle reaches to the west of Liupan Mountain.

Three classes, according to the composition, could be defined, i.e. earth flow, mud-rock flow and water-rock flow. Earth flow, very thick and high viscosity, consisting of loess, some silt and rock debris, mainly occurs in the loess area. Mud-rock flow, consisting of clay, sand and rock blocks, mainly develops in the bed rock area and distributes in the banks of Weihe River between Baoji and Tianshui and in the Zuli River basin. Water-rock flow, apart from water, consists of large hard rock blocks and boulders, and distributes in the north slope of Hua Mountain in Shaanxi Province. The volume of those three types of earth flow could be varied from several to millions of cubic meters. They obstruct excess, destroy farmland and silt up river course and water pool, and often cause severe loss of life and property.

Avalanching is frequently seen in the Yellow River basin. Generally speaking, where there are land sliding and earth flow, there is avalanching mostly. The concentrated distribution of the avalanches could extend tens of kilometers.

Landslide, earth flow and avalanching are extremely developed in the loess area, and they are the main source of sediment in the Yellow River.

Earthquake Damage

Altogether 6 strong earthquakes (M≥6), recorded in history or monitored through instruments, occurred in the basin between the year of B.C 780 and 1976. Those earthquakes took place in the area between Yinchuan and Tongwei, in Fenwei graben and in Dari in the north of Bayankala Mountain. 1920 Haiyuan M=8.5 seismicity made Lanni Valley in Xijie County jammed up, creating a series of paternoster dammed lakes. 1929 Bikeqi M=6 earthquake in Inner Mongolia resulted in ground water spraying, ground fracturing, landslide and collapsing.

A detailed investigation was carried out for the activities of 15 major faults in the water-transfer area. It was found that 4 of them show movement since Holocene. Left handed relative horizontal displacement rate varies from 3.3 to 17mm/a. It is impossible for the water transfer project to completely avoid the area with seismic intensity of 9. Hydraulic structures may be subject to seismic damage even fracture failure.

The long dykes along the banks of the lower Yellow River may also be subject to seismic failure due to the locally loosening and softening of the dykes foundation. The statistics show that the dykes cracking occurred because of 6 intensity earthquake, the dyke toe water spraying and sand boiling, and the dyke fracturing, even sliding because of 7 intensity earthquake.

Most of 12 large-sized reservoirs completed or under construction in the Yellow River basin are located in the strong seismic zone. There is a possibility of reservoir induced event, and in fact, RIS already occurred in some of the reservoirs. It is predicted that all the reservoirs could induce M=2.0 event. Generally speaking, M≥4.5 reservoir shake may cause certain failure and M=2.0 earthquake is strong enough to arouse panic among people settled in the reservoir area. Not only the direct damage of reservoir induced seismicity is great, some secondary calamity such as bank failure is also astonishing.

Loess Collapsibility

The middle, low mountains, plain and hill in a vast area from Zhengzhou to Lanzhou and from Great Wall to Qinling Mountain are almost continuously covered by loess. Based on the geological and geographical environment when the loess generated, four categories could be defined, i.e. plateau, valley plain, piedmont and intermontane basin. On the basis of the relative collapsing coefficient of loess, there are five regions: 1) slight collapsing region in the Huang River basin and from Tongguan to Tai mountains; 2) moderate collapsing region between Wei River to Fen River; 3) moderate-strong collapsing region scattered over in the west of Shanxi and the north of Shaanxi; 4) strong collapsing region in the east of Gansu and 5) extremely strong collapsing region in the west of Gansu. Loess in natural condition is often non-compacted, under certain pressure, elastic deformation is very limited, but the compaction deformation is great, which takes the forms of compressive and collapsing deformation. The loess, after soaked, collapses quickly, which is harmful to structures to varying degree.

Landslide

Landslides are extensively distributed in the strong upwarded plateau in the west and in relatively upwarded hilly plateau in the middle in the upper-middle reaches of the Yellow River. The loess slides are concentrated on the banks of river valley in the loess plateau, the edge of the loess tableland and gullies. When impervious layers such as clay, claystone or bedrock surface suffering from subsurface erosion or softening are dissected, loess collapsing and lateral relaxation may create large loess slide over large area. The bedrock slides are concentrated in the bedrock steep banks where the valley gradient is steep. When the gradient of a bank slope is steeper than the dip angle of structural plane including dip angle of intersection line of fissure planes, especially there is a weak plane, or ground water table is raised due to changes of hydrogeological condition, or when an earthquake occurs, a bedrock slide of varying volume is often generated. Very harmful landslides existed from Yehuxia in the upper reaches to Xiaolangdi in the lower reaches, some of them are natural, but most of them are man-made.

Land Subsidence and Ground Fissure

In Shanxi-Shaanxi-Henan provinces where there is loess cover, ground fissures occurred before the end of Ming Dynasty (1368-1644 A.D.), and some of them were associated with earthquakes, and some were naturally formed. From 1915 till now, ground fissures occurred one after another in the Fenwei graben including counties and cities of Xian, Weinan, Lantian, Huayin, Hancheng, Zhouzhi and Jingyang, etc., among which those in Xian were mostly developed. After 1977 Caotan earthquake, the ground fissures were somewhat developed, and so far 10 fissures are moving. Continuously or discontinuously, these fissures strike to NNE, dip to S at a dip angle between 75 and 90°. There are three opinions regarding to the genesis of fissures, one believes that the fissures are tectonic nature; one considers that it is a result of ground sinking caused by dropping of artisan water funnel; one holds that the fissures are controlled by tectonic stress, but their development are accelerated by drawdown of ground water.

Within the Xian city, the sinking area reached $181.3 km^2$ from 1959 to 1983. The subsidence was severely aggravated, the sinking rate was greater than 86mm/a, even 112mm/a locally from 1980 to 1983.

The aforesaid problems have brought various damage to industrial structures, communication and residence.

Desert Intrusion and Earth Desertification

In Tenggel, Kubuqi and Maowusu deserts where it is arid, precipitation is very limited, loose sand is scattered over a vast area, vegetation cover is sparse, rainfall and condensation water are infiltrated into the ground, the wind with speed over 4m/s could appear more than 220 times, "sandstorm" duration could last as long as 10 days in a year and the depth of wind abrasion is 1-3cm. Since a large quantity of sand carried by wind enters directly into the river course, which is over 30% of the silt transported by the river. Because the sand dunes close to the banks of the Yellow River and its tributaries are 10 to 100m higher than the river level, frequently occurred sand skating and collapsing are one of the main sources of river coarse sand and solid load. In the wind gap, sand dunes move as large as 10m/a annually, so far Maowusu Desert has southward intruded into the loess plateau 5 to 40km. Those have constituted a serious threat to subsistence and residence of herdsmen and peasants, and to the normal operation of large-sized communication facilities in both industry and agriculture.

Soil and Water Loss

The soil and water loss in the Yellow River basin is grave. There are a number of contributing factors for this. About an area of 480,000km² in the Yellow River basin including loess plateau, basin, desert highlands, middle-low hill and plateau mountain are suffered from various kinds of erosion, such as hydraulic erosion, wind abrasion, thawing-freezing hydraulic erosion, and artificial erosion caused by dumping waste soil and rock, mining and road building, particularly, landslides, earthflow and collapsing resulted from the combined effect of hydraulic force and gravity. Most areas in the loess plateau have an erosion modules of more than 0.5 (over 10,000T/a km²). There are two regions suffering the most serious water and soil erosion. One is the north main stream of the Yellow River and in the Jing-Luo River basins between Hekouzhen and Longmen, occupying an area of 121,000km², and producing 1.24 billion tons of silt annually which is 67% of the silt produced in the Yellow River basin. The other with an area of 29,000km² and the annual sand yield of 210,000,000T, 11.7% of the silt in the Yellow River basin is in the upper reaches of Wei River, Zuli River basin and Tao River basin. The soil and water loss does not only affect the development and utilization of the natural resources in the basin, but also is the origin of the silt in the Yellow River.

Heat Damage

The long deep-buried tunnels of West Route South-to-North Water Transfer Project cross the hot spring concentrated zone in Qingshui area. This zone, consisting of 12 hot spring sub-zones, is 120km long and 30km wide. The maximum temperature of the spring is 51°C, the maximum flow rate is 7.3L/s and the average flow rate is 1-3L/s. For another alternative, the tunnels will cross the Nigulou hot spring zone in Dari area. It is believed that the temperature of the surrounding rock mass of the tunnels could be as high as 53-68°C locally in the geothermal anomaly region. Evidently it is difficult to carry out construction in such high temperature.

Karst

Striped pure carbonate rock formations in the Yellow River basin are scattered over in the Qinghai-Shanxi-Inner Mongolia-Shaanxi-Ningxia-Henan-Shandong provinces, and distributed in the Yehuxia, Qingtongxia, Wanjiazhai, Longkou, Tianqiao, Ganzepo, crossing section of East Route South-to-North Water Transfer Projects, etc. completed or under planning along the banks of the Yellow River. The geological time of these formations are predominantly middle-upper Cambrian, middle Ordovician and middle-lower Ordovician. The formations within the Qinghai Province are mainly in Permian and middle-upper Devonian systems, and partially in Carboniferous system, middle-low series of Permian and Triassic systems, and in Proterozoic deposits.

In carbonate rock outcropped area, karst is developed. The rock fissures become wide and deep because of karst erosion and weathering, thus karst landform such as solution channel and groove are developed. The development of karst erosion along fault zone, joint, fissure and bedding plane below ground create karst holes, light holes and underground rivers. As a result of connection between infiltration of precipitation and surface water and fault zone, natural collection field or underground water access are created, which changes the dynamic conditions of the ground water and brings about severe reservoir leakage leading to engineering geological and hydrogeological problems such as foundation settlement and collapsing, even menacing the normal life of the local residents.

Lower River Course Accretion
The Yellow River flows eastward from Mengjin to Bohai Sea. The deposits in the river bed and banks consists of deltaic, lacustrine and marine materials. The grain size of the deposits becomes coarser from surface to depth and from east to west regularly. ·

In the three-thousand years history before the founding of the People's Republic of China, the lower course of the Yellow River migrated many times, and the dykes breaching had occurred over 1,500 times. The sediment carried by flood extended over 200,000km² region northward to Hai River and southward to Huai River, which slowed down the sedimentation in the main course and made the accretion in the banks and in the river course concurrently. After the founding of New China, dykes were built to force the river flowing within them in order to guarantee the safety of millions of people. As a result, 300,000,000 tons of silt was accumulated in the narrow river course annually, and the lower Yellow River suspended over the banks, even 7m higher than the ground locally. Therefore, there is a potential risk of dyke breaching in the lower reaches, which is the hidden trouble of Chinese people.

DISCUSSION ON IMPROVING THE ENVIRONMENTAL GEOLOGY.

Through the above analyses on 5 environmental geological conditions and 11 environmental geological factors, it is believed that the critical environmental geology issues are the shortage of water resources and loss of water and soil. In order to increase the water quantity, to store up and discharge silt, the Yellow River Conservancy Commission has successively launched water and soil conservation, water impounding and sediment retaining works, and investigation, planning and design of West Route South-to-North Water Transfer Project. Those work have made obvious progress individually. However the artificial erosion becomes severe since the people are not doing their best to protect the environment during the development and utilization of the natural resources in the Yellow River basin, consequently, although much work force has been put into the conservancy engineering works, the effect is not evident. In order to reduce sediment and increase water quantity, the following observations are provided for discussion.

Distribution and activity of the regional faults, parameters of ground motion and prediction after effects and measures should be studied. For the active faults crossing section of long tunnels mentioned above, besides the necessity of carrying out the study on bad geological phenomenon resulted possibly from earthquake and its after effects and the corresponding treatment measures, in-situ monitoring of three dimensional activities of the active fault walls also need to be studied, so as to resolve a number of significant problems such as shearing of structures.

In order to utilize and transform the frozen soil, and to guarantee safety operation of various kinds of engineering facilities, characteristics of the frozen soil must be thoroughly understood so as to define the relationship between the engineering activity and frozen soil, to control and regulate the changes occurred in frozen soil, and to minimize the damage resulted from frequent freezing-thawing.

The sediment in the Yellow River originates from five high tablelands of Gonghe county in Qinghai Province, etc., granite distributed area of Qinling, Yin and Luliang mountains, and loess area of Huangpuchuan, Gushanchuan, Tuweihe, Jinghe, Zulihe and Huluhe rivers. First of all, effective afforestation in water and soil conservation works should be strengthened, so as to increase vegetation cover, fix the silt and improve the ecological environment. At the same time, large tectonic settlement basins should be utilized to store up silt for creating farmland, and depressions exploited to improve the soil by desilting, and effective dam and reservoirs made use of for retaining debris and silt.

Silt carried by wind is one of sediment sources in the Yellow River. It is arid and windy in the desert highland of the northern basin. The climate in this region is continuously scaling the height of arid, and southward intrusion of desert is increasing. The damage resulted from wind abrasion and silt is becoming increasingly striking. Therefore, for the purpose of sand fixation wind break and desert transformation, windbreak forest belt should be continuously planted and grassland be enlarged.

With the development of mining, communication, water conservancy, hydropower, construction material and township enterprises, a large amount of waste soil and rock, waste material and sewage from industry, and urban garbage have been poured out into the river, which largely increased the sediment and polluted the water resources. Particularly, denudation of forestry and grass heavily counteract the good effect of water and soil conservation works. Therefore, the legal system must be established and improved, environmental impact of project construction should be assessed and monitored, and education of water and soil conservation should be strengthened. Those measures must be put into practice, which is beneficial to the coming generations.

In view of the environmental geology in the Yellow River basin, investigation, monitoring and prediction of the various environmental issues should be carried out, which include ground water pollution, landslide, earthflow and collapsing, earthquake and reservoir induced seismicity, land subsistence and ground fissure, soil salinization and desertification, and karst leakage of reservoirs. Those aimed at preventing from pollution of environment, destruction of natural resources and occurrence of disasters, furthermore providing basis for project construction and design of environment protection.

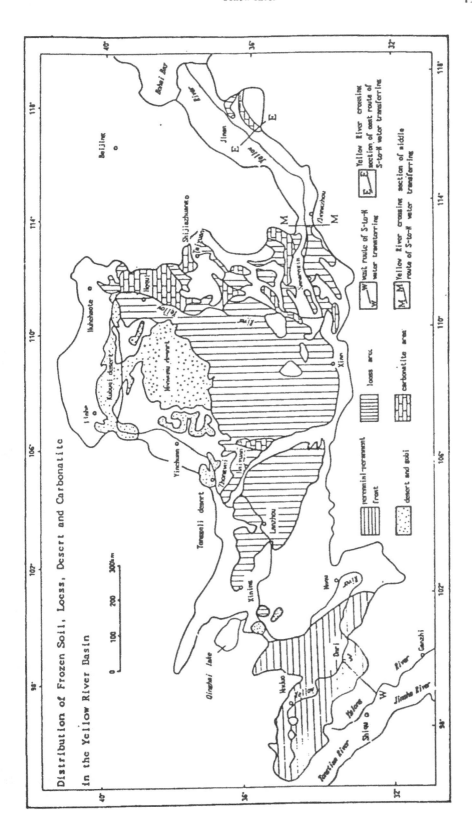

Distribution of Frozen Soil, Loess, Desert and Carbonatite in the Yellow River Basin

Proc.30th Int'l.Geol.Congr.,vol.24,pp.196-200
Yuan Daoxian (Ed)
© VSP 1997

Feature of the Geoscientific Environment in the Area of Poyang Lake and the Strategy of its Development and control

LU ZHONGGUANG

Jiangxi Provincial Bureau of Geology and Minerals. P. R. China

Abstract

Poyang lake is the largest Freshwater lake in China, its catchement area equals approximately to the administration area of Jiangxi province. One of the features of geoscientific environment in the area is the large annual variation of the water level. Another is large amount of sand and silt,and the high tractive - suspended ratio in the upper and middle reaches. Because the material flow and the energy flow consisted mainly of water and silt, it has linked the mountain - river - lake (M. R. L.) into an intergrated ecological system. Therefore, a strategical program has been laid, that is "to manage the lake,the river must be managed;to manage the river, the mountain must be managed". The international society shows a great interesting and concern for the M. R. L. project. The world's hope could be perceived from the M. R. L. project; and Jiangxi has been stepping into the world from the M. R. L. project.

Keywords: Poyang Lake, Annual Variation, Manage, Ecological System, Mountain-River-Lake (M.R. L) Project

INTRODUCTION

Poyang Lake is the largest freshwater lake in China, its catchment area equals approximately to the administration area of Jiangxi province.

Poyang Lake collects the water from five rivers as Gangjiang, Fuhe, Xingjiang , Raohe and Xiushui and pours the water into Changjiang River (that is, the Yangtze River) by one north mouth, Hukou. Its total water catchment area is 162, 225 square kilometers, 96. 8 % of them are located in Jiangxi province.

Jiangxi province covers an area of 166, 974 square kilometers, 94.09% of them are within the catchment area of the Poyang Lake.

In the hilly area of south-east part of China, the natural water catchment area is somewhat identical to the division of administration region, here is a typical example. That's why we say the development and control of Poyang Lake area is very much important to the economic construction and development in Jiangxi province.

The Poyang Lake water area, 3,210 square kilometers at high water level, is the largest take in the middle and lower reaches of Changjiang River, as well as the largest fresh water lake in China.

The middle and lower reaches of Changjiang River is the area where the freshwater lakes concentrate. Apart from Poyang Lake, there are Dongting Lake in Hunan Province, Hongzhe Lake and Taihu Lake in Jiangsu province and Caohu Lake in Anhui province. Although the catchment area and total annual runoff of Dongting Lake are larger than those of   Poyang Lake, the Poyang Lake surface area and its storage are larger than those of Dongting Lake.

The Feature of Geoscientific Environment in Poyang Lake Area
One of the features of geoscientific environment in the area is, as we call it, one line in dry
season and one tract in flood season" because of the large annual variation of the water level.
According to the observation data in the last 50 years, the difference between highest and lowest water level in many years is up to 15. 7 9 meters; the maximum annual variation is 14. 04, the minimum annual variation is 9. 59 meters.
As a result, the large variation of water level brings about a large variation of the lake surface area and its storage.   When the water level is lower at 9 meters, the lake surface area will be 216. 62 square kilometers; and when the water level is higher at 20 meters, the lake area will be 3218. 29 square kilometers.   The lake volume will be 460 million cubic meters at the lower level; and 25,800 million cubic meters at higher level, that is 56 times of the volume at lower level.
As a result, such a great variation causes the geoscientific environment landscape of, as we call it, "one line in dry season and one tract in flood season".
Why there is so much variation in water level?   From the viewpoint of geology, there are two contributing factors:
Starting from the geological age of its formation, by evolution and development, Poyang lake has been forming an inflow-outflow river-connected lake today.
Since the Quartemary, by the differential uplift and descent activities of the three faults of NE, NNE, NW strikes, the subsiding region of the lake has been formed, thus, a sunken land has been provided for the formation of the lake.   During the late Pleistocene, a large water surface lake has not been formed; there was only a sunken land and a river valley in their original state.
In the last glacial period in Quaternary, that is, the Wurm glacial stage (or,the Lushan glacial stage) , the sea level was 120 - 130 meters lower than that of today.   The ground surface slope in the reaches of Changjiang River, during that time, was many times bigger than the water surface slope in Changjiang River of today.   The upstreamward erosion had formed a deep river channel with minus 40 meters near Jiujiang; therefore, the water falls and drops down between the sunken land and the river channel of Changjiang.   The original river in the sunken land could drain away very smoothly without any obstruction, so there was no water stored up then.
In the Quaternary post -glacial period, climate turned to warm, the Holocene marine invasion curried.   Resisted by the high sea level, the River beds had become accretion and the water level of Changjiang River was uplifted.   Therefore, the water drop between the sunken land and Changjiang River decreased and even disappeared.   Not only the water inflow into the Lake from the five rivers increased due to the upstreamward erosion, but also the water could not drain away easily because of the disappearing of the water drop; so the water accumulated in the sunken land and formed the Lake.   From that time on, Poyang Lake, under the balance between the water level of Changjiang River and inflow rate of the five rivers, evolved, developed and formed a typical inflow-outflow river-connected lake.
The geographical location determines the large amount and concentration of the annual rainfall, and that results in large amount and concentration of total annual runoff.

Poyang Lake area is located in the subtropical high pressure area close to 30 degree north latitude. In the middle Eurasia, north Africa and the Mediterranean area on the same latitude as Poyang Lake, the annual rainfall are all below 600 millimeters. While, as the Poyang Lake area locates on the eastern coast of Eurasia and is bestowed by the southeast monsoon, the annual rainfall is up to 1, 470 millimeters. And only from April to September, the rainfall can be up to 1, 020 millimeters, taking the 69. 4% of the total annual rainfall. As a result, the annual runoff of the catchment area is as high as 145, 7 billions cubic meters. It equals to three times that of the Huanghe River and to four and half times that of Huaihe River.

The inflow rate of the Lake depends on the incoming water of the five rivers; whereas the outflow rate depends on the difference of water level between the Changjiang River and Poyang Lake. In normal years, the flood seasons of the River and the Lake are staggered and there is no problem with each other. But sometimes the flood peaks of the River and the Lake are encountered with each other (high possibilities in July), in this case, the Lake water can not be drained away easily, so it runs rampant.

Another feature of geoscientific environment in the catchment area is large amount of sand and silt, and the high tractive-suspended ratio in the upper and middle reaches.

The coarse sand rolling on the stream beds is called " tractional sand", the fine sand suspended in water and moving for-ward with water current is called "suspended load", the ratio between them is called tractive -suspended ratio. The tractive-suspended ratio in trunk stream of Changjiang River is 13. 6 percent; while the ratio in upper and middle reaches of Gangjiang River at Wan'an hydroelectric station is over 28 percent. As a result, a large amount of silt and sand is accumulated locally in erosive area of upper and middle reaches; the carrying and draining capacity of sand and silt is smaller than that of the normal rivers. Consequently, a wide and shallow river valley has been forming in the upper and middle reaches; in some part of the river, even an abnormal phenomena such as the "hanging river" (the river above the ground) occurred.

The reason why this phenomena occurred: since the Mesozoic, in this area, the crust upheaval and the intensive and extensive differential block faulting had formed a series of block-faulted uplifts and basins, large or small, with strike NNE, and the red rock - - the continental deposit consisted mainly of coarse clastics had accumulated in these basins. At the same time, a large number activities of intrusion and extrusion had formed numberous granite mass appearing on ground surface that makes up nearly one to thirds in the area of upper and middle reaches. In the relative stable geonomical environment in present times, the weathering crust had been preserved and accumulated. Its thickness is from 5 to 60 meters, usually 10 to 20 meters. Under the conditions of topographic cutting, flowing water scouring and vegetation destruction, the fully weathering zone often forms the collapse of hillock, promoting the water loss and soil erosion.

I have to point out that whether it is the red rock or the granite weathering product, they have a common feature that the coarse material takes a larger proportion; and they are the source of the sand and silt material in the stream, and will have the influence directly on the tractive-suspended ratio.

Mountain-River-Lake(M.R.L) project, a sustainable development strategy
In order to transform the unfavorable natural condition in the production and human life, and to bring into full play the advantage and potentials of natural resources, a strategical program has been laid, that is," To manage the lake, the river must be managed; To manage the river, the mountain must be managed". This is because the material flow

and the energy flow consisted mainly of water and silt have linked the Mountain - River - Lake into an integrated ecological system. If we try to harness or control the lake separately from the mountains and rivers, or the other way round, it is not possible for us to reach the ideal goal as a sort of sustainable development.

Jiangxi provincial government has accepted this strategical program, and set up a special organization as " MRL Development & Harness Commission". An outline of the general program has been worked out and it has been approved by the Jiangxi provincial people's congress. The program has been listed in the agenda of China's 21 century development initiative. The people are determined to abide by the program to do the harness and development one term of government after another, one generation after another.

Up to now, this program has been put into effect for over ten years. A primary achievement has been made. For example:

Mountain: Tree planting to prevent and control of water loss and soil erosion.

In the mountain area, an initial success has been made in the compaign of tree planting (afforestaton) to make the wind-break forest and sand fixation in each individual watershed area.

For more than ten years, the rate of the forest coverage in the whole Mountain River - Lake area has been increased from 35 percent up to 54. 6 percent. There are many poor mountain villages which used to suffer the water loss and soil erosion and be very backward in economy. And nowadays, they present an encouraging scene of green mountain, fertile farmland and well-off people.

River: Programmatic development by escalation. The rivers will be brought under control and developed step by step based on the program.

Gangjiang River is the largest river among the five rivers. Its catchment area takes up half of the total catchment area of Poyang Lake. Wan'an Hydro-electric power station has been completed, that is also a project item in this program. In the near and mid-term future, the Xiajiang and Taihe project, a higher-step in the program, will be launched. At a specified future date, the Dayu mountain - the watershed divide between Zhujiang River and Changjiang River - - will be tunneled through to build a Gang - - Yue Canal so as to link up the south north water transportation in China.

Lake: Bring the Lake under the control so as to make the radical change of the situation of "one line in dry season, and one tract in flood season. "

The area close to the Lake is a region where the politics, economics and culture are concentrated in Jiangxi province; and is a key area of harness and development in the program. It is planned that a key water conservancy project will be built to bring the lake under control like most of large lakes in China and in the world. The annual variation of water level will be controlled between 20. 5 - 16 meters (the maximum variation: 4. 6 meters) so as to make the radical change of the unfavorable situation such as flood disaster, difficult water transport, unreasonable land use, harmful snail fever and so on because of the "one line in dry season, and one tract in flood season".

There are many options in the construction of the control project such as the complete control or separate control options. Not only the cost of the project is high, but also it involves a lot of things and relation in the upper and lower reaches as well as in both banks of Changjiang River. Now a prefeasibility study linking with the research of the influence upon Poyang Lake by the construction of Three Gorge Project is under preparation by the departments concerned.

Before the construction of the project, some aquatic products farming has been made on a large-scale waters.

There are a lot of wild tracts of lake beaches and grassland islets or sand-bars in the Lake area. They are the propagation places for oncomelania, the second host of blood fluke as well as the source for snail fever epidemic. Nowadays, part of the beaches and islets or sand-bars have been remade into a ecological agricultural villages.

CONCLUSIONS

The World's Hope Could Be Perceived from the M.R.L. Project; and Jiangxi Has Been Stepping into the World from the M.R.L. Project

The international society shows a great interesting and concern for the M. R. L. project. They set a high value from viewpoint of ecological economy and provide a friendly assistance in sustainable development.

Since the MRL project got the UN aid in 1990, MRL office have conducted the exchange and cooperation with tens of international organizations and countries, such as Japan, Britain, America, Canada, Germany, Netherlands, Swiss, Israel, Thailand as well as

UN, EEC, the World Bank and etc.

The chief representatives of UNDP and FAO in China expressed that MRL project had achieved satisfied results, which can become a model for other areas to follow, and called on that the wealthy countries of the world should learn such creative spirit from the people of Jiangxi.

In march 1994, the Chinese Government declared that the M.R.L. project of Jiangxi was selected into the first priority projects of "21st century agenda of China". In the round table conference of international high - level "21st century agenda of China" held in Beijing, various international organizations and countries showed great interest in the project and intention of cooperation.

That is why we say "the world's hope could be perceived from the M. R. L. project; and Jiangxi has been stepping into the world from the M. R. L. project".

Proc. 30ᵗʰ Int'l Geol. Congr., Vol.24, pp201-209   201
Yuan Daoxian (Ed)
© VSP 1997

Geochemical Character of Selenium Fluorine and Human Health

LI JLAXI,   HUANG HUAIZENG,   LIU XIAODUAN,   GE XIAOLI
Institute of Rock and Mineral Analysis, Ministry of Geology and Mineral Resources, Beijing, 100037, CHINA

WU GONGJIAN
Lithosphere Research Center, Chinese Academy of geological Sciences, Beijing, 100037, CHINA

Abstract

The abundance characteristics of selenium and fluorine in stream sediments, soils, shallow underground water, grain and human hair are studied across China.   It shows that some environmental factors such as Eh, pH and the element composition of rocks affect the migration and distribution of fluorine and selenium.   And we come to the conclusion that there is a relationship between some diseases such as Keshan disease, Kashin-beck disease and liver cancer disease and the combination of elements such as Se, F, etc.

Key words: selenium, fluorine, environmental geochemistry

INTRODUCTION

Our project on "Regional Geochemistry and Its Application on Agriculture and Life Science" started in 1991 and will come to an end in 1996.   During the research, we have not only made use of the data available, but also have selected some area to carry out studies on geochemistry, geochemistry of shallow underground water, soil geochemistry, and the soil and other conditions suitable for some special agriculture products.   We have also conducted investigation into human food nutrition, mortality rate of malignant tumor, and trace elements in human hair.

The composition of the source rocks was controlled by the palaeo-environment and palacogeodynamics, and was independent of the modern environment.   Stream sediments are transitional products formed during the geological mega-recycling.   Water acting as a carrier is an important link between the bedrocks, stream sediments, and soils. The weather, landscape, organisms and element provenance together with the pathways of migration of elements in various media, control, therefore, the geochemistry of the environment .   Macroscopically, changes in environmental geochemistry came on account of the behavior of elements and their microscopic variation.   The difference in regional geochemical environment as reflected by abundance of elements and their mode of occurrences in the stream sediments, soils and waters is restricted by the overall effect of various factors.   However, in respect of environment, the most important factors are pH and Eh.   The pH value in China changes significantly in different geographical zones. China's territory can be divided into four regions from southeast to northwest, and the pH value increases   gradually northwestwards.   It is nearly consistent with the weather zonation of the continent. Longitudinally, it varies from wet to arid zone, and latitudinally from hot to cold zone northwards. The changes of pH in each zone are controlled also by regional bedrocks, landscape, and hydrological conditions. The constituents and forms of elements vary in different geological and geographical units. China can be divided into the following four regions, from the northwest to the southeast, based on the regional distribution of elements:

The northwest geochemical region in China (inland region).

The Daxinganling-North Tibet plateau geochemical region (plateau region).

The northeast Sanjiang-southwest Sanjiang geochemical region (Sanjiang region).

The southeast coastal geochemical region (coastal region).

Below is a general discussion on features of regional geochemistry and their relationship to human health, taking only selenium and fluorine as examples.

Distribution and Migration of Selenium and Its Relationship to Human Health

Distribution of selenium

Soil selenium: The national background selenium content is 0.29 mg/kg. The heterogeneous distribution of selenium in the crust is closely related with the bedrocks. The average selenium content in the red soil, laterite soil, yellow soil and rice field soil of the coastal region is 0.23 mg/kg. The selenium content changes with landscape. It is higher in the natural soil and nonirrigated farmland in hilly area, but lower in the rice farmlands. In the Sanjiang and plateau regions, there is a low selenium belt, with an average selenium content of 0.1 mg/kg . The belt is characterized by yellow brown soil and dark soil in the northeast, soft soil and gray limy soil in the loess plateau, purple soil and red brown soil in Sichuan, and the grassy marshland soil in east Tibet. However, the selenium content in the Enshi district in Hubei province and Ziyang district in Shaanxi province is rather high within the belt. In the inland area, the selenium content increases gradually. The selenium content in the dark limy soil, brown limy soil and wilderness soil is 0.19 mg/kg. The content of soil selenium varies away from coastal region because of the charge in weather and landscape. Selenium tends to migrate to the alluvial plain, flood plain and inland basin from the hilly and mountainous areas.

Water selenium: The selenium content of water is much lower than that of soil. In the dry inland region, the selenium content of water is higher, from 0.1 to 6.1 µg/L, but the water selenium in the drainage basin of river decreases to 0.08 - 0.25 µg/l.

Selenium in grain: The selenium content of grains, except for some special grain species, depends mainly on available selenium in soil. The selenium contents of corn, rice and wheat in the inland uncultivated soil are 0.052 mg/kg, 0.091 mg/kg, and 0.128 mg/kg, respectively. Those in the brown and yellow brown soils of the plateau-Sanjiang regions are 0.016 - 0.039 mg/kg, 0.021-0.04 mg/kg, and 0.019-0.061 mg/kg, respectively. In the coastal region, the selenium contents of grain grown on laterite soil are 0.071 mg/kg for corn, 0.077 mg/kg for rice and 0.061 mg/kg for wheat.

Food selenium: In the cities the grains are supplied from different sources. But the situation is different in the rural area, because the food are made from the local grains. Therefore, the average intake level of selenium depends either on food supply or the degree of the economic development. Generally, it can be divided into three categories: high, medium and low, depending on the degree of economic development. The average daily selenium assimilation is 52.3 µg for urban residents, 42.0 µg for people in township, and 36.7 μg for rural residents.   In view of regions, the average daily selenium assimilation is 57.4 μg in Xinjiang, 32.7 - 39.2 μg in plateau and Sanjiang regions, Sichuan basin, loess and Inner Mongolia plateau, and Songliao basin.   The daily assimilation of selenium is 55.8 μg in Beijing, and 53.2 μg in Tianjin.   In the coastal region, the daily assimilation of selenium is 43.9 - 53.0 μg in Shanghai, Jiangshu, Zhejiang, Fujian and Guangdong, but only 34.0   μg in Guangxi.

Selenium in human hair: From data available, we find that the selenium content of human hair in different geochemical regions varies from northwest to southeast.   It is 0.244 - 0.422 mg/kg for residents living in inland areas of gray limy soil, wilderness soil and grassy marshland soil, and 0.064 - 0.080 mg/kg in the plateau-Sanjiang regions where dark brown soil, black soil, brown soil and purple soil are developed.   The selenium content of the hair of people suffering selenium deficiency disease is significantly lower than that of normal one.   In the coastal region, the hair selenium content of the residents living on laterite soil, yellow soil, and yellow brown soil areas is 0.293- 0.500 mg/kg.

Migration of Selenium:
Selenium is a trace element.   Which is commonly dispersed in the structure of sulfide minerals.   It can also form selenides, especially in the coal measure strata, because as a reductant, the organic rock tends to reduce the selenium from the circulating water.   Eh and pH are the two most important factors controlling the solubility and the chemical status of selenium minerals.   In the coastal region, where the pH is around 5 in dispersed weak oxidation-reduction conditions, HSe^- is the main form of selenium in soil solution during the decomposition of selenium-bearing minerals.   In strong reducing environment, however, the Se^0 and Se^{2-} selenides are formed from HSe^- in the solution.
They are absorbed by the colloid [$Al_2O_3 \bullet SiO_2 \bullet nH_2O$] and either precipitate as an in-situ residue , or are transported to other places.   The absorption-desorption mechanism controls the solubility of selenium.   In the Sanjiang-plateau regions with the pH around 7, selenite is the main form. $HSeO_3^-$ is dominant when the pH is low, and SeO_3^{2-}, is dominant when the pH is high.   The former can form a strong absorption coordination compound together with $Fe(OH)_3$, and the latter forms the basic iron selenite $Fe_2(OH)_4SeO_3$ with iron hydroxide $Fe(OH)_3$,   immobilized in sediment and soil.   Most of them can keep in the place as insoluble selenite.   Amoung them $MnSeO_3$ is the most stable.   In the inland region with pH over 8.5, as the oxidation stay increases, selenium is oxidized to more soluble selenate, and its hydrate ionized to $HSeO_4^-$ and SeO_4^{2-}.   This desorption is supported also by our analysis made so far.   In Sanjiang region, in Mufu town, Enshi county, the selenium content is 45 mg/kg in the stone coal exposed on the hill top, but decreases to 18.5 mg/kg in the nearby soil, and to 7.3 mg/kg at the hill-side 450 meters away from the hill top.   The selenium concentration in the nearby water is less than 0.010 mg/L.   No obvious dispersion aureole is formed there, because of the weak migration of the selenite dominant in the soil.   In inland water, soluble Se accounts for 47.8% - 100% of total selenium .   Although it can migrate as selenate in quality solution, the quality meteoric water is insufficient for it to be removed.   Therefore, most selenium remains in-situ in primary minerals.

Selenium and Human Health
The selenium disorder diseases bear an obvious regional character in China.   In the inland region, where the selenium level is normal, no selenium disorder diseases are reported.   In the plateau Sanjiang regions, selenosis often takes place in the high

selenium areas, whereas Keshan disease (KD) and Kashin Beck disease occur in the low selenium areas. Malignant liver tumor occurs in the low selenium area in the coastal region. The selenosis which occurs in Enshi, Hubei province of Sanjiang region, has caught public attention since 1958. Before 1958, the vegetation cover was well developed, which protected the loss of soil and restrained the release of selenium. After 1958, the vegetation cover decreased rapidly, and the stone coal rich in selenium was mined to make up the shortage of energy. As a result, selenium particle was widely dispersed in the air. Statistics shows that 70% of selenium in coal could be released during burning, resulting in a rapid increase (over 100 times) of selenium in the air. The local farmers often used the soil heated with smoke of stone coal as fertilizer. Moreover, stone coals are also widely distributed in the cultivated farmland, which cause another addition to soil selenium level. Because the area is in a weak acid to weak alkaline environment, the low soluble selenite is difficult to be dispersed. The selenium content of soil depends greatly on the amounts of soil fertilizer treated with smoke of stone coal and the scattered stone coal in the farmland. The concentration of selenium in water in this area is higher, but is far less than that in the soil. The soil selenium and water selenium are in an order of mg/kg and μg/kg level respectively. The concentration of selenium in grains increases with the soil selenium level in farmland, and is controlled by the level of available selenium in soil. Distribution of selenosis depends on the concentration and degree of dispersion of selenium. In 1958 - 1963, because of the decrease in rainfall, most of the water-logged fields were changed into the dry ones. In the strong oxidation environment, the insoluble selenite became soluble selenate due to the increase of pH which is easy to be absorbed by the plant. The assimilation capacity of pumpkin, Chinese cabbage and radish can be many of times that of rice and corn. Vegetables were used as the main food instead of rice by the local residents at that time (personal correspondence with Mao Dajun). Apparently, the organic selenium amino acid is easy to be assimilated by human body causing selenosis of the local residents. The incidence of Keshan disease is inversely correlated with the content of Se in grain and human hair. The total concentration of Se in soil is increases with the Keshan disease incidence .

In the low selenium zone of the plateau-Sanjiang regions, the contents of selenium in soil, grain, food, human hair and blood bear no direct relationship to the incidence of Keshan disease and Kashin-Back disease. Recent investigation in the Zhangjiakou area, one important factor is the concentration of available selenium in soil and selective absorption by plants. Study shows that the action of selenium is quite complicated. The biological function of selenium was affected by a lot of factors. Statistical analysisses have been made by the authors in respect of trace elements in human hair, corn, rice and wheat in areas where Keshan disease and Kashin-Beck disease take place, with reference to the data of the nondisease areas (Table 1). In general, the Keshan and Kashin-Beck diseases belt appear to be coincident with the low selenium assimilation level of food.

In respect of liver cancer, its incident rate decreases gradually from the coastal to the inland region. The distribution and occurrence of the elements which are controlled by external geochemical environment play an important role. A trial of adding selenium to foods of residents has been carried out since 1984 in Qidong County, Jiangsu province. The incident rate of liver cancer before the trial was 52.85 over 100 thousand people, and it decreased to 14.15 per 100 thousand people in 1992. The incident rate of the people who did not take part in the trial is still 50 - 64 per 100 thousand people. Apparently, the trial shows that selenium can help to reduce the incident rate in the high liver cancer area, it can also prevent the Hbs Ag-carriers from the development of liver cancer. It is also effective in prevention for the families with high incident rate of liver cancer. The

Table 1. Relationship of Keshan and Kashin-Beck Diseases with Micro Element Contents in Foods as Compared with Nondisease Areas

Sample	Disease	Elements Lower as Compared with Reference	Elements Higher as Compared with Reference
Human hair	KD	Se, Zn, Cu, Sr, Pb, V, Fe	
	Kashin-back	Se, Zn, Cu, Sr, Pb, V, Fe, Ba, As, Ti, Al	
Corn	KD	Se	Ba
	Kashin-back	Se, Sr, Ba, Ti	
Rice	KD	Se	
Wheat	KD	Se	Ba, Pb, As, V, Ni
	Kashin-back	Se	Ba, V, Pb, Ti, Ni

statistics on the trace element contents in human hair after the standardization on age and the mortality rate of malignant tumor for 144 counties show that the liver cancer is closely related with Se, Cu, P, Fe and Mn, and also Ca, Sr, Al and Pb. According to the gradual regression module setup, the liver cancer shows a negative relationship with Se, Cu, P, Fe and Mn, and a positive relationship with Ca, Sr, Al and Pb. In the coastal region, the selenium level in soil is lower, and the Al level is higher; the available Cu is lower, the available P, Fe, Mn is higher. This coincides with the above analysis. There is, however, an exception in Shanghai where the personal daily selenium assimilation is of 53.0 µg, but the mortality rate of liver cancer is 24.49 per 100 thousand people, which is the highest in the country. The cause of cancer is complicated, and its incidence is related with many factors and different pathogens. There are many problems that are not very clear about the relationship of the factors that affect the selenium concentration in human body. Now, it has been found that As, Cd, Hg, Ag, Co and Zn can act higher in concordance or discordance with selenium. Because the stable concentration of micro elements in human tissue is very low, and their state is variable, so any slight change will affect the absorption and the metabolism of selenium.

Distribution and Migration of Fluorine and Human Health

Distribution of fluorine

Fluorine in stream sediments and soils: The average concentration of fluorine in both stream sediments and soils takes a saddle shape from coastal to inland regions. The low fluorine background is located in the plateau region, whereas fluorine concentration in the Sanjiang and coastal regions is higher, and in the inland region is moderately higher. The national average concentration of fluorine in stream sediments is 470 mg/kg, and the background in soil is 453 mg/kg . These figures are close to fluorine level of the source rocks. Abundance of fluorine depends on the original fluorine in the bed-rocks under different geological conditions. The concentration of soluble fluorine increases from southeast to northwest in both dry and wet farmlands. The variation in ratio of soluble to total fluorine shows the same tendency. It is different from the regional distribution of the total fluorine in soils, and bears different geochemical significance. The concentration of soluble fluorine in dry soils (and the ratio of dry soil soluble fluorine to total fluorine) in the coastal, Sanjiang, plateau and inland regions are 0.72 - 0.74 mg/kg (0.11 - 0.14); 1.33 - 1.71 mg/kg (0.41 - 0.42); 2.51 - 5.39 mg/kg(0.87- 1.87); 4.43 mg/kg (1.17) respectively. In the coastal, Sanjiang and plateau regions, the concentration of soluble fluorine in paddy fields (and the ratio of paddy field soluble fluorine to the total fluorine) is 0.28mg/kg(0.09); 0.38 - 0.54mg/kg(0.17); 4.05mg/kg(1.12) respectively.

Water fluorine: The dissolved fluorine is diluted because of the sufficient meteoric precipitation in the coastal and Sanjiang regions. So the fluorine concentration is greatly decreased in these areas. In the arid and semi-arid inland and plateau regions, however, the dissolved fluorine is easy to be concentrated because evaporation is twice rare of meteoric precipitation. As a result, the fluorine concentration decreases gradually from northwest to southeast in general. For example, the fluorine concentration in the shallow ground water is less than 0.5 mg/l in the coastal region and the fluorine concentration remains less than 0.5 mg/l in the Sanjiang, North China Plain to the north of Huai river and the most parts of the northeast China, it becomes more than 1 - 2 mg/l in the Bohai Bay and in areas to the west of the Harbin-Shenyang. The high fluorine water appears in a quite large area in the north of the inland -plateau regions, with its concentration being higher than 1-2 mg/l in general, and more than 2 mg/l in some particular areas. It is worth noticing that the concentration of fluorine in the water of hot springs is, without exception, higher. These springs are scattered all over the country, although their effect on human health is limited. Their fluorine concentration varies greatly from 1.10 mg/l to 16.3 0 mg/l.

Migration of fluorine

In addition to forming fluoride minerals, F- can also substitute for OH- in hydroxyl layer silicate minerals. The fluorine-bearing minerals are represented mainly by fluorite, apatite, mica, and cryolite. Fluorite and apatite are the common accessory minerals in igneous rocks, and also appear among sedimentary minerals. Mica is a main rock-forming hydroxyl layer silicate mineral . The above three kinds of minerals are often encountered in soil. Cryolite is rarely found in rocks, but mostly in soil.

The field investigation disclosed that, 10 species of fluorine, i.e. F^-, HF_{AQ}, $BF(OH)_3^-$, CaF^+, MgF^+, MnF^+, AlF^{2+}, AlF_2^+, AlF_3, AlF_4^- exist in the shallow underground water with pH between 7 - 8 in the North China plain. The activity of fluorine ion accounts for 79% - 96% of the total fluorine ion concentration. The activity of magnesium fluorine complex and calcium fluorine complex is 3.1% 19.2% and 0.3% - 3.0% of the total fluorine concentration, respectively. In the alkaline aqueous solution, OH^- reacts with Ca^{2+} to produce free fluorine ion ($CaF_2 + 2OH^- \rightarrow Ca(OH)_2 \downarrow + 2F^-$). Dynamic experiment on migration and accumulation of fluorine in the soil shows that free F^- concentration in water increases with the increase of pH. The adsorption capacity of OH^- decreases in the acid aqueous solution with the decrease of OH^- concentration. Under this condition, few Ca^{2+} can be taken from CaF_2 to replace F^-. Small amounts of released F^- might form soluble complexes of FeF^+, FeF^{2+}, FeF_2^+, AlF^{2+}, AlF_2^+, $BF(OH)_3$, BF_4^-, MgF^+, CaF^+, SiF_6^{2-}.

Different types of cations, anions and complexes might constitute different geochemical types of water. Active Ca^{2+} has a very strong control over chemical activity of fluorine in water. This accounts for the low fluorine concentration in HCO_3^--Ca^{2+} water and high fluorine concentration in HCO_3^- - Na^+ water in the North China plain, Conversely, apatite can be dissolved into water which contains CO_2 or organic acid. The increase in pH favors the precipitation of phosphate. Fluorine not only can immobilize the phosphorus in water solution, but also can replace the OH^- of the apatite, converting it to stable fluorine apatite $Ca_5(PO_4)_3F$ to form colloids of phosphate with carbonate apatite and hydroxyl apatite suspended as tiny particles in water.

For the fluorine in soil, there is a more complex physical chemical process. The soil colloid can not only absorb F^- and fluoride molecules, but also take part in reaction with F^-, resulting in destruction of the mineral crystal lattice and formation of new minerals, such as cryolite (Na_3AlF_6), malladiite (Na_2SiF_6), sellaite (MgF_2) and pleysteinite $Al_2PO_4F_2(OH) \bullet 7H_2O$ etc. These reactions are controlled by pH and Eh. In the acid soil, fluorite is deposited cryolite, and fluorine is released from apatite and mica. In the alkaline soil, however, the reaction proceeds in the opposite direction. In general, at the pH 5.0 - 6.5, soil has a strong capability to fix fluorine, and aluminum hydroxide $Al(OH)_3$ has a strong capability to absorb fluorine.

Fluorine and Human Health

There are various causes of endemic fluorosis in China. First of all, the disease is caused by the fluorine-bearing water. This type of disease is most common in China, which accounts for 80% of the total. Disease of this kind happens mainly in the basin, lower terrace and at the low-lying junction between the plain and alluvial fan, especially, in arid and semi-arid areas . Statistics shows that the correlation coefficient of incidence of dental fluorosis and concentration of fluorine in water is up to 0.9 - 0.97 ($p<0.01$). Endemic fluorosis can be caused also by other reasons including coal burning, mineral mining and also tea drinking, etc.

Except for pollution of foods caused by roasting and coal burning, this type of diseases can rarely be caused directly by foods. In the non-disease affected regions, such as in Guizhou and Hunan provinces, the average content of fluorine in grain is 4.75 - 21.10 mg/kg, which exceed apparently the content of the same kind of grains in the disease affected regions. This implies that the absorption of organic fluorine changes with the physical-chemical state of different composition. Experiment on rat shows, that the absorption rate of organic fluorine of bone meal and fish protein concentration (FPC) is less than that of inorganic fluorine , because 99% of the fluorine in human and animal bodies are in the inorganic state. Medical research shows, that fluorine mainly assumes a form of ion dispersion and percolation from the stomach into the blood, hence the fluorine ion is especially easy to be absorbed. The absorption rate of the soluble fluorine in drinking water can reach 95%, but CaF_2, Na_3AlF_6, and KBF_4 is difficult to be dissolved in the gastric juice at lower pH, so they are difficult to be absorbed. Spencer et al., conducted an equilibrium experiment on fluoride. They have increased the assimilation level of Ca, P and Mg, and then determined the fluorine concentration in urine and faeces There is no significant change in urine fluorine concentration, and the increase of fluorine in excrement is also limited. If aluminum hydroxide $Al(OH)_3$ is added to the meals, the fluorine in faeces increases obviously, but the fluorine in urine decreases. The absorption rate of fluorine decreases from 96% to 41% when the two experiments are compared. Iron has a function to help the stomach to absorb fluorine.

In the alkaline environment of the inland-plateau regions, the concentration of F, Ca, Mg is higher, whereas the concentration of Al, Fe is lower. Fluorine is easily absorbed by the human body when the F⁻ concentration is higher, and the Al complex concentration is lower. Conversely, in the acid environment of Sanjiang and coastal regions, although the concentration of iron increases rapidly, the concentration of F, Ca and Mg decreases. The ability of iron to accelerate the absorption of fluorine in the stomach is far less than that caused by fluorine ion and Al complex. The underground water chemical equilibrium model of Xingtai, Hebei province, shows that the total fluorine concentration (Fz), the activity of fluorine ion (α_F.), the activity of fluorine magnesium complex(α_{MgF}⁻) and the activity of fluorine-calcium complex (α_{CaF+}) are all positively correlated with the endemic fluorosis at the 0.01 significance level. The ratio of α_F/α_{MgF+} and the incidence rate of endemic fluorosis is negatively correlated at the 0.05 significance level. It means that the larger the α_F. and α_{MgF}⁻, the higher the incidence rate of fluorosis. Generally speaking, the role of Mg, Ca and Fe in accelerating the absorption of fluorine is far less than that caused by fluorine ion and aluminum complex. Because the activity of the fluorine ion is restrained and the activity of Al complex increases rapidly, it is suitable for fluorine to deposit and then be excreted by stomach and intestines. Therefore, to the south of Yangzi river, except for the cases of fluorosis caused by water contamination in mining areas and hot spring spots, no large scale water source related fluorosis occurs. It can be seen that although the rocks, stream sediments and soils are rich in fluorine, sometimes hundreds of times higher than the fluorine concentration of drinking water standard (1mg/l), it does not appear to cause fluorosis. It is the complex geochemical environment, mineral types and concentration of major and trace elements, including organic composition and its state and coordination that controls the incidence and degree of endemic fluorosis. During combustion of the fluorine-bearing stone coal, it is apparently harmful for people to absorb the released HF, SiF_4, H_2SiF_6 gases and NaF, $AlF_3+Na_3AlF_6$ dust. However, the degree of harm depends on the fluorine concentration in the source material and the ventilation condition of the coal burning stove.

CONCLUSION

From the inland region through the plateau and Sanjiang regions to the coastal region, the abundance of selenium and fluorine in stream sediments and soils assumes a high-low-high saddle shape. The concentration of fluorine and selenium in water varies accordingly. As an important factor of the environment, the Eh increases gradually from southeast to northwest. The physicochemical conditions control the migration and distribution of fluorine and selenium. Analysis of concentration of selenium and fluorine and their mode of occurrence in water and foods indicates that, apart from F⁻, there are also HF, $BF(OH)_3$⁻, CaF^+, MgF^+, MnF^+ and AlF^{2+} complexes in the shallow underground water of plateau-Sanjiang regions under weak alkaline to weak acid conditions. In the alkaline regime of the inland region, the concentration of the fluorine ion increases in shallow underground water. In the acid conditions of the coastal region, fluorine complex is the main form of fluorine in shallow underground water. It is inorganic fluorine that causes fluorosis through water drinking and coal burning. Fluorine is easy to be absorbed by the human body if it exists in the ionic form . The endemic fluorosis depends not only on the total concentration of fluorine, but is also closely related with the activity of the fluorine ion, fluorine-aluminum complex, fluorine-magnesium complex and fluorine-calcium complex. In the high Eh region, selenate is the most important mode of occurrence of selenium. Selenite is dominant in the medium Eh region. In the low Eh region, selenium exists only as element in form of selenide. Human body can mainly absorb the organic form of selenium, insufficient or excessive intake of selenium can cause diseases. Comparative analysis of micro element concentration in grains and human hair in Keshan disease, Kashin-Beck disease and nondisease areas. The pattern shows that Se, Zn, Cu, Sr, Pb, V and Fe are obviously lower in Keshan disease and Kashin-Beck disease areas, but Ba, Mn, Ti, Ni and As only significantly lower in Kashin Beck disease area. Se is a major factor that causes Keshan disease and Kashin-Beck disease. This point of view is also verified by the low selenium content in foods. Element combination that controls incidence of diseases might be different in different regions. The mortality rate of liver cancer shows a tendency to decrease gradually from the coastal region to the inland. In the high liver cancer incidence area, the incident rate is negatively correlated with Se, Cu, P, Fe and Mn, and positively correlated with Ca, Sr, Al and Pb. Although selenium is insufficient both in Sanjiang plateau and along the coastal regions, the combination patterns of elements are different due to the environmental and geographical conditions. As a result, Keshan and Kashin-Beck diseases happen mainly in the Sanjiang-plateau regions, whereas the liver cancer diseases mainly in the coastal region.

Acknowledgment

Experiment results and statistic data from programs of the research project of Science and Technology Committee of China "Regional Geochemistry and Its Application in Agriculture and Life Sciences" are cited in this article and Institute of Rock and Mineral Analysis, British Geological Survey collaborative research project. The authors sincerely thank the members of the project and their hard work.

REFERENCE

1. 全国环境监察总站. 中国土壤元素背景值. 北京: 中国环境科学出版社. 87(1990)
2. 寥自基. 微量元素的环境化学及生物效应. 北京: 中国环境科学出版社. 102-123. (1992)
3. 王子健. 硒形态与环境行为、生物效应关系的研究. 生态环境中心论文. 18(1991)
4. Johnson C., Ge X. et al., British Geological Survey Technical Report,wc/96/52(1996)
5. Drishnamachari K A V R (美)著. 朱莲珍 主译校. 人和动物的微量元素营养. 青岛: 青岛出版社, 297-339 (1994)
6. 王云剑, 孙玺, 陈英杰等. 氟骨症--X 线诊断图析. 北京: 中国环境科学出版社. (1990)

Proc.30th Int'l.Geol.Congr.,vol.24,pp.210-217
Yuan Daoxian (Ed)
© VSP 1997

Heavy Metals pollution in Anzaly Lake sediments, Northern Iran.

Soleyman Kousari

Department of Geochemical Exploration, Geological Survey of Iran. P.O. Box 13185-1494 Tehran-IRAN

Abstract

A geochemical statistic parameters have been done base on the analytical data of a previous geochemical exploration in Anzaly lake to identify the rate of heavy metal concentration and the primary sources of the lake sediments pollution. The enrichment factors as the coefficient of pollution of heavy metals show that the concentration values of Cu, Pb, Zn, Cr, Ni, TiO_2, and Fe_2O_3 are higher than their mean values, especially in eastern part of the lake.

There are several main rivers and channels in eastern part of the lake which carry the contamination water from the industrial and the populated zones, so the sources of pollution of this part of the lake could be mainly an anthropogenic origin.

The rank correlation of Cu, Ni, Co, Cr...... indicate a close relation with the ultrabasic and basic rocks of the high land where located on the western and southwestern parts of the lake. Consequently the source of concentration of these elements could be geogenic origin.

Comparing the coefficients of pollution of elements in Anzaly lake and the most 24 polluted lakes in Sweden, indicate that the degrees of pollution in Anzaly lake for Cu, and Cr are 3 times higher than the Sweden ones, but the contamination of Pb is lower than the Sweden polluted lakes.
Keywords: Heavy metals pollution, enrichment factor, coefficient of pollution. Geogenic and anthropogenic sources.

Introduction:

Anzaly lake is one of the largest back shore lake, where has been formed by the progressive of sand spits along the Iranian caspian beach, in late Pliestocene time, near the Ghazian -Anzaly harbour. (Fig.1)

In late 1986, a lake sediment geochemical exploration has been carried out for titanium investigation. (S. Kousari, and F. Azarm 1986) in 1992, once more, the spectrometric analytical data of the lake, reevaluated for distinquishing the rate of metal concentration in the lake bottom sediments.

The area is one of the populated and industrialized zone, where more than 20 large and medium size of villages and several towns and cities are developed all around the lake. So, most of the waste and garbages of the houses and factories are put directly into the rivers and lake water.

Obviously, the heavy metals concentration and pollution are associated with geogenic and anthropogenic sources. The main aim of this study is to distinquish the background values of elements and also the rate of heavy metals pollution by comparing with the degree of pollution in 24 polluted lakes in Sweden.

Lake Description

Anzaly lake is small and shallow with the average depth of 1.5m upto 3m. Several rivers and channels are distributed mostly in eastern part of the lake, so the rate of input matterials are higher than the western part. Every year more than 1,000,000 Tons of loaded materials are carried into the lake mostly by the eastern rivers and channels. Consequently, the eastern parts of the lake were changed into dried lands, most of these dried zones were converted into the farms and housing lands before the raising of sea level in recent years.

The loaded materials are essentially composed of soluble organic and nutrient materials and unsoluble sediments . (Table No. 1) (F.Moztarzadeh and etc. 1984). Most of organic and nutrient materials are used by plants, so, the growing rate of plants are too high. Many western parts of the lake became as a real swamp or dense marsh, but the eastern part of the Anzaly lake because of the highly energy of water current. still saves the characteristics of a real lake.

The extent of Anzaly lake in spring and late winter seasons is amounted to 400 km^2 , but in dry and hot seasons, reduced up to 240 km^2 .

Geologically, the lake has been formed by the developing of several sand spits on the caspian coastal plain where deltaic deposits are the main part of the lake basin. (Fig. 1)

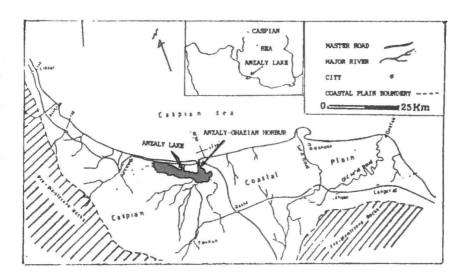

Fig. 1. Simplified geological and geographical map of study area.

Methods and Materials.

As it was mentioned, 52 bottom sediment samples were selected from the first 10 cm of top sediment layers of lake to identify the elements composition.

All the samples were analysed by the emission spectrometry for 8 major oxides and 13 trace elements. To find any genetic relation among elements, rank correlation has been calculated (table 2) and to idetify the rate of pollution the following equation and the statistic parameters were used.

1. Enrichment Factor = actual concentration
(pollution degree)   background value

No.	Materials	Ton/Year
1	Nutrient Loaded	700000
2	Chlorides	186700
3	Sulfides	89900
4	Phosphate	213500

Table 1: Aneual Loaded Soluble materials into Anzaly lake

	Fe2O3	TiO2	Ni	Co	Cr	Zn	Pb	Cu
Fe2O3	1.00							
TiO2	0.62	1.00						
Ni	0.77	0.53	1.00					
Co	0.70	0.86	0.73	1.00				
Cr	0.42	0.65	0.36	0.66	1.00			
Zn	0.17	0.14	0.26	0.23	0.20	1.00		
Pb	0.20	0.27	0.38	0.38	0.26	0.52	1.00	
Cu	0.61	0.36	0.77	0.52	0.28	0.55	0.43	1.00

Table 2. Correlation Maerix in Anzaly lake.

In order to find out the degree of pollution, the results of the enrichment factors of Anzaly lake should be compared with some other polluted and unpolluted lakes as references. By this way, the mean value of the enrichment factors of each element for all samples are divided by the same mean value of the enrichment factor of elements of the polluted and unpolluted lakes in Sweden. The results are shown in table 3.

Elements	Mean 1	Mean 2	Mean 3	CP1	CP2	References
Cu	112	40	37	1.1	3.3	M. Wallsten and 2.
Pb	17	79	63	1.25	0.27	Dressie 1993.
Zn	207	219	181	1.21	1.14	Dressie 1993.
Cr	191	67	61	1.10	3.13	Dressie 1993.
Ni	84	81*				J. Ingeric and Etal. 1993.

Mean 1 = Mean of Elements in Anzaly lake

Mean 2 = Mean of Elements in Polluted Swedish lakes

Mean 3 = Mean of Elements in Unpolluted area in Swedish lakes

CP 1   = Coefficient of Pollution in Swedish lakes

CP 2   = Coefficient of Pollution in Anzaly lake

*   = Ni Grade of Pollution in Dalalvan River Sediments (Sweden)

Table 3: Comparing of Pollution in Anzaly and Swedish lakes

The relative degree of pollution of the different elements are classified into 4 groups. Background value is less than mean value of each element. content. The third class of pollution range from 1 to 2 folds of background, the second from 2 to 3 and the first class more than 3 folds of background value.

Results and conclusions

Based on th 8 distribution maps, of elements, the following conclusion have been revealed:

1. The distribution of heavy metals contamination show that the eastern part of Anzaly lake is highly polluted for Cu, Pb, Zn, Cr, Ni, TiO_2 and Fe_2O_3. There are several rivers and main channels in the eastern part of the lake which carry the heavy metals and toxic materials from the industrial zones and populated areas, so the source of pollution for this part of the lake could be mostly an anthropogenic origin.

2. The results of the correlation matrix (table 2) indicate a highly geogenic relation between Ni, Cu, Fe_2O_3, MgO, TiO_2, Co, Ga, and Mo. These elements have been distributed over than 2/3 of the lake area. A wide sequence of basic and ultrabasic rocks are existed in the western Pre-Pliestocene high land, therefore the main sources of these high grade elements specially in eastern part of the lake, could be the ultrabasic country rocks, but in the eastern part of the lake, besides the ultrabasic rocks, the anthropogenic sources should be another factor for increasing the grades of the heavy metal elements.

3. The maximum contents of Pb and Zn both are more than 1000 ppm and the highly contaminated zones are clsoe to the Anzaly - Gazian harbour and this is because of the exsistance of the industrialized areas in these cities.

The downtown of Gazaian and Anzaly cities are conected directly to the north eastern part of the Anzaly lake. This area are too populated and there are a lot of car reparing centers, garages and bus

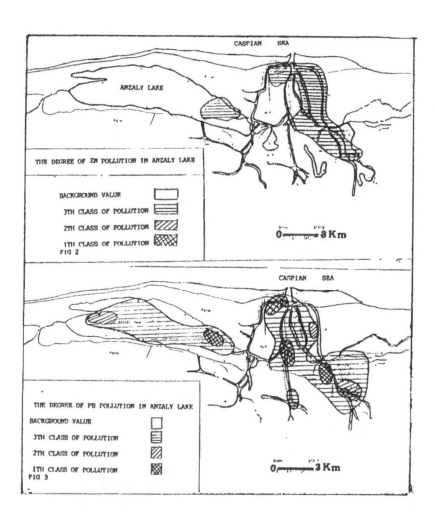

Fig. 2-9: The distribution of Heavy metals Pollutions in Anzaly lack, are shown in gigs 2 to 9.

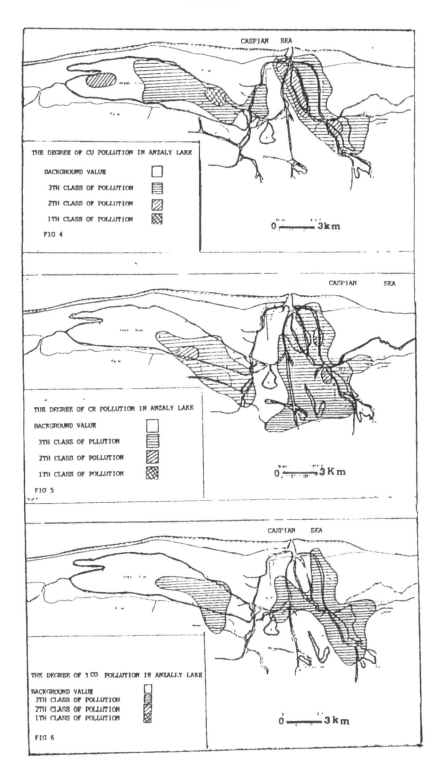

THE DEGREE OF CU POLLUTION IN ANZALY LAKE

BACKGROUND VALUE

3TH CLASS OF POLLUTION

2TH CLASS OF POLLUTION

1TH CLASS OF POLLUTION

FIG 4

0 ——— 3km

THE DEGREE OF CR POLLUTION IN ANZALY LAKE

BACKGROUND VALUE

3TH CLASS OF PLLUTION

2TH CLASS OF POLLUTION

1TH CLASS OF POLLUTION

FIG 5

0 ——— 3 K m

THE DEGREE OF 1 CO POLLUTION IN ANZALY LAKE

BACKGROUND VALUE
3TH CLASS OF POLLUTION
2TH CLASS OF POLLUTION
1TH CLASS OF POLLUTION

FIG 6

0 ——— 3 k m

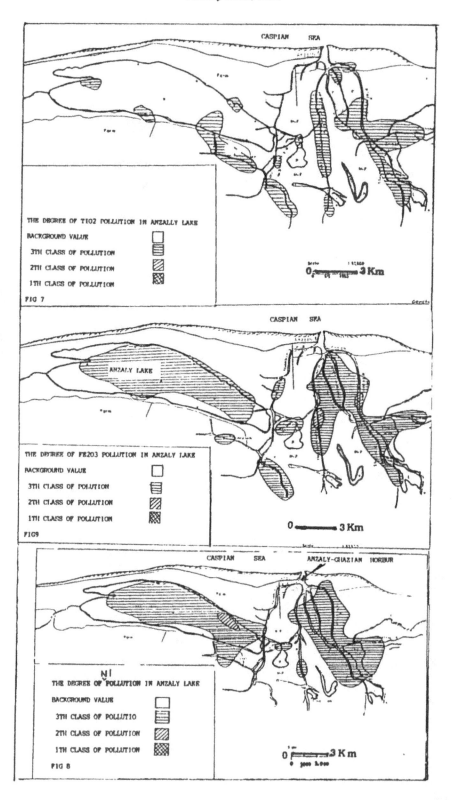

THE DEGREE OF TIO2 POLLUTION IN ANZALLY LAKE
BACKGROUND VALUE
3TH CLASS OF POLLUTION
2TH CLASS OF POLLUTION
1TH CLASS OF POLLUTION
FIG 7

CASPIAN SEA

0 _____ 3 Km

THE DEGREE OF FE2O3 POLLUTION IN ANZALY LAKE
BACKGROUND VALUE
3TH CLASS OF POLLUTION
2TH CLASS OF POLLUTION
1TH CLASS OF POLLUTION
FIG9

ANZALY LAKE

CASPIAN SEA

0 _____ 3 Km

THE DEGREE OF POLLUTION IN ANZALY LAKE
BACKGROUND VALUE
3TH CLASS OF POLLUTIO
2TH CLASS OF POLLUTION
1TH CLASS OF POLLUTION
FIG 8

CASPIAN SEA   ANZALY-GHAZIAN HORBUR

0 _____ 3 Km

terminals, where could contaminate directly both air and water. There is a main channel that cut accross the down town and carries town and carries the huge waste materials and garbages, specially the exhausted materials from the car reparing centers and bus terminal into the Anzaly - Gazian harbour. By this way the lead and zinc pollution in this part of the harbour are directly related to the anthropogenic sources. (Figs. 2.....9)

4. Comparing the Coefficient of pollution of Anzaly lake and the most 24 polluted lakes in Sweden reveals the degree of pollution in Anzaly lake is 3 times higher than the Sweden ones for Cu, and Cr. Except in Anzaly - Gazian harbour zone, the contamination of lead is lower than the Sweden polluted lakes (Table 3).

Acknowledgments

The author wishes to thank Engeneer F. Azarm and M.J. Shamsa for their great cooperation during the geochemical investigation in Anzaly lake, and tanks also are due to the Enginner V.Madelat and Mrs Eskanary for reading and editing the text. I am grateful to Engineer S.Mottaghi for his computer processing and maps preporations.

References

1. J. Ingri, C. Ponter, B. Ohlander, R. Lafvendahl and K. Bostrom 1993. Envrionmental monitoring with river suspended mat. Case study in the River Dalalavan, Central Sweden. Applied Geochemistry No. 2.

2. S. Kousari and F. Azarm 1984. Geochemical Exploration in Anzaly lake (Northern Iran). Geological Survey of Iran.

3. W. Lux. 1993. Long - Term heavy metal and As pollution of soils. Hamburg. Germany . Applied Geochemistry No. 2.

4. F. Moztarzadeh, B. Dehzadeh, M. Rajabi, Z. Mohamadi 1983. A research for phisico - chemical parameters in Anzaly lake. Research Foundation for Energy and Applied materials.

5. M. Wallsten and Z. Dressie 1993. An investigation of metal contents in lake sediments, country of Uppsala, Sweden. Applied Geochemistry No. 2.

Proc.30th Int'l.Geol.Congr.,vol.24,pp.218-228
Yuan Daoxian (Ed)
© VSP 1997

Distributing Pattern of Iodine and Selenium in the Daba Mountain Area, Shaanxi, P.R. China

LUO KUNLI   YAN LIDONG   ZHANG QI   GE LINGMEI

Department of Geology, Xian Mining Institute, Xian, Shaanxi, 710054, P.R. China

Abstract

In the mountain area, especially in the area that the bedrocks are predominantly exposed, such as Daba mountain area, how is the relation between trace elements in the bedrock and in the soil, grain and shallow groundwater? What is the degree of the geological character and the trace element of the bedrock impacting on the trace element content and its distributing pattern in the soil, grain and shallow ground water and what is the pathway through which the trace elements in bedrock enter the soil, water, grain in Daba mountain? And how the geographical and geological character effect on the trace element contents and distributing patterns in the soil, grain and the shallow groundwater? These questions have been studied in a preliminary way.

Through a systematic investigation on the geological and geographical environment, the contents and distributing patterns of the trace element, especially of iodine, selenium and fluorine in the bedrock, soil, crops, spring and surface water in the Daba mountain area, Shaanxi, P. R. China are studied[1]. We found that iodine content in this area is unbalanced, and the most area is seriously deficient, especially in the area of carbonic slate and limstone. The iodine content (5—25ug/L) in the spring water is closely related to the lithologic character and iodine content of the bedrock, but iodine content in the soil, crops and surface water are very lower than many other places of China and slight related to the lithologic character and iodine content of the local bedrock. The selenium contents in most Daba mountain area are abundant, especially in the area of lower Cambrian and lower Silurian carbonic slate. The selenium content in local bedrock range from 10mg/kg to 20mg/kg, even in some bedrocks the content can reach up to 300mg/kg. The selenium contents in maize, potato and rice range from 0.1mg/kg to 0.6mg/kg which is 200 times that of the same crops grown in many other places. Selenium contents in the crops are closely related to selenium content and lithologic character of bedrock. The fluorine content in spring water and surface water don't excess 1mg/L, only concentrated in the soil and air where the stone coal is mainly used for civil and industry. The there soil and air were polluted by fluorine which is released from Paleozoic stone coal which is abundant in fluorine (1000—2000mg/kg) when they are burning.

Keywords: distributing pattern, iodine, selenium, fluorine, endemic deseases

INTRODUCTION

The Daba Mountain in south Shaanxi province, P.R. China is south part of the Qinling mountain and is a boundary of North China and south China. In Daba mountain, most people have a lower standard of living. It is also one of the poverty area in P. R. China too. Meanwhile, the endemic cretinism is widespread in this area, and the cretins are average 10% of the whole local population[2], some places even more. The selenium poisoning and fluorosis are occured in a few places of this area. These illnesses are very harmful to the resident's health and obstruct the economic development of this district.

A lot of work about the trace element contents and their distribution patterns in the strata in Daba mountain has already been done by the geological scientists. Agricultural scientists and biologist have done many research work on trace elements content and distribution in soil, vegetation, water system, animal and human body. In the past, however, geologists put emphasis on research of the regional geological characteristics, such as the strata sequence, the composition of rocks, the depositing environment, the regularity of ore forming, and so on. Agricultural scientists and biologists focus on the trace element contents in the soil, vegetation and water system as well as animal and human body. Few put these two reseach fields together or study their interior relationship.

During recent years, we have studied the strata and their accompanying ore deposits and associated elements in the South Shaanxi, and the trace element contents and their distributing pattern in the bedrocks, soil, crops, water and their relationship with the distributing pattern and the incidence rate of the cretinism, flourosis, selenium intoxication are researched. We found that the distribution and the incidence rate of the endemic diseases is closely related to the geographical and geological characters as well as trace element contents of the bedrocks, soil, crops and water in this area. In this paper we report some results of our study on the geological factors and trace elements, especially on the iodine, selenium and fluorine distributing pattern in this area.

MATERIALS AND METHODS

The samples of typical rocks were collected according to the lithologic character and geologic time sequence of the bedrocks. In the places where the bedrocks have the same lithologic character and were depositing in the same time had collected one group samples.

The soil samples were collected from the different zones according to the sequence and lithologic character of the local bedrock too; crop samples were collected in the same way as in soil.

Water samples were stored in clean plastic bottles (4 liters), HNO_3 was then added in the bottles, PH was adjusted to $1-2$ and all water had analysed in 6 days since collected.

All samples were analysed by Northwest Geological Testing Center of Northwest Geological Research Institute of CNNC, a first class analytical laboratory of CNNC which has been accredited by China National Metrology Bureau and CNNC.

Sampling principle
The reasons of that all our samlpes were collected based on the unit of the same lithologic character and the time sequence of the bedrock is as the follows: the Earth surface where human being live on consists of various sediments and rocks that were deposited and formed in the different period of geologic history, and

they are the sources of the soil and deposits which we live on.

But in each time phase of the geologic history, the different regions of the Earth, the condition of astronomy, landform, climate and parent materials were different, and those differences caused the differences of the physical and chemical properties of rocks or sediments that formed in each time phase. So those differences would result in differences of the physical and chemical properties of the soil, water and crop in the regions where different sediments and rocks occur. When the rocks that have a resemble lithologic character and deposited in same time phase which indicate that these rocks had a similar parent material and depositional paleoenvironment, that is, the rocks have a similar mineralogical composition and chemical composition, so it is only needed to collect one group samples and is enough to express character of all this rocks, no matter how large area of this bedrocks distributing.

GEOGRAPHICAL AND GEOLOGICAL ENVIRONMENT

Natural geographical environment
Daba Mountain in the South of Shaanxi province is great and high, the hill slopes are very steep and the ravines are deep. The large and small surface runoff and gills are very developed and torrent, soil is barren with average thickness about 20 —30cm. The average altitude is between 1000 and 2000 metres above sea level. The height difference of peak is very great, ranging from 500 to 1500 metres on the average from the foot of a hill to hilltop. Rainfall is comparatively concentrated and annual precipatation is 750—1500mm. There are many little gills and rivers across the area and the stream current is torrent. The Daba mountain area belongs to intensive denudation zone. Because the QinLing mountain was across the north of this area, cold front has less influence on the area. The temperature in winter is higher than that of the eastern plain which lies in the same latitude as this area. There have poor transport facilities and only have one railway line and a few highway line in thie area.

Strata and bedrocks
The major strata that distributed in this area are lower Paleozoic strata, they are mainly carbonic slate, limestone and detrital rock and intruded by a large number of basic igneous rock—diabase and trachydiabase.

Cambrian system is the most widespread strata in this area. It is well development and very thick in this area. Due to the influence of the recurrent fault and fold, this set of strata is multi—repeated outcrop. The low part of lower Cambrian is mainly carbonic slate and siliceous carbonic slate and intercalated with stone coal and many thin beds of diabase. The upper part of lower Cambrian is micritic limestone. In the middle—upper Cambrian system are mainly limestones and intercalated with calc—slate, carbonic slate and stone coal. The Ordovician system are limestones and calc—carbonic slate and interbeded with 2—3 layers of stone coal. Silurian system are mainly fragmental rock and carbonic slate, and intercalat many thin beds of bioclastic limestone and in some places intercalated with a large number of basic igneous rocks—trachyte, volcanic breccia and tuff. The distribution of Ordovician system and Silurian system, same as that of Cambrian system, are

Table 1. The trace element contents of Paleozoic strata in Daba Mountain (mg/kg)

stratigraphic unit	rock type	sample number	Cu	Zn	Pb	Cr	Ni	Mo	V	Co	Ag	Li	Be	La	Yb	Y	Zr	As	P	Ba	Sr	Mn	Ti	Cd	Se	Ge	F	Hg
	clarke value		63	94	12	110	12.3	1.3	140	25	0.08	21	1.3	39	2.7	24	130	2.2	1200	390	480	1300	6400	0.3	0.08	1.4	450	0.089
	siliceous slate	4	42	150	18	271	38	6.2	198	8.7	0.05	46	3.2	50	3.3	23	350	6.5	580	1300	120	32	6700	0.2	4	2.0	1010	6
	trachydiabase	3	86	139	8.9	89	59	9.1	76	2.0	0.05	69	1.70	82	4.6	40	460	5.2	1200	3600	265	740	6100	1.70	0.3	0.8	610	/
S$_{(1-2)}$	sandstone	5	25	100	16.2	180	50	1.0	73	19	0.03	36	1.8	38	2.7	21	260	3.3	390	1300	201	440	4900	0.2	0.4	1.0	565	1.5
	silty slate	2	30	40	20.2	250	60	2.3	120	14	0.05	45	2.5	35	3.0	23	260	7.0	410	1003	756	280	7100	0.08	1.6	1.5	780	2
	carbonic slate	8	140	500	46	301	52	0.6	358	3.1	2.1	25	6.1	90	6.1	45	470	40.2	920	6020	160	261	6900	1.80	5	1.20	1280	8
O	marl	3	26	36	1.8	60	20	0.6	42	81	0.1	16	1.6	30	1.3	17	75	2.2	290	1360	280	401	1890	0.1	1.2	0.76	750	30
	cale—slate	4	56	101	9.3	85	26	1.2	75	21	0.08	76	2.8	36	1.8	16	73	3.1	401	790	210	910	3260	0.1	2.2	1.8	801	19
	diabase	2	75	/	/	86	/	/	86	20	/	/	/	/	/	/	/	2.8	/	3900	860	/	8200	1.3	0.06	1.3	420	/
	marl	4	36	38	7.8	30	15.6	1.2	98	10	0.05	25	1.2	20	1.0	10	20	3.2	364	450	812	320	1200	0.1	4	0.5	1020	12
	cale—slate	3	40	35	6.2	32	20	2.0	120	8.2	0.05	26	1.6	22	1.6	13	36	3.0	480	690	450	210	1400	0.2	2.2	0.8	460	10
	limestone	5	26	28	4.6	24	12	1.0	101	7.8	0.05	20	0.5	20	0.5	8	18	2.10	550	390	950	98	910	0.1	0.6	0.5	450	16
Ɛ	diabase	2	70	/	/	72	/	/	88	20	/	/	/	/	/	/	/	2.6	/	4100	910	/	1.2%	1.2	0.8	1	360	/
	carbonic slate	8	276	1440	63	366	650	22	1800	10	5	56	3.2	36	4.7	38	140	20	3200	8500	460	820	3600	15.0	12	2	1110	8
	siliceous slate	10	340	1300	56	280	41	42	1200	7	2	29	3.0	3.0	4.0	30	120	16	1260	6260	350	624	3031	6.1	20	1.5	890	6

Note: All samples analysed by the Northwest Geological Testing Center of Northwest Geological Research Institute of CNNC. Se—sample treatment: $HNO_3+H_2SO_4$, deter method: atomic absorption fluorescence, absolute error: 0.0003%; Reporting limit: 10^{-9}; I, F—sample treatment: alkali fusion, deter method: ISE & distillation. Reporting limit: 10^{-9}; Pb, Zn—HCL+HNO_3, AAS; Cd—HCL+HNO_3, GAAS; Hg—HCL+HNO_3, AFS.

multi—repeated outcrop as zonal distribution too.

DISTRIBUTING PATTERN OF IODINE AND SELENUIM

Bed rocks

In this area bedrocks are predominantly exposed. Table 1 is the content of some trace element of Paleozoic strata in Daba mountain. Siliceous slate and carbonic slate of the lower Cambrian and lower Silurian have an anomalous value of selenium, fluorine, phosphorus, vanadium, zinc and barium. The selenium content in the carbonic slate is about $10-20mg/kg$; in the limestone and marl is about $1-2mg/kg$; in the basic igneous rocks and sandstone is about $0.05-0.08mg/kg$, and in silty slate is about $1.5mg/kg$.

Table 2 is iodine contents of bedrocks and soil as well as potato in their distributing area.

Table 2.   Iodine contents of bedrocks and soil as well as potato in disease area, Daba mountain

	←₁		←₂:			O		S					
	carbonic slate	limestone	marl	limestone	calc—slate	carbonic slate	limestone	carbonic slate	marl	sandstone	trachy—lava	limestone	carbonic slate
bedrock (mg/kg)	2.6	1.3	1.9	1.2	2.1	3.0	2.2	2.8	1.2	1.8	0.9	1.1	1.8
soil (µg/kg)	43.5	22.2	32.4	22.2	32.0	42.2	21.6	31.1	21.0	31.0	87.6	21.1	31.1
potato (µg/kg)	15.8	14.6	15.2	14.3	16.1	16.1	15.2	16.8	16.0	14.8	98.0	13.8	14.0

Note: All samples analysed by the Northwest Geological Testing Center of Northwest Geological Research Institute of CNNC. I—sample treatment: alkali fusion, deter method: ISE & distillation. Reporting limit: 10^{-9}

The iodine content in the carbonic slate of the Cambrian is about $2.0-3.0mg/kg$; in the limestone is about $1.0-1.5mg/kg$ (Table 2); in basic igneous rock is about $0.5-0.9mg/kg$; in the limestone of Silurian system is $1.1mg/kg$; in the limestone of Ordovician system is $2.2mg/kg$; in the diabase and trachydiabase are about $0.8-0.9mg/kg$, which are the lowest in this area.

Soil

In this area mountain slopes are mainly planting fields. The soil in this place is very thin about $10-30cm$ thick, and are mainly podzolic soil and peat soil. Generally they are slope washes and talus of the Paleozoic limestone and microclastic rock that outcropped and distributed in the locality.

Table $2-5$ are the analytical results of the iodine, selenium and fluorine contents of bedrocks and soils in their distributing area. From these table we can find that iodine content of soil in this area is very lower than average content in the soil of

China, and closely related to the lithologic character of the local bedrock. The selenium contents in soil is closely related with selenium content and lithologic character of the bedrock as well as landform characteristics[3], and is related with the thick of local soil(see table 3). The soil iodine content are mainly controloed by the lithologic character of the local bedrock. In the distributing area of basic igneous rock, the iodine content of the soil is about 120ug/kg; in carbonic slate area is about 50ug/kg and in limestone area is 20—30ug/kg. So the iodine content of soil in Daba mountain area is much lower than others area in China. Especially in the area of limestone, carbonic slate and microclastic rock, the iodine contents are very low.

Table 3.   The selenium content of rocks, maize and rice in the distributing area of Paleozoic strata in Daba Mountain ($\times 10^{-6}$)

	€₁				€₂₋₃							O		S						
	diabase	silicous slate	carbonic slate	stone coal	silicous slate	carbonic limestone	marl	stone coal	calo-slate	limestone	marl	carbonic slate	sandstone	carbonic shale	carbonic slate	silicous slate	trachy-lava	stone coal	limestone	phyllite
bedrock	0.08	16.4	10.6	40.1	20.1	4	4	30	2.2	0.6	1.2	5.8	0.4	6	5	4	0.3	30	0.2	0.8
maize		0.6	2.6	6.6	0.8	0.6	0.4	/	0.9	0.06	0.06	0.8	0.01	1.8	1.1	0.8	0.06	/	0.06	0.1
potato		0.08	0.4	0.8	0.1	0.1	0.08	/	0.2	0.01	0.04	0.04	0.01	0.8	0.4	0.2	0.06	/	0.04	0.06
rice		0.2	0.6	1.6	/	0.2	0.1	/	0.2	0.1	0.06	0.1	0.5	0.4	0.5	0.4	0.03	/	0.1	0.1

Note: All samples analysed by the Northwest Geological Testing Center of Northwest Geological Research Institute of CNNC. Se—sample treatment: $HNO_3 + H_2SO_4$, deter method: atomic absorption fluorescence, absolute error: 0.0003%; Reporting limit: 10^{-9}

Table 4.   The fluorine content of rock and maize in the distributing area of Paleozoic strata in Daba mountain area, Shaanxi, P.R. China ($\times 10^{-6}$)

	€						O			S				
	silicous slate	carbonic slate	diabase	limestone	calo—slate	marl	diabase	calo—slate	marl	silty slate	sandstone	trachy—lava	silicous slate	stone coal
bedrock	890	1110	360	450	460	1020	420	801	750	566	280	565	1010	1280
maize	0.5—4	1—6	<1	<1	0.8—1	1—2	0.5	0.8	1.2	1—10	3	2—3	0.5—2	1—8

Note: All samples analysed by the Northwest Geological Testing Center of Northwest Geological Research Institute of CNNC. F—sample treatment: alkali fusion, deter method: ISE & distillation. Reporting limit: 10^{-9}.

Crops

In this area main crops are potato and maize. The local residents, except the town residents, mainly consume local food. Table 2 shows the iodine content of potato in this area. From this table, we know the iodine content in the potato is positively related to the local soil iodine content, ranging from 13ug/kg to 90ug/kg. In this area iodine content in all potato is very low except those in the area of basic igneous rock which is a little higher. Two maize samples were collected from marl distributing area of Maobaguan group in the Cambrian system and two from carbonic slate distributing area of Wuxiahe group of the Silurian system. The iodine contents of them (dry weight) are about 10—40ug/kg, which are lower than that of the maize grown in many other areas. The selenium contents in the crops vary in different places of Daba mountain areas. Selenium contents in maize, potato and rice in the Daba mountain area range from 0. 1mg/kg to 6. 6mg/kg which are 3—200 times that of the same crops grown in many other places. Fluorine content of the maize in Daba mountain area is very difference and the content of the most maize don't exceed 1mg/kg except towns that are located at or near the stone coal mine where a great quantity of stone coal are burnt and released fluorine that polluted the local crops and air.

Table 5. Seleninim content of the maize, potato and rice in Daba mountain area, Shaanxi, P. R. China ($\times 10^{-6}$)

	location	Zihuang	Malui	Maoba	Shuang men	Gaotan	Wafang dian	Mixi	Huangu	Shuanan
	maize	0. 6	2. 6	0. 9	0. 06	0. 07	1. 1	1. 2	0. 8	6. 6
	potato	0. 08	0. 4	0. 2	C. 04	0. 06	0. 4	0. 26	0. 2	0. 8
	rice	0. 3	0. 6	0. 2	0. 06	0. 1	0. 5	0. 2	0. 4	1. 6
soil thick (m)	maize	720	<20	±20	720	720	<20	<20	<20	<20
	potato	720	±20	±20	720	720	<20	<20	<20	<20
	rice	30-50	30-50	>30	>30	>30	30-50	30-50	30-50	>30

Note: All samples analysed by the Northwest Geological Testing Center of Northwest Geological Research Institute of CNNC. Se—sample treatment: $HNO_3 + H_2SO_4$, deter method: atomic absorption fluorescence, absolute error: 0. 0003%; Reporting limit: 10^{-8}

Spring and surface water

The rainfall is very plentful in Daba mountain. Water system is well develped and there are many rivers and streams all over the region. Spring is also developed. They are main source of drinking and irrigating water for local residents.

Table 6 is the contents of iodine and other elements in the springs. The iodine content of spring is 5 — 25ug/L which is closely ralated to lithologic character of bedrocks. The iodine content in the springs that flow out from basic igneous rock area is the highest. It is 25ug/L on the average. The second highest is in the limestone, which is about $8-16$ug/L. The iodine content of spring in the limestone is closely related to structural location. If the spring originated in the limestone that lies in the deep fault, the iodine content of spring can reach 16ug/L and even more. If it is fissure spring or much far from deep fault, the iodine content is usually 8ug/L and even lower. The spring that originated in carbonic slate is lowest, which is 5ug/L.

Table 6.　Some trace element contents of spring occurrence in Paleozoic rocks in Daba mountain, Shaanxi

rock type		assay value $\rho(\mu g)$/L					assay value $\rho(mg)$/L					
		Se	Pb	Hg	Cd	I	Li	Sr	Zn	F	H_2SiO_3	free CO_2
$\in_2$	carbonic limestone	0.3	6	<1	0.4	14	0.003	0.29	0.021	0.28	6.1	9.7
$\in_1$	carbonic slate	1.9	10	<1	1.4	5	0.002	0.12	0.040	0.24	0.9	11.1
O_1	marl	0.1	4	<1	0.2	5	0.001	0.26	0.019	0.20	1.7	19.4
S_1	thachy lava	0.4	10	<1	0.3	25	0.005	0.45	0.016	0.42	8.0	2.8
S_1	carbonic slate	0.4	9	<1	0.2	5	0.002	0.38	0.014	0.42	12.5	19.4
$\in_2$	limestone	0.25	5	<1	0.3	16	0.003	0.25	0.02	0.25	8.9	10.6
$\in_1$	silicious slate	6.0	9	<1	2.0	6	0.002	0.20	0.10	0.30	6.8	22.1
S_1	trachy—lava	2.0	7	<1	0.3	20	0.002	0.45	0.10	0.45	11.9	6.57
$\in_2$	diabse	0.1	4	<1	0.3	25	0.01	0.55	0.019	0.4	9.0	3.0
$\in_2$	limestone	0.2	4	<1	0.5	16	0.001	0.30	0.01	0.20	10.1	12.5

Note: All samples analysed by the Northwest Geological Testing Center of Northwest Geological Research Institute of CNNC. Se—sample treatment: $HNO_3+H_2SO_4$, deter method: atomic absorption fluorescence, absolute error: 0.0003%; Reporting limit: 10^{-9}; I, F—sample treatment: alkali fusion, deter method: ISE & distillation. Reporting limit: 10^{-9}; Pb, Zn—HCL+HNO$_3$, AAS; Cd—HCL+HNO$_3$, GAAS; Hg—HCL+HNO$_3$, AFS.

We also analyzed the iodine content of several main river in this area (Table 7). The iodine contents of analyzed samples are very low, all of them don't exceed 1ug/L, much lower than the standard of drinking water.

By analyzing, we can find that the iodine content in the soil, crops and spring is closelly related to the lithologic characters of bedrocks. In this area the iodine content in the widespread limestone and carbonic slate is deficient seriously. It is one of the most deficient regions in our country. To this area more attention should be paid by Chinese government.

So, this district is very short of iodine especially in the area of carbonic slate and limestone, which is also the serious area of cretinism, the cretins are commonly 10% of the whole residents in this area. The selenium content is abundant especially in the area of carbonic slate of lower Cambrian and lower Silurian in this district.

Table 7.   Iodine and flouride contents of some main rivers in Daba mountain

	Hanjiang river	Renhe	Dadaohe	Haop- inghe	Heishui- he	Mopan- gou
I(ug/L)	0. 8	0. 25	0. 9	0. 6	0. 8	0. 2
F(mg/L)	0. 4	0. 28	0. 3	0. 58	0. 2	0. 2
Sampling location	Ziyang city	Bajiao	Yuechi	Haoping	Shuang hekou	Renjia village

Note: All samples analysed by the Northwest Geological Testing Center of Northwest Geological Research Institute of CNNC. I—sample treatment: alkali fusion, deter method: ISE &. distillation. Reporting limit: 10^{-9}

ENDEMIC DISEASES DISTRIBUTING IN THIS AREA

The endemic diseases such as cretinism are widespread and very serious. The incidence of the endemic cretinism is about 10% of the whole local population and in some places is higher, such as the dull—witted deaf, short and low—intelligence. The endemic fluorosis is not very widespread in this area, but is very serious in towns such as Haoping town in Ziyang county. The selenium intoxication is only occured in the Suanan of Ziyang county. These diseases, in particular, cretinism have seriously infected the health of the residents, reduced the quality of population and limited economic development in this area.

On the relation to the distribution and cause of the endemic cretinism the predecessors have done a lot of work. They believed cretinism is an endemic desease which emergenced in the endemic goiter region[4]. The study on the distributing pattern of the endemic cretinism in middle and north zone of Qinling mountain was limited to Ningshan, Foping and Zhouzi in the north and the middle of Qinling mountain. However, very few scientists are involved in studying the endemic cretinism in the South Qinling (Daba Mountain). If so, is only in a preliminary way.

We have done field geological work along Ziyang — Songping, Ziyang — Huangu, Maobaguan — Wanyuan, Ziyang — Wafangdian, Wafangdian — Hongchunba, Gaoqiao — Liuhe, Shuangmen — Jieling, and so on in the Daba mountain area, and have preliminarily studied the distribution and incidence rate of the endemic cretinism in different regions. We found that the distribution and incidence rate of the endemic cretinism is closely related to iodine content in the drinking water and food in this area.

In the remote mountain area, especially in the Paleozoic carbonic slate and limstone distributing area, the incidence rate is highest, for example, Renjia village, which is 8km to the west of the Ziyang town, mainly distributed carbonic slate, carbonic —siliceous shale and sandstone, the incidence rate of the endemic cretinism is very high. Almost every family has the foolish and low—intelligence person. The incidence rate is about $10-15\%$. The villagers mainly drink the water of the Renhe stream, its iodine content is 0.25ug/L only, and mainly consume locally produced food. The incidence rate is the second highest in the limestone distributing area if people mainly drinking the spring water. The lowest incidence rate is in the towns, where people mainly consume the food which is produced in other places and drink the spring water that rises from basic igneous rock such as Ziyang town. That is, the incidence rate of the endemic cretinism in Daba mountain area is closely related to the iodine content in the food and drinking water in this area.

For the convenience of planting crops, most of the mountain villagers in Daba mountain area live on the slopes, and generally have to drink local creek water and spring water, and to consume the local produced food (because this area is a poverty district, the people except the towns, have not money to buy crops that grow in other places). The iodine content of surface runoff is very low (see table 7). For instance, the iodine content of Renhe water, which is the drinking water for the residents of Renjia village, is 0.2ug/L on the average, and it is very low. But the iodine content of the springs is comparatively higher, especially the springs that orinqinated from the basic igneous rock and the limestone near the deep fault. Therefore, although the iodine content in the soil and crops is low in this area, if the residents drink the spring water occuring in the igneous rock, more iodine would be absorbed than those drinking the spring originating in the carbonic slate and surface water, and the suffering rate is lower. The geological and geographical environment of Ziyang county in the carbonic slate area of Cambrian system and Silurian system is similar to that of Renjia village in the west of Ziyang town. But in the Ziyang town residents drink the spring water being originated in the diabase and trachydiabase in the north of Ziyang town and eat the food produced in the other places. The iodine content of this spring is 25ug/L. So the suffering rate of Ziyang towns is very low, which is about 0.2%.

In a summary, the incidence rate of the endemic cretinism in Daba mountain area is high, the distributing area is wide. But it is not all mountain area having the endemic cretinism, and the suffering rate isn't same in everywhere of Daba mountain area.

So it is certain that the primary factor of widespread cretinism in this region is that the environment is seriously short of iodine. All these need more study in the future.

The incidence rate of fluorosis in this area is not balanced and only concentrated in the place where the stone coal is mainly used for civil and industry. The main reason of the fluorsis in this area is that corn and air were polluted by fluorine which is released from Paleozoic stone coal which is abundant in fluorine when they are burning.

Proposal

The residents living in this district shall mainly drink the water of springs that rise from the basic igneous rock and not drink the surface water and spring water that rise from the carbonic slate, and mainly eat the crops that are growing in the basic igneous rock area or the other places where iodine is not short. Although the iodine intake from drinking water is little, compared with that from the food. So it

is not enough to add iodine only by drinking the water and shall popularize the iodine salt in this region and sometime shall add more iodine to the common iodine
salt according to the quantity that different residents needed in Daba mountain
area.

Acknowledgements

All our field and laboratory work during 1990—1995 were suppoted by the Coal
Industry Ministry of P. R. China and the Xian Mining institute. I would like to
express my gratitude to Prof. Yuan Daoxian of the Institute of Karst Geology in
Guilin, P. R. China for his encouragement and his excellent suggestions through
the final editing stages of the paper.

REFERENCES

1. Luo Kunli, Gelinmei etc. *Accompanying and associating ores of Paleozoic black shale and coal, Shaanzi province.* Press of Northwest university. Xi´an (1994).
2. Editorial board of Ziyang county annals. *Ziyang county annals.* San Qin Press. Xi´an (1988).
3. Luo Kunli, Qiu Xiaoping. Analysis on selenium —rich crops in Ankang district, Shaanxi— the case of Ziyang county. *Journal of Natural Resources*, No. 2, 68 — 72 (1995) (in Chinese)
4. Song Guangshun and Wang Shaohan, *Bviromental medicine.* Tianjin Scientific and Technical Press. Tianjin. 188—196(1987).

Proc. 30th Int'l. Geol. Congr., vol. 24, pp. 229-240
Yuan Daoxian (Ed)
© VSP 1997

Heavy-Metal Contamination in the Grand Canal, People's Republic of China

HUANXIN WENG[1] and CHARLIE Y. XU[2]

1. Department of Earth Sciences, Zhejiang University, Hangzhou 310027, P.R. China
2. Department of Geological Sciences, University of Illinois at Chicago, M/C 186, Chicago, IL 60607, USA. E-mail: yuping@uic.edu, Fax: 1-312-413-2279.

Abstract

Many regions of the Grand Canal in the People's Republic of China have been highly contaminated with heavy metals and organic contaminants from both industrial and municipal sources since the 1950s. The awareness of this environmental problem by the Chinese government in the 1980s prompted an investigation into the sources of heavy-metal discharges to Hangzhou section of the canal and their impacts on local water resources. This report describes the distribution of some heavy metals (As, Cr, Cd, Cu, Ni, Pb, and Zn) in river water, river sediments, and soils adjacent to the canal. The results show that these metals are largely accumulated in the surficial river sediment, whereas heavy-metal concentrations in solution are elevated only in several locations. The only exception is Zn with its concentrations are extremely high in both solution and the surficial sediment, where Zn appears to have nearly reached its sorption capacity. Overall, the order of contamination level in the canal is Zn > Cu > Pb > As > Cd. It is speculated that Cu distribution has been significantly affected by its binding to organic matter in the surficial sediment. The study indicates that the surficial sediment in the Grand Canal likely serves as a sink to anthropogenic contaminants consisting of heavy metals as well as organic compounds.

INTRODUCTION

Heavy metals in aquatic and soil environments have received increasing attention in recent years due to public awareness of environmental issues related to human health and agricultural productivity. As a result, more stringent regulations have been established in industrialized countries toward controlling heavy metals in drinking and natural waters. Developing countries have been left behind in terms of awareness of environmental problems. In the 1980s, research was initiated and funded by Chinese environmental agencies and local governments to investigate the extent of contamination in the Grand Canal. The Grand Canal was built in the 14th century by connecting existing canals [1]. It runs from Hangzhou in the south through Beijing in the north, connecting Yangtze River and Yellow River (Fig. 1). In addition to being a channel for flood controls, the Grand Canal (*Da Yun He* in Chinese) was once a chief waterway of transport to maintain

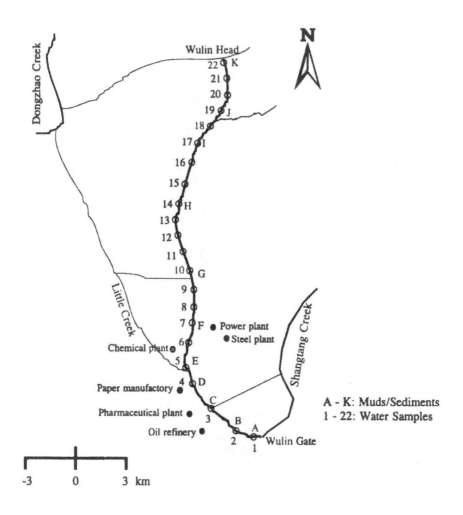

Figure 1. Schematic diagram of the study area with sampling locations.

the supply of grain from the southern provinces to the capital in imperial times [2]. In the mid 19th century, the canal was severely damaged and blocked in the northern sections [1]. The Hangzhou section of the Grand Canal runs from Wulin Gate, Hangzhou city in the south through Wulin Head, Deqing county in the north (Fig. 1). This section is 23.6 km long, 100 m wide, and 2-3 m deep in average. The gradient of the river bed is small and water flows normally very slowly from south to north. This section of the canal was once a source of drinking water. However, direct emission of industrial and municipal wastes into the canal since the late 1950s has polluted the easily accessible water source. Several major industries, such as steel and chemical plants, have contributed a significant quantity of contaminants into the canal (Fig. 1). The canal also receives polluted runoff from the urban area of Hangzhou city, industries, and farm land from south to north.

Hangzhou lies within a subtropical region of China. The Hangzhou section of the Grand Canal is mainly recharged by precipitation and surface runoff. Because of the small flow rate, advective transport of contaminants is very limited. The water is mainly HCO_3-Cl-Ca-Na type in chemistry with several locations dominated with HCO_3-SO_4-Ca-Mg [3]. The study area is covered with a layer of Holocene sediment up to 15 meters deep [4]. The sediment consists of fine sand and clay from lacustrine and other deposits. Using Fe and Mn as indicators, Wu and Weng [4] showed that the interaction between the river water of the Grand Canal and groundwater was minimal. The lack of interaction was attributed to the low hydraulic conductivities (K) of the Holocene sediment where K was estimated to be in the range of 10^{-8} - 10^{-4} cm/s [4]. In this report, we will evaluate heavy-metal contamination of the water body and river sediment throughout the section.

EXPERIMENTAL SECTION

Sampling
Samples of the bed sediments were taken at eleven locations by drilling four to six meters deep into the river bed (A - K in Fig. 1). At each location, three cores across the canal opening were taken. At the same locations, water samples were taken 50 cm beneath the surface. At each location, bottles of water were drawn from three spots across the canal and mixed to obtain a composite sample. Water sampling was conducted bimonthly for the period between May 1983 and March 1984 and a total of 22 samples were obtained during each sampling event (No. 1 - 22 in Fig. 1). pH, Eh, temperature and transparency of each water sample were measured on site. The water samples were filtered and the solid residues on the filters were weighed and recorded as the total suspended matter (TSM). In addition to the sediment and water sampling, soil samples in the study area adjacent to the Canal were taken sporadically to obtain background concentrations of heavy metals.

Sample Analyses
Filtrates from the water samples were analyzed using the following methods. Concentrations of K, Na, Mn, and Zn were determined directly with flame atomic absorption spectroscopy, whereas concentrations of Cu, Pb, and Cd by the graphite furnace using an AA180-50 Atomic Absorption Spectrophotometer (Hitachi Ltd., Japan). Total Ca and Mg were measured by the EDTA titrimetric method [5]. Ammonia (NH_4^+) was determined by the Nessler method, Cl by the potentiometric method, SO_4^{2-} by either the turbidimetric method or the gravimetric method, and HCO_3 by acid-base titration [5]. Fe^{2+} and total Fe were measured by the phenanthroline method, As by the silver diethyldithiocarbamate method, total phosphate by the ascorbic acid method, and total organic nitrogen by the Kjeldahl method [5].

The core samples of bed sediment and soil in the area were acid digested and analyzed according to the methods listed above. Each core was analyzed for its heavy-metal contents according to the depth: <0.3 m (surficial sediment, or mud), 0.3 m, 1.0 m, 2.0 m, 3.0 m, and 4.0 m. The organic matter (OM) contents of the surficial sediments were measured by a wet oxidation method in which an excessive amount of the $K_2Cr_2O_7$-H_2SO_4 mixture was used to oxidize the organic matter at 185 °C, followed by the titration of the excessive $K_2Cr_2O_7$ with $FeSO_4$ [6].

MINTEQA2 computer code [7] was used for the equilibrium calculations of the water chemistry in the Grand Canal. No precipitation of solid phases was allowed in the calculations, and the saturation indices (SI) obtained were used as indicators to the saturation status of each metal in the solution.

RESULTS AND DISCUSSION

The water chemistry of major and heavy-metal elements is shown in Table 1. The pH values and the concentrations of major elements represent the water samples taken in November 1983, whereas the concentrations of heavy metals are the average values of six consecutive bimonthly sampling events conducted between May 1983 and March 1984. Seasonal variations in heavy-metal concentration were found to be insignificant. Overall, the water chemistry is characterized by Ca (mean 42.3 mg/L) as the dominant cation and HCO_3^- (mean 147.6 mg/L) as the dominant anion. The calculated ionic strength ranges from 2.96×10^{-3} to 1.25×10^{-2} M. The measured Eh of water at 50 cm depth is in the range of 0.68 to 0.71 V, suggesting an oxic environment in the upper water layer.

The heavy-metal concentrations and organic matter (OM) contents of the surficial sediments and of the adjacent soils are listed in Table 2. The total heavy-metal concentration is the summation of As, Cr, Cd, Cu, Ni, Pb, and Zn in weight (ppm). Because no size-fractionation studies were initially planned for metals in solution, the distribution of each metal in soluble and colloidal forms is not determined in this report. The metal concentrations in water samples taken with the available filtration method may have included metals in soluble and colloidal forms. However, the total suspended matter (TSM in Table 1), ranging from 27 to 97 mg/L, suggests little solid suspension, presumably due to an insignificant flow rate in the canal.

Water in the Grand Canal
The major-element composition of the canal water resembles that of the Yangtze River. Both contain elevated Ca and HCO_3^- compared to the average values of major rivers in the world [8]. However, their concentrations vary geographically as shown in Table 1. In the vicinity of the Wulin Gate (Fig. 1), water is enriched with most of the major chemical species. Samples taken at sites 3 and 4 have the highest ionic strength (0.012 M). Runoff from Hangzhou's metropolitan area and discharge from the nearby oil refinery and other plants may have contributed to the high ionic strength, which gradually decreases toward Wulin Head. The sample at site 22 contains much lower concentrations of most aqueous species. The chemical composition of this water most likely represents water not affected by industrial pollution.

Table 1. Chemical composition of water in the Grand Canal (Hangzhou section).

Sample Site	Unit	1	2	3	4	5	6	7	8	9	10	11	12	13	14	15	16	17	18	19	20	21	22
Temperature	Deg C	16.2	15.8	19.8	16.3	19.5	19.7	19.8	21.7	21.5	21.4	20.8		20.8	18.7	18.7		19.5	19.5	19.1	18.3	18.5	18.8
pH		6.6	6.8	7.0	6.9	6.6	6.6	6.9	7.2	6.7	6.8	6.7	6.9	7.6	7.0	7.0	7.1	7.0	7.0	7.0	7.0	7.0	7.0
Na	ppm	51.5	55.5	89.0	88.0	50.0	42.0	25.6	24.8	22.1	22.1	20.8	20.8	17.3	35.0	35.4	35.4	21.8	23.2	21.2	21.2	21.8	10.4
K	ppm	14.1	15.0	14.4	14.6	7.4	8.6	4.4	4.2	4.4	4.2	4.2	4.0	3.4	6.0	6.0	6.4	4.4	4.0	4.4	3.6	3.6	2.2
Ca	ppm	57.3	57.1	70.3	64.1	49.3	51.1	40.1	40.1	34.1	39.1	39.1	34.1	34.1	43.1	43.1	30.1	31.7	34.3	31.5	31.5	33.9	21.6
Mg	ppm	10.8	8.4	12.5	12.5	9.7	10.2	8.5	9.7	8.5	8.5	6.7	7.3	7.9	9.7	10.3	9.7	7.2	6.9	7.8	7.8	6.9	5.5
NH_4	ppm	2.60	2.60	3.00	2.60	0.90	1.10	0.80	6.0	6.0	6.0	5.5	5.0	8.5	2.8	2.8	3.6	3.2	4.5	5.5	4.0	4.0	1.6
Cl^-	ppm	87.9	85.4	157.4	157.4	75.2	56.4	90.0	78.7	42.5	39.0	37.2	35.5	31.9	56.7	60.3	35.5	33.7	35.5	33.0	34.0	34.0	15.2
SO_4^{-2}	ppm	72.0	81.1	46.1	58.4	61.3	56.8	42.0	41.2	41.2	42.8	34.2	40.7	41.2	50.2	44.8	27.2	28.0	35.0	32.0	32.0	32.0	16.0
HCO_3^-	ppm	219.7	228.8	262.4	164.8	189.2	166.6	131.2	131.2	122.0	119.0	115.9	129.1	106.8	140.3	149.5	106.8	109.8	109.8	112.9	109.8	109.8	73.2
TFe	ppm	2.0	2.0	1.5	1.5	4.2	3.2	2.2	1.6	1.6	1.6	1.8	1.6	1.4	2.4	2.0	1.3	1.3	1.3	1.4	1.4	1.6	0.7
Fe^{2+}	ppm	2.0	2.0	1.5	1.5	3.2	2.2	1.6	1.1	1.2	1.1	1.3	2.0	0.0	2.0	1.0	0.7	0.7	0.7	0.7	0.7	0.8	0.2
Fe^{3+}	ppm	<0.1	<0.1	<0.1	0.2	0.2	0.2	0.6	0.5	0.4	0.5	0.5	0.3	1.4	<0.1	<0.1	0.3	0.6	0.7	0.7	0.7	0.8	0.5
TMn	ppm	0.264	0.264	0.453	0.434	0.491	0.547	0.26	0.25	0.26	0.28	0.26	0.26	0.23	0.47	0.46	0.26	0.29	0.3	0.3	0.3	0.3	0.16
TSM	ppm	62.0	540	331	300	338	391	478	426	857	660	594	524	515	975	316	790	278	480				
TKN	ppm	0.61	1.14	0.72	0.9	1.42	1.38	1.42	1.94	1.2	1.49	1.46	1.77	1.92	1.43	1.6	2.02	2.8					
Tot P	ppm	0.287	0.183	0.0335	0.271	0.34	0.0733	0.431	0.555	0.72	1.187	0.747	0.899	0.72	0.57	0.929	1.889	1.031	2.774				
As	ppb	3	40			98	28	13	2	4	2	2	2	3	3	9	19	4					
Cd	ppb	0.53	0.59	0.25	0.4	0.68	0.35	0.94	0.72	0.12	0.6	0.8	1.2	0.7	1.03	0.68	1.2	1.01	0.8				
Cr (VI)	ppb	<1	<1	<1	<1	<1	<1	<1	6	<1	<1	<1	2	4	<1	<1	<1	<1	4				
Cu	ppb	9.9	9.3	12.1	10.2	8.9	16.2	11.9	11	15.5	27.4	88.2	37.2	9.2	17.3	30.1							
Pb	ppb	8.6	8.8	9.3	9.3	9.8	8.9	9.8	12.8	14.8	16.8	10.8	6.5	9.1	2.6	16.5	5.6	8.5					
Zn	ppb	245.5	158.8	313.5	364.5	413	32.8	108.3	106.2	202.6	76.3	341.5	210.5	193	77.8	256.3	112.8	115.8					
Ionic Strength	(10^{-3} M)	9.65	9.83	12.5	12.06	7.86	8.5	7.91	5.77	5.77	5.16	5.26	5.04	6.41	4.63	4.32	5.01	4.53	4.78	2.96			

TSM = Total Suspended Matter, TKN = Total Kjeldahl Nitrogen, TFe = Total Fe, and TMn = Total Mn.

Table 2. Heavy-metal content of surficial sediment (mud) in the Grand Canal (Hangzhou section) and of soils in the area. Units are ppm (parts per million) unless otherwise noted.

Sample Site	OM (%)	As	Cr	Cd	Cu	Ni	Pb	Zn	Total
A (1)*	20.6	20.01	176.5	0.65	428.9	58.9	22.4	678.8	1386.2
B (2)	9.57	12.06	104.3	0.49	117.7	37.7	16.5	239.6	528.4
C (3)	4.76	12.81	76.4	0.58	89.0	34.8	41.7	206.7	462.0
D (4)	8.85	65.77	107.8	6.91	772.3	76.7	81.8	1644.2	2755.5
E (5)	5.55	23.37	87.3	1.02	115.3	41.2	74.1	1263.7	1606.0
F (7)	6.96	23.12	84.0	2.92	128.3	44.4	104.9	2922.2	3309.8
G (10)	5.45	19.27	94.5	1.57	97.4	41.1	29.7	2442.6	2726.1
H (14)	6.25	25.05	100.6	1.56	81.8	41.8	62.6	3132	3445.4
I (17)	3.88	13.09	89.1	0.45	46.9	32.5	36.2	1155.9	1374.1
J (19)	2.11	10.3	75.1	0.38	26.6	27.4	25.5	336.8	502.1
K (22)	1.85	7.5	64.1	0.27	25.2	22.4	15.5	401.8	536.8

Heavy-metal concentrations in soils adjacent to the Grand Canal (Hangzhou Section)

	As	Cr	Cd	Cu	Ni	Pb	Zn	Total
Sample #	78	68	80	80	81	78	73	
Mean	7.86	69.29	0.301	26.47	36.3	32.12	82.74	255.1
Minimum	1.33	46.81	0.201	13.6	22.45	19.85	11.8	
Maximum	15.25	89.19	0.455	38.72	49.18	51.26	147.39	

* The number in the parenthesis is the corresponding water sample site.

Distribution of heavy metals

Figure 2 shows the concentrations of dissolved heavy metals in the canal. Although the canal has been heavily polluted with industrial discharges, elevated concentrations of dissolved heavy metals occur only sporadically at a few locations, with the exception of Zn. However, the heavy metals are almost three orders of magnitude or more enriched in the surficial river sediment (depth <0.3 m) than in water body above (Fig. 3). The metal concentrations display a general order of Zn > Cu > Pb > As > Cd in both water and sediment. Figure 4 depicts the total concentrations of these heavy metals in the surficial sediment at all sampling locations. Compared to that of the soils in the study area (Table 2), the heavy metals are about two (site C) to 14 times (site H) more concentrated (Fig. 4).

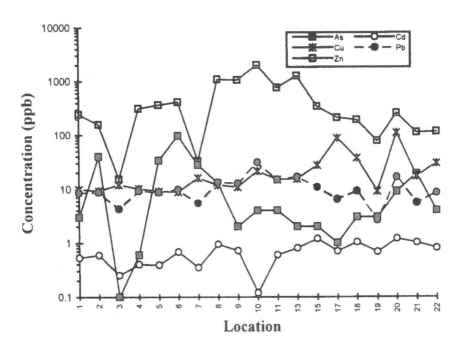

Figure 2. Heavy-metal composition in solution in the Grand Canal (Hangzhou section).

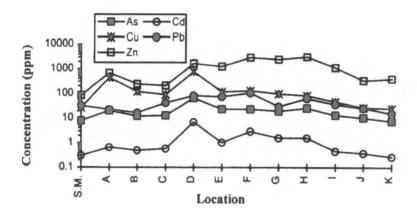

Figure 3. Heavy-metal contents of surficial sediment (<0.3 m depth) in the Grand canal. S.M. = soil mean concentrations in the study area.

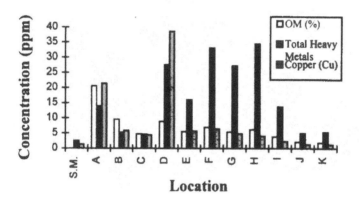

Figure 4. Comparison of organic matter (OM) content with total heavy-metal and Cu concentrations in surficial sediment. S.M. = soil mean concentrations in the study area. Heavy-metal and Cu concentrations are plotted with 100 and 20 times reduction, respectively, to fit into the diagram.

Heavy-metal contamination in the canal has been largely attributed to discharges from nearby industrial sources. The Environmental Protection Agency of Zhejiang province estimated that the daily discharge of heavy metals consisted of 1.538 kg Cu, 0.526 kg Pb, 0.139 kg Cd, 0.966 kg As and 2028 kg Zn [4]. These discharge rates seem to agree with the metal concentrations in both water and sediment. The large emission of Zn to the canal leads to elevated Zn concentrations in water and extremely high Zn contents in the surficial sediment. The metal distribution in Fig. 4 indicates two major pollution areas: site A and the area from site D to site I. Site A at Wulin Gate, the mouth of the canal, receives significant runoff from Hangzhou's urban area and waste streams from nearby manufacturing plants. Heavy metals and organic pollutants adsorbed on the surfaces of aerosol and road dust are carried into the canal through precipitation and runoff. A preliminary investigation revealed that a significant amount of Pb, Cd, and Zn were accumulated on dust particles in Hangzhou's streets. The area between sites D and I appears to be highly influenced by discharges from adjacent industrial operations (Fig. 1). In comparison, the area north of site I is much less contaminated with heavy metals. In this area, concentrations of dissolved phosphate are higher which is probably related to the farming practice in the northern part of the study area.

Figure 5 illustrates the distributions of heavy metals with depth in river sediments at sites A (Wulin Gate) and K (Wulin Head). Other sites show similar vertical distribution of heavy metals in river sediment. The surficial sediment (mud) is highly enriched in all the heavy metals. The heavy-metal concentration in deeper sediment decreases dramatically to a level equal or smaller than the mean heavy-metal concentration of soils in the study area. The concentration in the upper 0.3 m sediment is large at many locations in the south where major industrial and municipal pollution occurs the most. At site H, high Zn concentrations extend down to at least 0.3 meter depth at site H (Fig. 6). The surficial sediment stores large quantities of Zn emitted from industrial sources in the study area. The concentration profiles in Figures 5 and 6 represent an historical influx of heavy metals to the canal. The deeper (0.3 m or below) sediment records natural processes by which soils and sediments in the area were weathered and carried to the river bed prior to 1950's. The fact that the heavy-metal contents in the deeper sediment are slightly lower than that of the soils in the area may suggest a loss of these metals due to transport and dissolution during the weathering of the soils. The surficial sediment documents the significant anthropogenic input of the heavy metals since the 1950s when population growth and industrial development were accelerated at an extremely high rate in China. In the past decades, surficial sediment has served as a sink to heavy metals and organic contaminants.

Organic matter (OM) contents in the surficial sediment correlate positively with the distribution of heavy metals, especially with Cu (Fig. 4). We suspect that the elevated OM contents in the southern part of the canal are largely a result of anthropogenic input. The organic matter in several Chinese river sediments was found to have more nonpolar characteristics than their nearby soils, based on the greater K_{oc} (partition coefficient) values of carbon tetrachloride or 1,2-dichlorobenzene to the sediment OM [9]. Erosion and diagenetic processes of soils might have caused the decrease of polarity of organic matter in river sediments. Alternatively, an addition of hydrocarbon contaminants such as

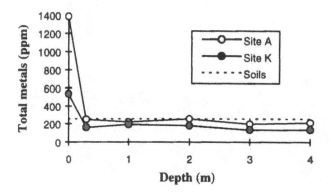

Figure 5. Vertical distribution of total heavy-metal content (excluding Cd) with depth in bed sediment of the Canal. The dash line indicates the mean heavy-metal content of soils in the study area.

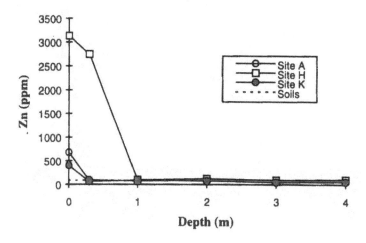

Figure 6. Zinc distribution with depth in bed sediment of the canal. The dash line is the mean Zn content of soils in the study area.

oil to the sediments would also increase the K_{oc} values and the resulting organic matter would become more nonpolar [10,11]. The discharges of the oil refinery and other industries adjacent to the Grand Canal (Fig. 1) provide ample opportunities for such organic contamination. The high correlation between heavy metals and OM contents (Fig. 4) suggests that the heavy metals and part of the organic matter in the surficial sediment many have come from the same contamination sources, either independently or through some degree of metal-organic complexation. Organic matter adsorbed on sediments has been reported to retard metal transport [12]. The positive correlation between Cu and organic matter in Fig. 4 is consistent with their strong mutual binding [13.14]. Because no major Cu sources exist north of site H (14), it is quite likely that Cu fixed in the surficial sediment in the south underwent a desorption or remobilization process by which Cu was released back to the solution and transported along the canal. This hypothesis explains the pattern of Cu distribution throughout the Hangzhou section of the canal.

SUMMARY AND CONCLUSIONS

This investigation of the Grand Canal reveals that the Hangzhou section of the canal has been highly polluted by heavy metals from both industrial and municipal sources. These metals are largely accumulated in the surficial river sediment. Despite the contamination, heavy-metal concentrations in solution are elevated only in several locations except for Zn whose concentrations are extremely high in both water and surficial sediment. The study shows that the surficial sediment in the Grand Canal serves as a sink to heavy metals as well as anthropogenic organic contaminants.

REFERENCES

1. Geelan P.J.M. and Twitchett D.C. (1974) The Times Atlas of China. The New York Times Book Co., New York.
2. de Crespigny R.R.C. (1971) China: The Land and its People. St. Martin's Press, New York, 235 pp.
3. Weng, Huanxin (1989) On the Accumulation of Arsenic in the River Mud and its Environmental Effect. Journal of Zhejing University, 23, 49-56 (in Chinese with English abstract).
4. Wu Z. and Weng H. (1987) On the elevated concentrations of iron and manganese in the shallow groundwater along the Grand Canal (Hangzhou section). Acta Scientiae Circumstantiae (Environmental Science Bulletin), 7, 279-287 (in Chinese).
5. Greenberg A.E., Clesceri L.S. and Eaton A.D. (1992) Standard Methods for the Examination of Water and Wastewater, 18th Edition. American Public Health Association.
6. Nelson D.W. and Sommers L.E. (1982) Total carbon, organic carbon, and organic matter. In: A.L. Page et al. (Eds.) Methods of Soil Analysis. Part 2. Chemical and Microbiological Properties. Am. Soc. Agron., 9, 539-579.
7. Allison J.D., Brown D.S. and Novo-Gradac K.J. (1991) MINTEQA2/PRODEFA2, a geochemical assessment model for environmental systems: version 3.0 user's manual. U.S. Environmental Protection Agency, Athens, Georgia, EPA/600/3-91/021.
8. Faure G. (1991) Principles and Applications of Inorganic Geochemistry. MacMillan, New York, p. 442.

9. Kile D.E., Chiou C.T., Zhou H., Li H. and Xu O. (1995) Partition of Nonpolar Organic Pollutants from Water to Soil and Sediment Organic Matters. Environ. Sci. Technol., 29, 1401-1406.

10. Boyd S.A. and Sun S. (1990) Residual Petroleum and Polychlorobiphenyl Oils as Sorptive Phases for Organic Contaminants in Soils. Environ. Sci. Technol., 24, 142-144.

11. Sun S. and Boyd S.A. (1991) Sorption of Polychlorobiphenyl (PCB) Congeners by Residual PCB-Oil Phases in Soils. J. Environ. Qual., 20, 557-561.

12. Haas C.N. and Horowitz N.D. (1986) Adsorption of cadmium to kaolinite in the presence of organic material. Water Air Soil Pollut., 27, 131-140.

13. Tessier A., Campbell P.G.C. and Bisson M. (1980) Trace metal speciation in the Yamaska and St. François Rivers (Quebec). Canadian Journal of Earth Sciences, 17, 90-105.

14. Nriagu J.O. and Coker R.D. (1980) Trace metals in humic and fulvic acids from Lake Ontario sediments. Environ. Sci. Technol., 4, 443-446.

Proc. 30th Int'l. Geol. Congr., vol. 24, pp. 241-253
Yuan Daoxian (Ed)
© VSP 1997

The Impact of Excavation and Landfilling on the Hydrogeology of a Clay Pit

D. HALLIDAY*, J.D. MATHER* AND J.B. JOSEPH**

* Geology Dept. Royal Holloway, University of London, Egham, Surrey, TW20 0EX, UK.
**Shanks & McEwan Ltd, Woodside House, Church Road, Woburn Sands, Bucks, MK17 8TA, UK.

Abstract

Extensive quarrying of the Jurassic Oxford Clay in the area south and west of Bedford, UK, has resulted in the presence of a series of disused brickpits. Current demand for waste disposal sites has led to the development of several of these pits as landfill sites. The current study addresses the impact which the removal of large volumes of clay, and the subsequent use of several of the pits as landfill sites, has had on the local hydraulic regimes of the pits.

Water bearing strata beneath the Oxford Clay have low permeabilities of 10^{-07} to 10^{-09} m/s, which, coupled with relatively flat lying piezometric surfaces, means that any contaminants which may escape from the landfills will spread extremely slowly. Groundwater quality is naturally poor, with salinities of the order of 500 to 1000 mg-Cl/l. To date no conclusive evidence of widespread leachate migration exists. However, study of piezometric data and groundwater quality analyses suggests that prolonged periods of clay extraction have had the greater impact on the local hydraulic regimes of the pits. Potential flow directions between the two water bearing units beneath the pits appear to reverse on the down gradient side of several of the pits. This is considered a consequence of the old quarrying practice of digging drainage trenches into the Kellaways Sand to collect surface run off. Pumping of the collected water also served to lower groundwater levels in the Kellaways Sand.

Keywords: Oxford Clay, landfill, hydrogeology, hydraulic conductivity.

INTRODUCTION

The Peterborough Member [1] of the Jurassic Oxford Clay Formation in the UK has been extensively exploited for the purpose of brick manufacture. Large scale excavation has resulted in the presence of a series of voids which are flooded, back filled or remain open. In Buckinghamshire and Bedfordshire a number of these disused brickpits extend from Calvert in the south west to Elstow in the north east (Fig. 1).

These pits are in much demand as landfill sites for the disposal of domestic, commercial and industrial wastes, due to the natural containment properties of the clay, in conjunction with excellent road and rail links to the area. However, excavation of large volumes of the Oxford Clay has materially affected the physical, chemical and hydraulic properties of the formation. Field investigations at the Bletchley landfill have been conducted in an attempt to understand the hydraulic regimes of the pits and how the excavation practices, followed by the placement of wastes of mixed origins, may impact on the local groundwater environment.

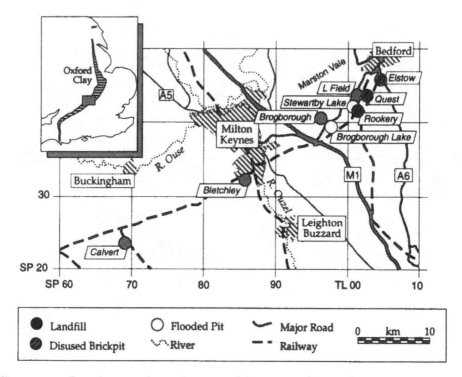

Figure 1   Distribution of sites between Calvert and Elstow. Inset map shows the outcrop of the Oxford Clay in southern England.

BLETCHLEY SITE INVESTIGATION

Shanks and McEwan. a major UK waste management company, have been operating a landfill site at Bletchley since 1986. The landfill is located in Buckinghamshire, UK (Fig. 1), and occupies worked out clay pits, up to 25m deep, excavated into the Stewartby and Peterborough Members [1] of the Oxford Clay (formerly known as the Middle and Lower Oxford Clay). The site is divided into areas A, B, C, and D (Fig. 2). Areas A and B are filled and restored and filling is currently taking place in area C. Parts of area C (C3) and area D are excavated but unfilled and are separated from the rest of the site by a bund of replaced clay constructed to carry a surface water ditch across the site. As of 1994, the remaining void space was 13.1Mm³ which gives a corresponding lifetime of 30 years at current filling rates.

Over the past two years monitoring boreholes have been drilled in pairs around the perimeter of the site to penetrate the two principal water bearing strata beneath the landfill. Each pair comprised one borehole completed in the Kellaways Sand and one in the deeper Blisworth Limestone. A total of four pairs were drilled with one additional borehole in the Kellaways Sand (Fig. 2). Drilling was carried out using rotary air flush techniques. Completion was with 110mm (O.D.) HDPE casing, screened opposite the water bearing unit under investigation. Slotted casing was in the form of a multi-layer, non-metallic filter mesh well screen with a slotted base pipe covered with a 150 micron geotextile wrap. The open area averages 7%.

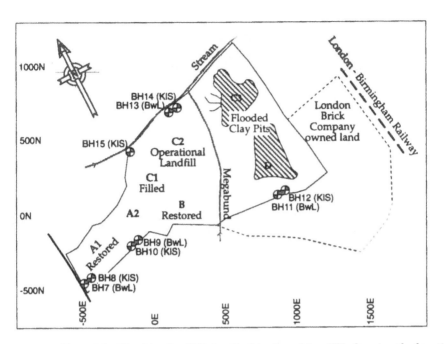

Figure 2   Plan of the Bletchley landfill site, Buckinghamshire, UK. showing the location of the groundwater monitoring boreholes installed in 1994 (site grid is calibrated in m). The site covers an area of 93ha, of which 28ha are landfilled and 10ha restored 55ha are quarried but not backfilled.

Data from the core logging operations at Bletchley were used to supplement existing data from the large number of investigatory and monitoring boreholes drilled over the past few years by Shanks & McEwan in the Marston Vale and Calvert (Fig. 1).

GEOLOGY

The geological succession in the area ranges from the Middle Jurassic to Recent (Fig. 3). All the landfill sites are located in the Oxford Clay Vale of Bedfordshire and Buckinghamshire and are underlain by a series of interbedded clays, clay rich sands and limestones. Regionally the strata dip gently (2-3°) towards the south east.

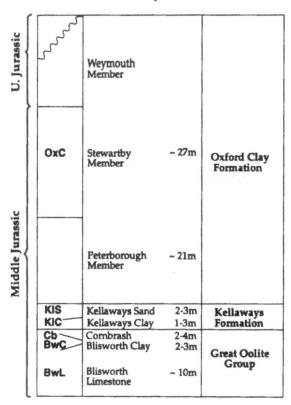

Figure 3   Generalised geological section for the Marston Vale, south and west of Bedford (modified from [2]).

Core logging at Bletchley showed that the Kellaways Sand, immediately beneath the Oxford Clay, is a mid greenish grey clay and silt rich sand varying in thickness between 1.64 and 3.06m (Table 1). The lower water bearing unit, the Blisworth Limestone, consists of up to 10m of dense biomicritic limestone, with thin marl or mudstone interbeds. The two formations are separated by 10m of interbedded clays and limestones (Table 1).

HYDROGEOLOGY

In terms of groundwater resources the Jurassic strata beneath the Oxford Clay offer little or no potential, with very low yields and naturally poor quality groundwater. Of the potential water bearing strata beneath the research area the main unit is the Blisworth Limestone. However, its limited recharge area, lack of fracture permeability, gentle dip beneath clay cover and poor quality water mean that it has little economic potential.

Table 1. Generalised classification of the Jurassic strata beneath the Bletchley landfill, Bucks., UK.

Age	Formation		Description	Thickness (m)
Quat.		Glacial Till	Silty clay with an abundance of rounded erratics and selenite crystals.	up to 4m
Middle Jurassic	Oxford Clay Formation	Stewartby Member	Moderately calcareous, locally silty, variably shelly mudstone.	up to 22m
		Peterborough Member	Variably silty, bituminous, moderately fissile mudstones and shales. Fossil fauna are diverse and well preserved, allowing biostratigraphic zonation by ammonites [3]. Finely disseminated pyrite is present throughout.	up to 20m
	Kellaways Formation	Kellaways Sand	Clay and silt rich fine sand, wispily bedded in places. Variably shelly and intensely bioturbated. Localised calcite cementation forms 'doggers', hard lenticular masses of sandstone up to 3m in diameter.	up to 3m
		Kellaways Clay	Smooth-textured, slightly fissile mudstone. Upper part contains sand filled burrows, many of which are pyritised.	up to 2m
	Great Oolite Group	Cornbrash	Hard, medium grey, shell detrital limestone with a micritic matrix. Clay interbeds up to 0.4m thick.	up to 3m
		Blisworth Clay	Comprises two main lithologies; a lower green/grey calcareous mudstone, and an overlying unit of emerald green, mottled, silt clay.	up to 4m
		Blisworth Limestone	Medium to dark grey, shell detrital, limestone with occasional mudstone and marl interbeds.	up to 10m

The Kellaways Sand, which is comprised of clay and silt rich sands, is treated as the second water bearing horizon. The low transmissivity of the beds, coupled with the saline nature of the groundwater means that there is no potential for utilising the formation as a public, or other, water supply source. However, it is a potentially important horizon as far as monitoring the effects of the landfill on the environment are concerned.

The only other potentially water bearing stratum beneath the Oxford Clay is the detrital limestone of the Cornbrash. This has minimal significance however, due to its dense matrix (low primary porosity), infrequent incidence of fracturing (low secondary porosity) and limited thickness (less than 4m).

The Oxford Clay, Kellaways Clay and Blisworth Clay are effectively aquicludes, confining the water bearing formations (Fig. 3).

Hydraulic Parameters

Hydraulic testing, carried out on completion of the Bletchley boreholes, was conducted to assess the potential rates at which any contaminants could migrate from the landfill. Slug tests were chosen for this purpose. The tests involved monitoring the water level recovery in a borehole after the withdrawal or addition of a known volume, or 'slug', of water. Measurement of the rate at which recovery occurs allows the formation transmissivity to be calculated. Results of the slug tests were interpreted using the Cooper curve matching method which is suitable for screened or slotted piezometers that are open over the complete thickness of a confined aquifer [4].

Hydraulic conductivity values in the Kellaways Sand at Bletchley were found to be low, ranging between 10^{-07} and 10^{-09} m/s, and reflecting the clay and silt rich nature of the formation.

Available data from a number of other sources [5-7] for field tests conducted in the area between Calvert and Elstow (Fig. 1) are quoted in Table 2. The field tests used were also mainly rising head slug tests and gave comparable results to the tests conducted at Bletchley. A plot of all the data points gives a median value of 3.1×10^{-07} m/s with 30% of the values below 1×10^{-07} m/s and 30% above 1×10^{-06} m/s. In addition, laboratory determinations of both vertical and horizontal hydraulic conductivity quoted by Williams, (1985) [8] are generally one or two orders of magnitude less than those calculated from the field tests.

Table 2   Hydraulic Conductivity Data for the Kellaways Sand

Site or Area	Data Source	No. of Tests	Type of Test	Hydraulic Conductivity (Range in m/s)
Marston Vale	Klinck [5]	6	Rising Head	1.41×10^{-06} - 2.03×10^{-08}
Elstow	Klinck & Noy [6]	14	Mainly Rising Head	7.10×10^{-06} - 1.30×10^{-08}
L Field	Parry [7]	2	Falling Head	1.26×10^{-07} - 3.00×10^{-08}
Bletchley	Halliday (unpublished)	4	Rising Head	1.2×10^{-06} - 3.8×10^{-09}

Four rising head tests in the Blisworth Limestone, recently carried out at Bletchley by the authors, gave values of hydraulic conductivity comparable to the data for the Kellaways Sand.

It is clear therefore that, with a thickness of less than 5m and transmissivities of the order of 1.5×10^{-06} m/s, the Kellaways Sand will not yield water to pumping wells. This is confirmed by experience at the Bletchley site where investigatory boreholes could be pumped dry by a small pump in a matter of minutes but took in excess of 24 hours to refill.

Groundwater Movement

Over the past few years Shanks and McEwan have drilled a large number of investigatory and monitoring boreholes in the Marston Vale and around the Calvert and Bletchley sites to the south east (Fig. 1). In addition, UK Nirex Ltd drilled a network of boreholes around Elstow as part of their now abandoned plans to investigate the Elstow Storage Depot, 5km south of Bedford, as a potential repository for the shallow land burial of low and intermediate level radioactive wastes. Many of these boreholes penetrated the Kellaways Sand and some have been continued into the Blisworth Limestone and provide a network of points for measuring piezometric levels.

On a regional scale piezometric levels indicate that flow within the Kellaways Sand is from the south west towards the north east (Fig. 4). The piezometric surface is relatively flat lying in the Marston Vale with a gradient of about 1 in 350 or 0.003, but steepens towards the outcrop along the western side of the River Ouse Valley. This steepening could be the result of lower hydraulic conductivities in this part of the Kellaways Sand or a reduction in its effective thickness.

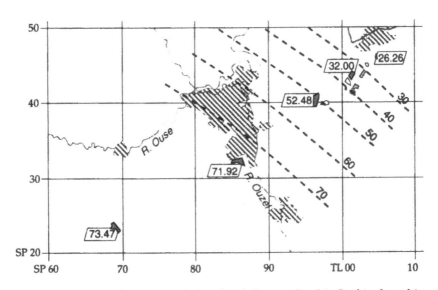

Figure 4   Piezometric levels recorded in the Kellaways Sand in Buckinghamshire and Bedfordshire, April 1995. Values quoted are averages for the numerous monitoring points surrounding the perimeter of each landfill site. Contour values are in metres above ordnance datum.

It is generally the case that the Kellaways Sand heads are above those in the underlying Blisworth Limestone, indicating downward flow from the sands into the limestone. Locally however, excavation activities have influenced conditions immediately down gradient at many of the pits, where the piezometric heads are reversed, and water can, in theory, flow upwards from the limestone into the sands. This is thought to be a consequence of the excavation practice of digging drainage trenches at many of the sites. These trenches were excavated into the Kellaways Sand to collect surface water running into the brickpits. In addition to removal of the collected surface water, pumping also served to lower piezometric levels in the Kellaways Sand, which drained into the pits. This also illustrates the impermeable nature of the Kellaways Sand, which produced very little water, and as far as the quarry operators were concerned was not obviously more permeable than the overlying Oxford Clay.

Groundwater Quality

The groundwater quality in the water bearing Kellaways Sand, which lies immediately beneath the Oxford Clay, is the most likely to be influenced by the presence of the brickpits and landfill sites. Groundwater quality in the Kellaways Sand is known to be poor, largely due to its high sodium chloride content. It has been suggested in earlier research that this salinity is the result of leachate contamination of the groundwater [9]. Whilst elevated salinity is ubiquitously associated with leachate pollution from landfill sites, historical data suggests the high concentrations of sodium chloride certainly predate any anthropogenic activities.

Woodward and Thompson [10] stated, as long ago as 1909, that;

> *"The Kellaways Beds yield water, but not as a rule of good quality. It is particularly saline in the area to the south and south-west of Bedford ..."*

For some wells they concluded that [10];

> *"This water is simply an impossible one to contemplate using as a regular supply, either for drinking or general domestic use. The amount of salt is enormous"*

A review of the available historic analyses for the Kellaways Sand groundwater at nine locations in Buckinghamshire and Bedfordshire shows that chloride concentrations range between 550 and 1345 mg/l, in all cases exceeding the current maximum admissible concentration in drinking water of 400 mg/l [10-13] These analyses were all carried out prior to brick clay extraction or landfill operations and highlight the naturally poor quality of the Kellaways Sand groundwaters.

Contemporary chemical analyses of the groundwaters beneath the Bletchley landfill have confirmed their continuing poor quality. Sodium, chloride and sulphate dominate the major ion chemistry in the Kellaways Sand. The high salinities (up to 1000 mg-Cl/l) are quite normal in gently dipping, low permeability, confined strata, several kilometres from outcrop. The elevated sulphate concentrations (500 to 2000 mg-SO_4/l) are most probably associated with the presence of pressure relief fractures in the Oxford Clay caused by the removal of large volumes of material in the recent past. These fractures provide the ideal environment for oxidation of the finely disseminated pyrite present throughout the formation [14] which comprises up to 2% of the rock mass . Oxidation of the pyrite will result in the release of sulphate.

Concentrations of selected major and minor constituents from the Marston Vale boreholes are quoted in Table 3, together with the current EC maximum admissible concentrations (MAC) for drinking water, which are included for comparative purposes. The Marston Vale boreholes were drilled between 1988 and 1991 to monitor hydrogeological conditions beneath the Vale. Concentrations of chloride, sulphate, sodium and potassium all exceed the EC MAC in both the Kellaways Sand and the limestones of the Great Oolite Group, confirming their continuing poor water qualities. High COD values result from the high content of organic material (up to 2%) in the Oxford Clay and Kellaways Sand.

Table 3   Contemporary concentrations of selected major and minor groundwater constituents in the Kellaways Sand and Great Oolite Group. Values given represent mean values for data recorded since 1988 in the network of boreholes in the Marston Vale, Bedfordshire(Fig. 1).

Determinand (mg/l)	Kellaways Sand	Great Oolite Group	EC MAC
pH (pH units)	7.6	8.0 (min 6.5)	8.5
BOD	4.0	1.0	-
COD	144	110	-
Ca	80	50	250
Mg	24	23	50
Na	684	790	150
K	16	18	12
Cl	834	970	400
HCO3	188	234	-
SO4	630	410	250
NH3-N*	1.00	2.67	0.5

* NH_3-N average values were calculated taking values less than the detection limit (0.01 NH_3-N mg/l) as zero.

Whilst it is not suggested that the presence of the landfill sites has resulted in no deterioration of the groundwater environment, the existing quality data is not consistent with plumes of migrating leachate, but can be more confidently associated with the excavation history of the pits.

LANDFILL CONTAINMENT

According to the newly published NRA groundwater vulnerability map covering Bedfordshire [15] the Kellaways Sand is classified as a minor aquifer, defined as;

> *"fractured or potentially fractured rocks, which do not have a high primary permeability, or other formations of variable permeability including unconsolidated deposits. Although these aquifers will seldom produce large quantities of water for abstraction, they are important for both local supplies and in supplying base flow to rivers".*

The waste regulation authorities also consider the Kellaways Sand a minor aquifer and, as such, modern waste disposal licenses require the installation of an engineered containment system requiring a minimum thickness of 1.5m of natural and engineered clay above the Kellaways Sand of which not less than 0.5m is native clay and not less than 1m is engineered to a hydraulic conductivity equal to, or better than, 1×10^{-09} m/s.

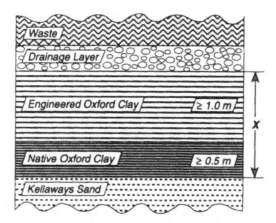

Figure 5   Schematic cross section through the base of an Oxford Clay landfill site, showing the presence of the engineered clay liner required by the regulators In practice the thickness of cell X may range from 1.5 to 5.0m.

The field work carried out over the last few years, together with an analysis of the existing data, confirms that the Kellaways Sand in the study area is a thin, fine grained clay rich sand. It has a low hydraulic conductivity, of the order of 10^{-07} m/s. Hydraulic gradients are low and water quality is naturally poor. It will not yield water to pumping wells and cannot be regarded as a minor aquifer.

With thicknesses of less than 5m and hydraulic conductivities of the order of 10^{-07} m/s the clay rich Kellaways Sand behaves more like an aquitard or aquiclude than a minor aquifer. In fact it corresponds well to the definition of a non-aquifer given in the Groundwater Protection Policy [16];

> *"Formations with negligible permeability. Only support very minor abstractions, if any".*

In the absence of local abstractions from the formation, the only argument for regarding the formation as a minor aquifer and insisting on its protection by engineered containment of the landfill sites, is that it may contribute to base flow in adjacent rivers or groundwater from it may flow into an adjacent aquifer. Due to the low hydraulic gradient and transmissivity of the bed any seepage into the River Ouse will not exceed more than a few m^3 per year distributed along an outcrop length of several kilometres. There are no springs issuing from the Kellaways Sand outcrop and a site visit failed to find any evidence of discrete discharges.

There is also an argument that water from the Kellaways Sand might enter the Blisworth Limestone, part of the Great Oolite formation (Fig. 3), as it has been demonstrated that the regional hydraulic picture is reversed around some of the pits. However, slug tests in the Blisworth Limestone have shown that hydraulic conductivities are comparable with those in the Kellaways Sand and water quality is generally even poorer.

CONCLUSIONS

Despite the widely held belief that the waste industry is responsible for wide scale pollution events, this is not confirmed by the statistics. Whilst there are documented cases of groundwater contamination by landfill the majority involve the local contamination of minor aquifers as a result of the previously accepted practice of operating dilute and disperse sites, which allowed the controlled release of materials back into the environment. These relatively isolated cases must be viewed in context of the much more common and widespread pollution incidents resulting from agricultural activities and leakages from underground storage tanks.

Research on the Oxford Clay landfill sites of Bedfordshire and Buckinghamshire has highlighted the importance of examining all possible pollution sources and anthropogenic activities, past and present, which may have resulted in a change in the groundwater quality. Existing groundwater quality and piezometric data have revealed that the act of excavating the Oxford Clay brick pits has had a greater influence on local hydrogeological environments than usage of the voids for landfill.

Risk arises when an activity such as landfill takes place at a particular location. The risk to groundwater can be assessed by taking into account the nature of the waste to be disposed of and the leachate which will be generated from it, the natural vulnerability of the groundwater and the scale of preventive measures proposed. It is suggested that in the case of the Kellaways

Sand the preventive measures which are installed are out of all proportion to any rational assessment of groundwater vulnerability arising from waste disposal.

Acknowledgements

The authors would like to thank the Shanks & McEwan staff at Head Office, Woburn Sands and the Bletchley landfill site for their assistance throughout the research period, especially Mr R Fluckiger and Mr D Bird for help during the site investigation. Miss N Ingrey is also thanked for her assistance with the hydraulic testing carried out at Bletchley.

REFERENCES

[1] Cox, B.M., Hudson, J.D. & Martill, D.M. 1993. Lithostratigraphic nomenclature of the Oxford Clay (Jurassic). Proceedings of the Geologists' Association, 103, 343-345.

[2] Barron, A.J.M., Horton, A. and Sen, M.A. 1992. Geology of the Marston Vale. British Geological Survey, Technical Report WA/92/23C. pp82.

[3] Callomon, J.H. 1968. The Kellaways Beds and the Oxford Clay. In: Sylvester Bradley, P.C. & Ford, T.D. (eds). The geology of the East Midlands. *Leicester University Press*, Leicester. pp264-290.

[4] Cooper, H.H., Bredehoeft, J.D. & Papadopulos, I.S. 1967. Response of a finite diameter well to an instantaneous charge of water. *Water Resources Res.*, 3, pp 263-269.

[5] Klinck, B.A. 1992. Marston Vale groundwater study, Bedfordshire: results of rising head slug tests. BGS Fluid Processes Group, Technical Report WE/92/19C.

[6] Klinck, B.A. & Noy, D.J. 1992. A groundwater flow model for the Marston Vale. British Geological Survey, Technical Report, WE/92/21C.

[7] Parry, R.H.G. 1972. Some properties of heavily overconsolidated Oxford Clay at a site near Bedford. Geotechnique 22, 485-507.

[8] Williams, G.M. 1985. Preliminary assessment of the geology and hydrogeology for the proposed site investigation at Elstow Storage Depot, Bedfordshire. Nirex Report No. 22.

[9] Phipps, O.C. 1993. The hydraulic regime and hydrogeochemistry of the Jurassic Kellaways Sand, Bedfordshire. MSc. Thesis. Univ. Birmingham. Oct 1993. (unpub).

[10] Woodward, H.B. & Thompson, B. 1909. The Water Supply of Bedfordshire and Northamptonshire, from underground sources: with records of Sinkings and Borings. *Mem. Geol. Surv. Eng. Wales*. HMSO. pp230.

[11] Jukes Browne, A.J. 1889. On a boring at Shillingford, near Wallingford on Thames. *Midland Naturalist*, 14, 201-209.

[12] Jukes Browne, A.J. 1889. The Occurrence of Granite in a Boring at Bletchley. *Geol. Mag.*, 7, 356-361.

[13] **Woodward, H.B. 1910.** The Geology of Water Supply. Edward Arnold, London, UK. pp339.

[14] **Milodowski, A.E. & George, I.A. 1985.** The petrography of the Jurassic core from the Harwell Research Site. Part 2: Kellaways Beds, Great Oolite Group and Inferior Oolite Group. BGS Report, FLPU 85-8.

[15] **National Rivers Authority (NRA). 1995.** Policy and practice for the protection of groundwater; Groundwater Vulnerability 1:100,000 Map Series, Sheet 31, Bedfordshire. HMSO, London.

[16] **National Rivers Authority (NRA). 1992.** Policy and practice for the protection of groundwater. National Rivers Authority, Bristol.

Proc. 30th Int'l. Geol. Congr., vol. 24, pp. 254-260
Yuan Daoxian (Ed)
© VSP 1997

Hydro-Fraise Diaphragm-Wall Encapsulating Bitterfeld Hazardous Waste Sites

LUTZ KRAPP

Bureau of Applied and Regional Geology, Birkenweg 30, 52080 Aachen, Germany

Abstract

Next to the chemical industries at Bitterfeld there are about 50×10^6 m³ of toxic wastes deposited in uncontrolled open pits. Chemicals from these wastes are infiltrating into two upper aquifers. At 55 m depths a natural clay-horizon has been encountered which appears suitable as a basal buffer. Remedial measures are necessary, however complete decontamination is excluded for economic reasons. As a feasible alternative, the encapsulation of the wastes by vertical diaphram walls reaching down into the basal clay layer in combination with a surface cover is recommended. The extreme requirements of these vertical walls, in terms of depths, chemical milieu during construction and long-term stability under complex chemical attack are not met by any standard technique. Two years of investigations on materials (clays, lignite, waxes, polymeric silicates) were necessary to establish applicable mixes in laboratory- and semi-technical scale tests. Presently the technological improvement of the necessary equipment (hydro-fraise) is in progress and construction of a pilot segment of the wall of about 1 km length is planned. Because of its innovative character and its assumed general economic-ecologic value for the containment of hazardous sites world-wide, this project has been selected as one major show-piece of the EXPO-2000 / Germany in the correspondency-area Bitterfeld-Dessau.

Keywords: fraise-diaphragm wall, hazardous wastes, Bitterfeld Chemical Industries, EXPO-2000

INTRODUCTION

Environmental contamination by the Bitterfeld Chemical Industries in Germany goes back to more than 100 years of industrial history on the basis of chlor-organic compounds. Townships, industrial sites, disposal sites, lignite open pits and agricultural use have developped simultaneously in this region without a general concept. Now soil and water are heavily polluted and, consequently, human health as well as ecology are affected.

There exist several hazardous land-fills in old open pits which contain millions of cubic meters of toxic chemicals, only controlled by groundwater drawdown in the surrounding lignite mines. Groundwater contamination extends over more than 50 km² and reaches down to about 50 m depth. Protective measures are necessary since mining activities have been stopped and the groundwater level will rise high into the land-fills. Transfer of highly polluted groundwater under township areas and into the recipient streams is inevitable.

GEOLOGY AND HYDROGEOLOGY

The rock basement in about 100m depth is covered by Tertiary sands, clays and lignite seams, and by Quaternary sandy gravels, tills, loess and valley fill (Fig. 1). Of special significance for the diaphragm project are

- marine clays (Rupel) of 10 to 25m thickness, which occur in about 50 to 60m depth
- mica sands of 20 to 30m thickness
- a lignite seam of upto 10m thickness being widely extracted
- top clays of a few m thickness
- river terrace sediments
- glacial tills and sands, partly filling deep eroded valleys and
- loess sediments outside the valleys.

The natural conditions have widely been changed by the lignite mining operations as well as by backfillings and hazardous land-fills.

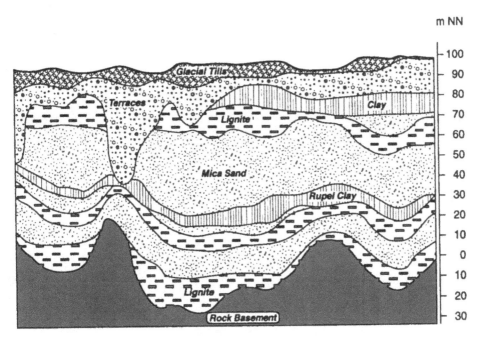

Fig. 1   Schematic geological cross section
           South of Bitterfeld

Groundwater has suffered a corresponding impact. The original flow in NE direction is
regionally altered by the great drawdown measures. The upper aquifer with its Quaternary
sediments is highly permeable (10^{-3} m/s). Due to the effects of the old erosion valleys and the
mining activities this aquifer is in direct hydraulic contact with the Tertiary mica sands (10^{-4} to
10^{-5} m/s).
The Rupel clays represent a natural horizontal buffer which separates the two upper aquifers
from the lower Tertiary sands and lignite seams.

HAZARDOUS LAND-FILL ANTONIE

The most critical waste site of the region is a residual open-pit filled in the last decades up to
the surface, partly with ashes and contaminated rubbles, but predominantly with toxic
chemical wastes as, forinstance, CHC, PCB, HCH etc. Intermittently, concentrated hot acids
have also been disposed into this pit, causing uncontrolled chemical reactions and deep
infiltration of dense non-aqueous phase liquids (DNAPL) including chlor-organic solvents
(Fig.2). This means that the wastes together with the supporting soils down to about 50m depth
must be considered as one contamination hot spot, continuously emitting pollutants if no
remedial measures are taken.

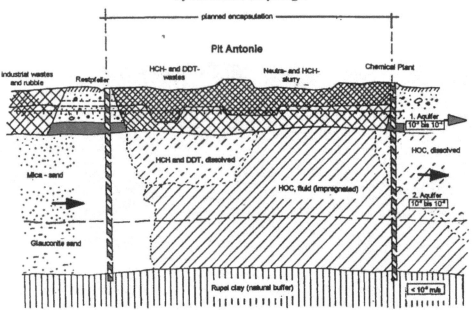

Fig. 2   Cross section Pit Antonie with DNAPL impregnation down to Rupel clay (GFE)

ALTERNATIVE STUDY

In 1990/91 a detailed alternative study of possible remediation techniques has been performed in order to find the optimum solution regarding effectiveness, technical feasibility and costs. The spectrum of considered alternatives ranged from

- hydraulic measures over
- biological in situ treatment and
- encapsulation to
- complete decontamination, i.e.
  extraction and thermal treatment

With respect to the future groundwater rise, the toxic and acid milieu and the costs, only encapsulation as a temporary protective measure has been considered feasible. Allthough the costs for thermal treatment have rapidly decreased in the past few years, decontamination is stimll out of reach and has to be postponed for a future clean-up.

The special requirements for a sealing wall under the prevailing conditions in Bitterfeld are:

- depth about 55m
- construction at high salt content of groundwater
- construction in an acid milieu
- overcoming of findlings, rubble and silicified layers
- required wall permeability < 10^{-8} m/s
- long-term resistance against heavy chemical attack.

Additional construction techniques will be necessary under buildings, traffic ways and pipelines.None of the usual diaphragm technologies meets all the requirements. Main obstacles are the great depth and the corrosiveness of cements under the extreme ly acidic conditions.

A very promising technology has been devellopped for groundwater problems in German lignite mines in the Lausitz. These mines are operated along the German/Polish border with a necessary groundwater drawdown of about 60m. In order to avoid a similar drawdown on the

Polish side, more than 10km of a special hydro-fraise diaphragm-wall down to 100m depth have been successfully built since 1970. The wall is cement-free and highly impervious. The technology is approved under normal groundwater conditions.

TECHNICAL CONCEPT

The construction procedure is shown in Fig.3. Starting with a 1m borehole down to the envisaged depth, a 0.4m pile is lowered into the hole. This pile is used as guidance for the hydro-fraise and for air-lifting of the fraised soil material. The fraise is moving up and down the pile, cutting a continuous trench of 1m width into the ground. During excavation the trench is filled with a clay suspension containing few additives. Subsequently, the trench is perpendicularly subdivided by pre-fabricated concrete slabs every few meters. During the cutting progress a filter cake layer of 8 to 10cm thickness devellops on both walls of the trench. This process requires a hydraulic head of about 10m of the suspension above the groundwater table.

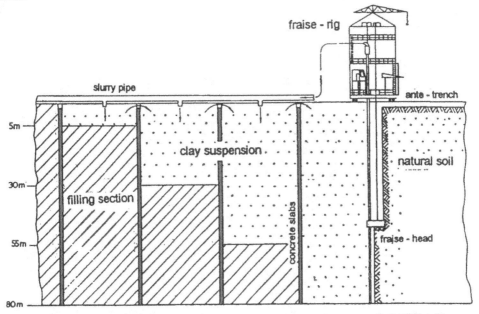

Fig. 3   Schematic construction procedure of hydro-fraise diaphragm-wall (LAUBAG)

The filter cakes consist of oriented and densely packed clay minerals. They represent the main sealing layers with K-values $<10^{-9}$ m/s. The residual trench compartments are back-filled with the excavated soils, producing a very mineral-rich earth layer. It consists of clay minerals from the suspension and of the back-filled soil. The K-values reach 10^{-7} m/s. Alltogether, we obtain a three-layered sealing wall of great hydraulic effectiveness.

Rock obstacles and others have been successfully tackled by cracking or blasting without removal of the rig.

Using rough and heavy mining equipment, progress in the past was slow (some m/d) and land-use very high (80 to 100m).

NEW EQUIPMENT AND MATERIALS

Modern hydro-fraise rigs are meanwhile available on the market. They can be applied in all normal soils and in semi-consolidated rocks. The fraises consist of a steel frame with two contra-rotating transmissions. The transmission axes are supplied with cutter heads adapted to the occurring soils. The fraise is mounted on a heavy crawler type excavator with an integrated hydraulic system. Lifting and lowering of the fraise head is accomplished by a rope winch. The contact pressure can be regulated by a hydraulic ram. The soil is continuosly fraised and fractured by the contra-rotating fraise heads, mixed with suspension and transported to the suction pump. The loaded suspension is afterwards recycled.

Penetration depths greater than 50m require an additional hose support. Working progress mainly depends on:

- cutting tools
- rock conditions
- contact pressure
- torque and rotational speed
- suspension flow per time and
- trench geometry.

Depth, progress, rotational speed, pressure and suspension flow can be automatically registered. This allows for corrections during performance. Compared with conventional technologies the new hydro-fraise permits greater depths, rapid progress of >30m/d, high quality performance, better controls, greater range of application in various soils and safe disposal of the excavated soil. The plant requires separate recycling and mixing machines.

For the existing walls in the Lausitz kaolin-rich clays have been transported from deposits in Northern Germany. Meanwhile a number of clay deposits in the vicinity of Bitterfeld has been investigated and tested in order to reduce transport costs. All other components (bentonites, sodium silicates, etc) are required in small quantities only and can be easily provided to the site. Dominating wall material in any case is the back-filled excavated soil.

RESEARCH PROGRAMMES

Since technical feasibility and long-term effectiveness of this type of wall are not yet confirmed under extreme chemical attack, several questionable influences and reactions have been tested in laboratory-, field- and semi-technical scale during the last two years.

Funded by DBU (German Environmental Foundation) a wide range of laboratory tests on mixtures with different components, additives and dispergators as well as different solution agents has been performed in order to establish optimum mixtures. The main interest concentrated on the effects of lignite wax, lignite resins and a locally produced polymeric silicate, named Poly-Quat. The programme comprised:

- testing of eco-compatible dispergators (DBI/IWS)
- testing of resistivity against contaminated groundwater, acids and alkalines (DBI/GBD/RWTH-LIH)
- testing of microbial resistivity of lignite wax under aerobic and unaerobic conditions
- testing of sufficient strength and water tightness (DBI/GBD/RWTH-LIH)
- testing of immobilization by lignite wax and Poly-Quat (BLZ/BVV)
- testing of additives in filter cake sedimentation (Apero/RB/DBI)
- assessment of corrosion effects (E.S./NUD)
- testing of resistivity against radiologic attack (UAW)

There are good indications that Poly-Quat additives will be useful in HPG mixtures and lignite-wax emulsion in fraise-diaphragm mixtures regarding long-term effectiveness.

Field- and semi-technical tests comprised:

- test-fields for horizontal sealing measures (Flowmonta)
- testing of new drilling and grouting techniques
- testing of regional clay deposits (GFE/RWTH-LIH).

Generally the results are positive or promising.

In order to get a representative number of filter cake samples, a special apparatus was used which produced filter cakes at different pressures and different percentages of additives. Optimum effects were achieved when adding about 5 percent of lignite-wax emulsion. The most obvious effects were greater thickness and faster devellopment of the filter cakes.

TEST SECTION

Prior to final approval, both the fraise-diaphragm- and the HPG-walls require a large-scale field test. A suitable test field has been selected close to the hazardous land-fill Antonie in order to provide the extreme groundwater milieu to be demanded for this test. The test walls will be some hounderd m long, about 55m deep and 1m thick. They will penetrate through all occurring strata, including hard drilling obstacles, mine tailings and open mine adits. The test section will be built along the western rim of the land-fill and can ,hopefully, be integrated into the final encapsulating measure.

QUALITY CONTROLS

The test walls have to be seriously controlled during and after performance. Because of the extreme conditions and the high long-term quality requirements these controls are considered very essential and will be performed with ultimate care. The envisaged control programme includes the standard wall testing procedure as well as a special control shaft with adits, control boreholes with different wall penetration and control boreholes into the basal buffer, all being equipped with modern monitoring devices (Fig.4). Continuous registration of working progress and long-term behaviour as well as easy access to all controll installations have been recommended, because this test wall will represent a very important progress in remediation technologies.

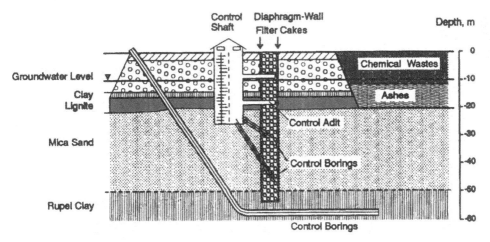

Fig.4   Envisaged control programme for diaphragm-wall test section (KRAPP)

The project has also been recommended as a special environmental show piece of the EXPO-2000 in Germany.

REFERENCES

1. H. Koerner et al, Umweltschutz-Pilotprojekt, Chemie AG Bitterfeld-Wolfen, project report (1991).
2. J. Hille et al (editors), Bitterfeld, modellhafte ökologische Bestandsaufnahme einer kontaminierten Industrieregion, 1. Bitterfelder Umweltkonferenz, Schadstoffe und Umwelt **10**, 325 (1992).
3. L. Krapp et al, Umweltgeologie beim Pilotprojekt Bitterfeld-Wolfen, Die Geowissenschaften **11:1**, 1-9 (1993).
4. L. Krapp, S. Bunge, K.H. Steinberg and T. Böhme, Schaffung des wissenschaftlich-technischen Vorlaufs zur Sanierung eines hochkontaminierten Chemiestandortes (Bitterfeld-Wolfen), summary project report, 132 (1994).
5. H. Weiss, L. Krapp, T. Böhme, R. Ruske, K.H. Steinberg, Informationsmaterial zu vorgeschlagenen Projekten der Altlastensicherung/ -Sanierung im Großraum Bitterfeld, Arbeitskreis 'Altlasten' des Landesbeirates EXPO-2000 (1996).
6. L. Krapp, Die Schlitzwandtechnologie als Sanierungsverfahren für die Bitterfelder Altablagerungen, 2. Bitterfelder Umwelttage, 6 (1996).

Proc.30th Int'l.Geol.Congr.,vol.24,pp.261-265
Yuan Daoxian (Ed)

Redox Condition in the Vadose Zone intermittently Flooded with Waste Water

Min Wang, Yongfeng Wu
Department of Hydrogeology, China University of Geosciences, Beijing 100083, P. R. China

Abstract

When the ground surface is intermittently flooded with waste water, a special and complicated redox condition would form in the vadose zone. This controls the migration and transformation of some pollutants, especially to nitrogen and organic compounds. Therefore a soil column test was set up to study this situation. The test consisted of three plexiglas columns with 15 cm internal diameter and 240 cm height and lasted more than 8 months. It was found that in the special case the whole column profile could be divided into three parts: gravitational saturated zone, capillary saturated zone and aeration zone. In the aeration zone, the value of Eh decreased to -500 mV during the flooding period and increased to 350-400 mV during the drying period. In the capillary saturated zone, the Eh value did not decrease but increase in the beginning of the flooding and then decreased. In the gravitational saturated zone, the environment always kept in the anaerobic state even in the drying period This result tells that the design and control of rapid infiltration systems for wastewater land treatment should put the emphases on the soil above 150 cm depth.

Keywords: redox condition, wastewater, groundwater pollution, vadose zone, aeration zone

INTRODUCTION

When the ground surface is intermittently flooded with waste water, for example, in the case of waste water irrigation or in rapid infiltration (RI) treatment systems for waste water or seasonal waste water river, channel and pond, a special and complicated redox condition can form in the vadose zone. This controls the migration and transformation of some pollutants during the infiltration of waste water, especially to nitrogen and organic compounds in RI systems [3-5].

To reveal the oxygen distribution in vadose zone and its relation to nitrogen removal, Lance, J.C. et al. [1-2] carried out a soil column test intermittently flooded with domestic effluent. Wu, Y., et al. [6] studied the environmental characters in RI systems intermittently flooded with brewery waste water.

However, up to the present, no one precisely knows the special redox condition in the common case of intermittent flooding with municipal waste water. Therefore, a soil column test was set up to study this situation.

EXPERIMENTATION

The test was carried out in the laboratory. Three Plexiglas columns with 15 cm internal diameter and 240 cm height were constructed. The bottom 8 cm was packed with coarse gravel to simulate the aquifer. Over the gravel, sandy loam or loamy sand or fine sand was packed to a depth of 192 cm. In the top part was 40 cm waste water obtained from an effluent source in the campus of China University of Geosciences. To get a stable input for the experiment, a 4 month monitoring to the source was made before the start of the test [5]. It was found that the water quality had a periodic change with daytime and the most stable values of the main concentrations appeared at 8:00 - 9:00. The water was then picked up on that time during the whole experiment. Its average concentrations for COD (Chemical Oxygen Demand), SS (Suspended Solid) and NH_4-N were 240, 160 and 62 mg/l respectively.

To measure Eh values, platinum black electrodes were separately installed in soil depth of 20, 50, 80, 110, 150 and 195 cm as the measuring probes. A calomel electrode was placed in soil depth of 10 cm as the comparing probe (Fig.1).

During the experiment, the hydraulic period was in the way of 5 days around for flooding and 9 days around for drying in which the flooding water was kept with a Marriotte bottle of 20 liters. Before the formal measurement, the columns were biologically cultivated about 4 months in the above hydraulic way. When Eh values were detected through the measuring probes, the amounts of DO (Dissolved Oxygen) in the soil solution were also measured in the same positions with a portable oxygen meter.

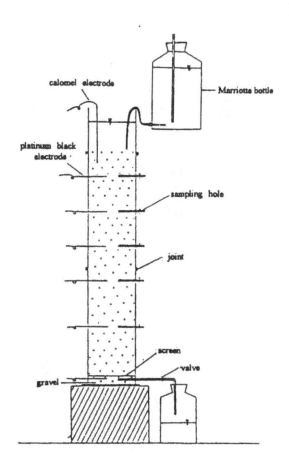

Fig.1 A plexiglas column used in the test

RESULTS

The typical changing of Eh, DO and pH was illustrated in Fig.2. It was found that with the beginning of the flooding, Eh values in the soil depth above 150 cm were all decreased and at last stabilized around -500 mV. The finer the soil was, the slower the decreasing rates of Eh values would be. When the operation came into the drying period, the Eh values in different depth (above 150 cm) were then increased in turn. The more the depth, the later the increasing period started. As long as the drying time was long enough, for example more than one day, all the Eh values could be recovered to 350 - 400 mV.

As to the depth below 150 cm, it seemed much different. The value of Eh detected in the depth of 195 cm was not decreased in the beginning of the flooding. On the contrary, it was first increased to reach a maximum value of -150 mV (sandy loam) or -50 mV (loamy sand) or 250 mV (fine sand) and then decreased to -500 mV. During the drying period, the values of Eh in the depth of 195 cm kept stable for sandy loam and loamy sand columns. However for the fine sand column it was increased and stabilized at -100 mV.

All the amounts of DO in the soil solution in different depth changed in a similar way with time. When

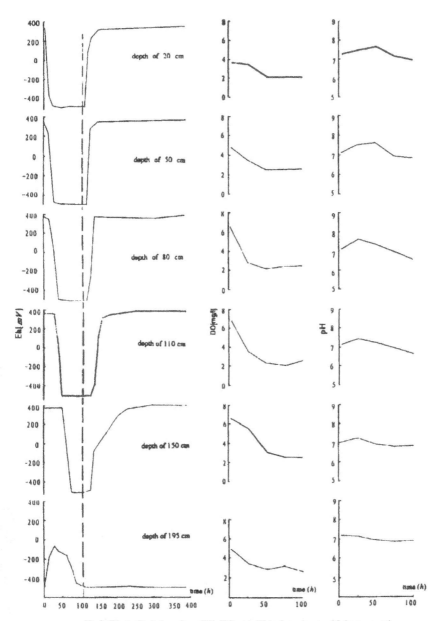

Fig 2 The typical changing of Eh,DO and pH in the column with loamy sand

the columns were flooded, they tended to decrease and keep at 2-3 mg/g at last. But the initial values of them in the beginning of the flooding were different. It was higher in the middle part of the columns but lower in the top and bottom parts.

Values of pH remained in the range of 7.0-7.5 during flooding period with no obvious change.

DISCUSSION

Since the simulated groundwater table was controlled in the soil depth of 197 cm, the whole column profile could be divided into three parts, that is, gravitational saturated zone, capillary saturated zone and aeration zone. This can be illustrated in Fig.3.

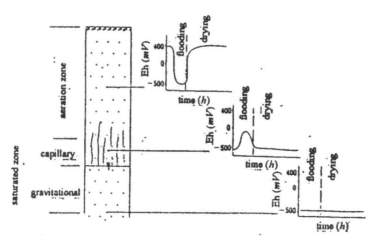

Fig. 3 Conceptual model in the vadose zone intermittently flooded with wastewater

In the aeration zone, as the column was flooded, the frontal wetting surface began to move downwards and to make the environment anaerobic (Eh decreasing); as the column was dried and gravitational water drained off, the Eh value was recovered to the state of aeration.

As the frontal wetting surface went on to move down, the oxygen in the top part of the column was driven and concentrated in the capillary saturated zone. This made the environment somewhat aerobic. Eh values increased. Only if the oxygen was exhausted, the Eh value would decrease.

In the gravitational saturated zone or in the top part of the groundwater body, the redox condition generally kept in an anaerobic state even if it was in the drying time.

It was clear that in the test the soil in the depth above 150 cm belonged to the aeration zone no matter what kind of soil the column was filled with. In the depth between 150-195 cm, there was at least part of it existing in the capillary saturated zone.

The above experimental results explained the reason why RI systems or geological porous media have the better ability to renovate most kinds of waste water. It is because of the three states of redox environment existing in the same time. Since the alternating of aerobic and anaerobic states mainly happens in the aeration zone, the design and control of RI systems should put the emphases on the soil above 150 cm depth.

Acknowledgements

The authors are grateful to Prof. Shen Zhaoli and Prof. Zhong Zuoxin who reviewed earlier versions of this paper. The authors wish to thank Prof. Jinag Jingcheng, Mr. Chen Liang and Miss Lu Xiaoxia for their helpful discussion and assistance in the experiment.

REFERENCES

1. Lance,J.C., et al., Oxygen utilization in soil flooded with sewage water, J. Environ. Quality 2:3 (1973)

2. Lance,J.C., Renovation of waste water by soil columns flooded with primary effluent, 52:3 J WPCF, (1980).

3. Mathew, k., Ground water recharge with secondary sewage effluent, Australian Water Resource Council, Technical Paper, No. 71 (1982).

4. Wang, M., Zhong, Z. and Wolff. J, The nitrogen pollution of groundwater by municipal waste water irrigation. Z. Wasser-Abwasser-Forsch, 25:1 (1992)

5. Wang, M., et al., Nitrogen removal in wastewater irrigation systems and its influence on ground water pollution, J. China University of Geosciences,6:2 (1995).

6. Wu, Y.,et al., Rapid infiltration treatment system for brewery wastewater, Proceedings of International Workshop on Groundwater and Environment, Beijing, Seismological Press (1992).

Proc. 30th Int'l. Geol. Congr., vol. 24, pp. 266-281
Yuan Daoxian (Ed)
© VSP 1997

Mapping Units on São Sebastião Geohazards Prevention Chart, Northshore of São Paulo State, Brazil.

PAULO CESAR FERNANDES DA SILVA
CRISTINA DE QUEIROZ TELLES MAFFRA
LIDIA KEIKO TOMINAGA   RICARDO VEDOVELLO
The Geological Institute, São Paulo State Secretariat of Environment
Av. Miguel Stéfano 3900, São Paulo - SP, 04301-903, Brazil

Abstract

Since 1988, the Geological Institute (IG/SMA-SP) has been performing multidisciplinary studies for land-use planning and environmental management in São Paulo State (Brazil), a region highly developed both urban and industrially. These studies provide earth-science useful information at the District, City and State administration levels to formulate and implement planning policies which include hazard mitigation, zoning regulations and control ordinances. The elaboration of the São Sebastião Geohazards Prevention Chart represented an attempt to reduce adverse environmental impacts (and disasters), as well as to reconcile many conflicting influences. Furthermore, one of the most meaningful governmental concerns is related to risk assessment applied to urban growth and development into a context of tropical environment.

The study consisted of a first regional evaluation on 1:50.000 scale, and a second stage on 1:10.000 scale, that is, a detailed engineering geological cartography work. The results now presented refer to the first stage. It included a full desk study using multipurpose background data; and field work associated with remote sensing techniques in which satellite imagery (Landsat, band 4) and aerial photographs (at three different scales) were the fundamental tools for analysis. In this paper a land classification system comprehending "basic physiographic units" (bpu's) is presented. A bpu is characterized by the evaluation of the relevant attributes or factors which should be elected during the performance of the study. The choice of the attributes and of their relevance is based on a combination of the following aspects : the scope of the work, the objectives, previous data basis and field work.

Keywords: land-use planning, land evaluation, physiographic compartimentation, environmental management, geological hazards, engineering geology, risk assessment.

INTRODUCTION

In 1994, the Geological Institute (IG/SMA-SP) and the Municipality of São Sebastião have signed an agreement for technical cooperation. The agreement included the elaboration of a geohazards prevention chart as well as consultancy and advising services to support the Mayoralty in outlining an implement and structural plan for the city. The work was carried out in accordance with the main concerns of local authorities which were highlighted as: i. formulation of public policies emphasizing the prevention of disasters and hazard mitigation. It would have to include immediate measures and remedial actions, with special attention to landslides and flooding events; ii. formulation of public policies in planning, as guidelines for urban growth and sustainable development as well as environmental protection and management.

Such sort of study implied in different scales of analysis and it should involve the identification of current and potential phenomena for later prediction and evaluation of geological hazards. Risk assessment should be based on the magnitude and on the frequency of geohazards in combination with the probability of damage (security of life, destruction of property, disruption of production).

The methodology of homogeneous terrain-unit for land classification [7, 10, 20] has been used by the IG/SMA-SP in previous studies performed in other parts of the São Paulo State with similar purposes. The area of each municipality was used as territorial reference for approaching and analysis. It is interesting to note that those studies could be recalled as geotechnical cartographic work in the usage of Thomas [17] or as applied environmental geological mapping in the sense used by Doornkamp *et al.* [3].

After achieving further experience on this type of study, the IG/SMA-SP has changed its approach and started considering portions of drainage basin as territorial reference . Another land classification system using *diagnostic units* was developed [8]. The characterization of these units was realised in the basis of land capabilities, land susceptibilities and natural ground constraints for use. The *diagnostic units* represent land portions with similar response to similar types of use, instead of the homogeneous terrain-unit which regards inherent ground properties,independent of their use.

This way-of-working has proved to be very feasible in mid to long term earth-science studies performed in the central eastern of São Paulo state. In that region the topographic gradients and amplitudes are not extremely high. The data basis is relatively dense showing more adequate scales, and the relationship between the types of land-use are less complex.

As the characteristics of the São Sebastião region are quite different from the ones found on precedent works we had to review our methods and land classification systems.

The Region of São Sebastião

The area of study is a peculiar coastal environment located in the northshore of São Paulo State (Fig.2). It discloses a remarkably contrasting landscape involving widespread ridges (commonly more than 800 meters high) associated to narrow smoothed relief areas (lowlands or coastal plains). The climate is humid tropical with frequent rainstorms at the summer. Many conflicting influences in land-use can be observed. The main urban area of São Sebastião is situated at the seaside, and it is washed by sheltered waters of the São Sebastião Channel between the island of Ilhabela and the continent. Due to this peculiar location, an important harbour with oil and gas storage facilities was set out there, and pipelines depart from this point to supply refineries at the neighbour regions. Navigation for fishing activities and nautical leisure is also permanent. All the region is a very important centre for tourism due to its beaches and sightseeing. High economic pattern housing contrasts to bad housing with absent or inadequate infrastructure, although both are sometimes built on hazardous areas.

Previous Work

During the period between the 30's to 50's decades, some geological and geomorphological studies concluded that regional faultings and diferential erosion would have been the possible origins of the Serra do Mar Ridge [*apud* 12]. In the 60's, Pichler & Rodrïgues [*apud* 12] and Machado Filho [*apud* 12] carried out important engineering geological works for slope stability control. Engineering geological works were also carried out by Hidroservice [*apud* 12] to construct the Paraibuna River Dam. Silva *et al.* [12] and Hasui *et al.* [6] recognised the regional geologic setting at 1: 250.000 and 1: 200.000 scales: it comprehends a predominant outcroping area of pre-Cambriam cristalline basement embodied by a major gneiss-migmatitic complex in association to granites, and minor portions of granulitic rocks. This basement was affected by small ultramafic to alkaline intrusions of post-Jurassic age. Restricted sedimentary fillings of quaternary age were also recognised. Suguio & Martin [16] studied the quaternary formations of the São Paulo State shore zone at 1: 100.000 scale. They concluded that most quaternary sediments had been reworked during the two late transgressive episodes. More recently Souza [14, 15] studied the current and sub-recent coastal dynamics at Caraguatatuba (neighbourhood of São Sebastião). Campanha & Ens [1] and Campanha *et al.* [2] contributed with studies of structural geology and morphotectonics in the region.

REGIONAL EVALUATION

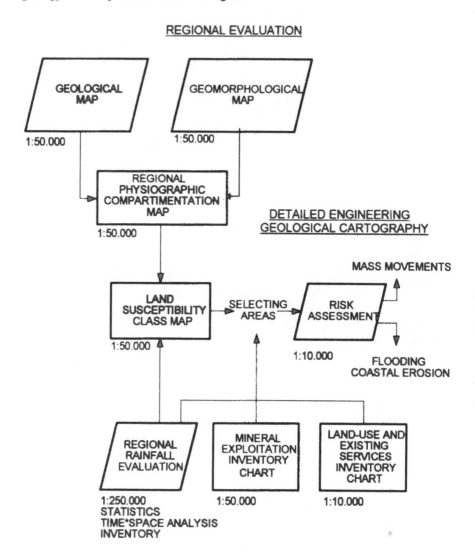

Figure 1. Flowchart showing research stages and products in the São Sebastião Geohazards Prevention Chart

PROCEDURES AND METHODS

The present study has been differentiated from the precedent ones due to many factors. As mentioned before, some of the reasons are related to the nature of the region itself, like geology and physiography. Other reasons could be considered as non-natural and related to the scope of the work. For example: the necessity to obtain fast results reconciling risk assessment with regional to local planning interests; singular geographic conditions generating a very complex and conflicting relationship between the current land-uses; insufficient and discontinuous background information requiring intensive desk studies to compile a data basis in adequate scales; demand for field work in areas of adverse natural conditions and access. The whole municipality's territory involves an extension of 580 square kilometers approximately.

The study consisted of two main research stages: **1.** regional evaluation on 1: 50.000 scale in order to produce a preliminary response under multidisciplinary approach; **2.** detailed engineering geological cartographic work on 1: 10.000 scale in which the main targets were centered on risk assessment with emphasis on landslides and flood (Fig. 1). The regional evaluation comprised a desk study with fully comprehensive collection of background data, remote sensing techniques and field survey work. Multisource background information was gathered and compiled with the help of computing facilities. Data bank included: topography, geology, geomorphology, mineral resources and exploitation, climate, rainfall, tidemarks, geodynamic processes, land utilization and existing services. Satellite imagery (1: 100.000 scale) complemented by aerial photographs (1: 60.000, 1: 45.000, 1: 25.000) were fundamental tools used in combination with previous data and site-specific data from field work. The results of this first stage of research were displayed in base products (as geological and geomorphological maps, as resource inventories) and also in intermediate products as the Regional Physiographic Compartimentation Map and the Land Susceptibility Class Map as shown in Figure 1. The data framework and products concerning the regional evaluation were applied to the next stage of detailed engineering geological mapping. They had also a further application by the Mayoralty in planning and environmental management.

A combination between the above so-called intermediate maps and Land Use plus Mineral Exploitation Inventories had supported the selection of areas for risk assessment in the second research stage. The ultimate product was a set of charts indicating geological hazards and classifying areas in degrees or classes of risk. An explanation note and user's guide come along with all this material.

REGIONAL PHYSIOGRAPHIC COMPARTIMENTATION

The physiographic compartimentation consists in identifying portions or land divisions on a satellite image (TM-Landsat, band 4) using hierarquical levels related to the morphological, and to the lithological and structural characteristics of the bedrock [18, 19]. For supporting and complementing the interpretation of the image, information from Geological and Geomorphological Maps were used and specific field work was executed in order to verify on-site ground conditions.

The Geological Map

The geological survey was applied in support to regional evaluation and to engineering geological cartographic work. The first scanning was carried out at 1:50.000 scale whereon two meaningful domains of bedrocks were recognised: cristalline basement and quaternary sediments. They were used as code reference for classification and for presentation of geological units (Figure 2). On cristalline basement (**Cb**), five relevant units were characterized under lithological and structural criteria as follows, they all strike according to the regional trend which is NE-SW to locally E-W. The main structural feature observed is the foliation, which is associated to regional shear zones. Despite its exiguous presence, folds are observed in metric scale. Brittle tectonics are represented by penetrative metric to sub-metric slickensides as well as joint systems. Quaternary sediments are unconformably overlying the cristalline basement rocks. These sediments are found either filling the coastal plains or accumulated at the footslopes. The classification of the quaternary sediment units is related to the predominant depositional environment, although the correlative ages are also taken into account. The code **Qc** represents a continental origin whilst the code **Qm** means that the sediments have a marine or a mixed origin. Age correlations were based on Suguio & Martin [16] and on Souza conclusions [14, 15] who associated the changings in sedimentation to sea level variations and coastal dynamics.

Figure 2. Location and geological map of the municipality of São Sebastião, southeast Brazil

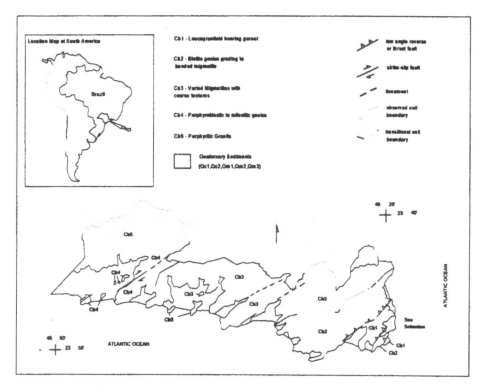

Leucogranitoid bearing garnet (**Cb1**) outcrops at the eastern portion of the studied area into two restrictly distributed stripes of 1 to 1.5 km wide. It consists of a very homogeneous leucocratic rock with fine to medium granitic textures, or locally grading to pegmatoid types. Essential mineralogy is formed by feldspar (microcline, pertite, oligoclase), xenomorphic quartz and garnet (almandine) with minor and variable amounts of biotite. Foliation is absent or incipient, except near the contacts with unit **Cb2** which is intercalated or hosting the stripes of **Cb1**. Foliations and contacts are associated to regional structures (thrust or low angle reverse faults) which is believed to have been generated during Neoproterozoic times (Braziliano Tectonic Cycle) by a transpressive transcurrent regime according to Maffra *et al.* [9].

Biotite Gneiss grading to Banded Migmatites (**Cb2**) - it is a widespread unit at the northeast - southeastern part of the studied area. Its boundaries are transitional to unit

Cb3 but they have a clearly tectonic nature in relation to unit **Cb1**, as highlighted before. The biotite gneiss is a dark coloured rock constituted by a fine grained matrix containing layers of biotite, plagioclase (albite - oligoclase) and anedric quartz. Another mineral constituent is garnet in sub-rounded and fractured crystals bigger than matrix texture. Foliation is represented by a strong schistosity which is determined by aligned grains of quartz and biotite. In the direction of unit Cb3, from east to west, the schistosity evolved to compositional banding under metamorphic segregation conditions. The new formed leucossomatic granitic portions - with quartz, feldspar and phylosilicates - are separated from biotitic mesossomes enriched in garnet. Thus the resulting rock is banded migmatite, due to its prevailing stromatitic structures. The foliation/schistosity follows the NE regional trend generally dipping to NW at medium to low angles. Therefore, it is sub-horizontal, gently dipping to NNE in the centre of the unit. Folds are more common in this unit. They are either observed in metric scale, semi-recumbent with one disrupted flanc or in slight undulations with sub-vertical axial plane.

Varied Migmatites with coarse texture (**Cb3**) - it represents the main unit which outcrops from the north-northeast extremity to the south where it underlies into the Atlantic Ocean. Its boundaries are transitional. The foliation corresponds to compositional banding characterized by alternating layers of gneissic mesossome and granitic to pegmatitic neossome. The coarse texture is a commom feature including different migmatitic structures as stromatitic, schlieren, scholen, oftalmic or locally agmatic (paleossomatic fragments surrounded by neossome). Gneissic mesossomes are formed by milimetric to centimetric layers of biotite intercalated to continuous or partial interrupted quartz and feldspar lenses (plagioclase, pertite, mirmequite). Neossomes present mainly interstitial to poligonized xenomorphic quartz grains, garnet aggregates, some potassic feldspar and sericite-muscovite. The foliation strikes to NE following the regional trend, and it dips to NW in low to high angles. Mylonitization features are locally observed within this unit.

Porphyroblastic to Mylonitic Gneiss (**Cb4**) - it comprises a single stripe 5 to 6 kilometers wide which outcrops at the southwestern part of the area. The unit shows transitional boundaries to unit **Cb3** but they are strongly tectonized when contacting unit **Cb5**. Most part of the unit was affected by strike-slip movements, therefore the resulting foliation is mylonitic, characterized by mineral elongation. The rock-type consists of a strong oriented fine grained matrix containing biotite, quartz ribbons, mirmequite, strenghtened microcline and eventually garnet. Porphyroblasts of plagioclase are outstanding in this matrix. Thin bodies of amphibolite as well as of hornblende gneiss are commonly intercalated in the unit. The foliation behavior is possibly associated to a major regional transcurrent shear zone mentioned as a positive flower structure by Campanha & Ens [1]. The foliation dips in high angles to SE near unit **Cb5** and it gradually changes its dips to vertical and medium angles to NW eastwards.

Porphyritic Granite (**Cb5**) - the unit corresponds to the highest topography in the área and it outcrops at the northwest extremity. The rock is a coarse grained type, essentially constituted by quartz, feldspar and biotite. The phenocrysts are potassic feldspar, strongly deformed near the contacts to unit **Cb4**, wherein the textures changed to *augen* terms. Milonitic foliation is only observed in association to these contacts.

Oher rock-types Some lamprophyres as well as basic to alkaline igneous intrusions are observed within the cristalline basement. They are often represented by sub-metric to metric dykes. On the other hand, a single lense of quartzite was also described in intercalation with unit **Cb3**.

Recent to sub-recent fluvial deposits (**Qc1**) - The fluvial deposits are associated to the wider coastal plains including the well developed drainage basins. They comprehend the accumulations of gravels, cobbles and pebbles in areas of higher steepness gradients. They correspond to badly sorted and mineralogically imature sandy deposits at low relief areas. The sub-recent fluvial terraces are locally and restrictly recognised as mixed gravel - badly sorted sand - dark plastic clayey deposits elevated 3 to 4 meters above the active stream level.

Slope associated deposits (**Qc2**) - The unit is composed by colluvionar ramps, talus or boulders, and dejection cones associated to valleys. The granulometry is characteristically bad sorted and sometimes unconsolidated. The main texture comprises cristalline rock fragments surrounded by sandy-silty-clayey mixed matrix. Rock fragments are angular or rounded with many different sizes. The thickness of these deposits varies depending on the local landform and slope steepness.

Activel beach deposits (**Qm1**) - According to Souza & Suguio [13] the beaches of the São Sebastião region correspond to three of the seven morphodynamic compartiments described at the State of São Paulo coastal zone. They are classified in regarding the following aspects: wave energy, morphological beach profile, grain size and type of sediments, depositional features and coastal transport. The majority of the beaches in the region are classified into dissipative and intermediate types, only one beach was recognised as reflexive according to Souza [14].

Sub-recent marine sand deposits (**Qm2**) - This unit is characterized as preserved surface beach crests. Its sediments occur at the backside of unit **Qm1** corresponding to the pleistocenic to holocenic sediments described by Suguio & Martin [16]. Locally, the deposits are considered as ancient marine terraces made up of well sorted quartzose sands.

Mixed deposits (**Qm3**) - The unit consists of sandy-clayey sediments with a likely fluvio-lacustrine origin occurring at topographic lows within the coastal plains. According to Souza [15] these sediments refer to paleolagoons covered by active floodplain sediments or colluvionar material.

The Geomorphological Map

This study aimed to delimitate and characterize the landforms, including aspects of morphometry as well as morphodynamic processes. The municipality territory was divided into four regional morphological domains which by their turn comprised units of relief or units of local landforms. The geomorphological study concluded that the units are strongly controlled by tectonic features as well as by bedrock lithology. The attributes used to characterize each unit were the following: local landforms, morphometric constraints as steepness and altitude, rock-type, morphodynamics including different processes such as creep, gullying, and so on. The resulting units are summarized in Table 1.

The use of BPUs - an operational land classification system

The regional physiographic compartimentation refers to a territorial zoning (the municipality, in this case) which is based on selected criteria or physiographic factors. The procedures for compartimentation involved i. the use of background information by means of compiled maps and inventories; ii. analysis of satellite imagery complemented by aerial photographs to recognise *"basic physiographic units"* (bpus) as a function of the selected criteria; iii. field work for bpu on-site characterization. The result was a Regional Physiographic Compartimentation Map (1:50.000) in which the territory of the municipality was divided into four hierarquical levels comprehended by the bpus.

Table 1. Geomorphological classification system for the region of São Sebastião

REGIONAL MORPHOLOGICAL DOMAIN	LOCAL LANDFORM UNIT	MORPHOMETRY	GEOLOGICAL UNIT (LITHOLOGY)	MORPHO DYNAMICS
Juqueriquere Plateau	**MMTp** Plateau Hills	Steepness: 10 - 30% Altitude: 580 - 660m	Cb2 Cb3	Low intensity of erosional processes. Some rilling, small ravines.
Paulistano Plateau	**MMH** Mountains	Steepness: 20 - 30% Altitude: 900 - 1100 m	Cb5	Susceptibility to the following phenomena: rockfall, soilcreep, planar landslides, gullying.
Serra do Mar Ridge	**Ea** Concave Scarp Face	Steepness: 20 - 30%, 30 - 40% and more than 40% Altitude: 100 - 760 m	Cb2 Cb3	High intensity of erosional processes. Soil creep, planar and rotational landslides, rockfall, block tilt, rilling and gullying. Torrential drainage processes with mass transport
	Ed Scarp face with fingered crestline	Steepness: 30 - 40%, 20 - 30% Altitude: 100 - 600 m	Cb2 Cb3 Cb4	High intensity of erosional processes as above
	Er Scarp face in rectlinear section	Steepness: 30 - 40% and more than 40% Altitude: 100 - 900 m	Cb1 Cb2	Predominantly processes of slab failure and rockfall
	MMTl Littoral or isolated residual hills	Steepness: 10 - 30%; 30 - 40% Altitude: 20 - 160 m	Cb2 Cb3 Cb4	Rilling, gullying, laminar erosion, soil creep and landslides
Coastal Plain	**FM** Fluvio-marine plain	Steepness: less than 2% Altitude: 4 -12 m	Qc1 Qc2 Qm3	Swamp-like areas and intermitent flooded areas. Stream erosional processes.
	M Marine Plain	Steepness: less than 2% Altitude: 1 - 8 m	Qm1 Qm2	Actual beaches under coastal dynamics. Ancient marine terraces affected by stream erosional processes

The selection of criteria or physiographic factors should be a function of the objectives and of the scope of the work. Under this perspective, the choice of factors becomes an operational decision to be taken along the work. Such selection evaluates the relationship between different environmental land components and the relevance of each component, in order to achieve a suitable presentation of results. The selected factors for characterization of the bpus were: a) bedrock lithology and geological elements as strikes, dips, geometry of faults and fractures; b) landforms and surface drainage pattern; c) slope steepness; d) geodynamic processes as erosion by rilling, soil creep, etc; e) soil and land cover conditions including thickness plus weathering. A simplified codification of the bpus is illustrated in Table 2. Thus, similar bpus codes are reflecting similar ground conditions. The basic physiographic unit (*bpu*) could be directly associated to the concept of *geoform* as used by Vedovello [19]: "geoform is considered as a portion of land wherein a determinated group or combination of landscape components occurs. It is a result from the action of environmentally exogenic factors (as climate, human influence), landscape evolution dynamics, as well as of intrinsic physiographic properties (as bedrock lithology)". In this way, the significance of each compartiment involved different conotations apart from the usage on other terrain classification or land-unit system [4, 7, 10, 11, 17, 20].

Table 2 - Summarized codification of the basic physiographic units (bpus)

REGIONAL MORPHOLOGICAL SECTOR	BEDROCK LITHOLOGY	LOCAL LANDFORM COMPARTIMENT	BASIC PHYSIOGRAPHIC UNIT
PLATEAU (P)	Granite (G)	Mountain (H)	PGH (1 to 3)
	Porphyroblastic Gneiss (P)		PPM
	Migmatite (M)	Hill (M)	PMM (1 to 3)
	Biotite Gneiss (B)		PEM (1 and 2)
ESCARPMENT (E)	Granite (G)	Concave Scarp Face (A)	EGA (1 to 3)
		Scarp Face with fingered crestlines (D)	EGD (1 to 4)
		Hill (M)	EGM
	Porphyroblastic Gneiss (P)	Concave Scarp Face (A)	EPA
		Scarp Face with fingered crestlines (D)	EPD (1 to 3)
		Hill (M)	EPM
		Littoral or isolated residual hill (I)	EPI
	Migmatites (M)	Concave Scarp Face (A)	EMA
		Scarp Face with fingered crestlines (D)	EMD
		Hill (M)	EMM
		Littoral or isolated residual hill (I)	EMI
	Biotite Gneiss (B)	Concave Scarp Face (A)	EBA
		Scarp Face in Rectlinear Section(R)	EBR
		Scarp Face with fingered crestlines (D)	EBD
		Hill (M)	EBM
	Leucogranitoid (L)	Concave Scarp Face (A)	ELA
		Hill (M)	ELM
	Quaternary Continental Sediments (C)	Colluvionar/Talus/Boulder Deposit (T)	ECT
LOWLANDS OR COASTAL PLAIN (B)	Quaternary Marine Sediments (M)	Colluvionar/fluvial deposit (F)	BCF
		Mixed fluvio-marine deposit (M)	BMM
		Marine terrace(T)	BMT
		Beach (P)	BMP

The first hierarquical level of compartimentation divided the municipality territory into three major sectors or physiographic domains: upland or plateau (P), hillslope or scarpment (E) and lowlands or coastal plain (B). The second level is concerned with the main attribute assigned at the geological map: the bedrock lithology. The unit boundaries are slightly different from those depicted in the geological map because physiographic constraints were applied (Table 2, letters L to **CT**). The third level of compartimentation added the analysis of different types of local landforms contained in the geomorphological map to the divisions established by precedent levels (Table 2, letters **A to I**) .

The fourth and last level allowed the land compartimentation in *basic physiographic units* (bpus) which are the smallest unitary division adequated for geotechnical characterization purposes. In this case, the emphasis was on the evaluation of land susceptibility to landslides and flood. The obtained bpus corresponded to the individualization of specific lanscapes under the selected factors mentioned above. For classification and presentation of the bpus (Figure 3), the three letters used in the codification represent, respectively, the three first levels of compartimentation. The numeric character refers to the fourth level which indicates particular local features, such as type of weathered land cover material, soil thickness, bedrock jointing density and others. Table 2 shows the unit codification format into a summarized reference, it does not contain a full description of each unit.

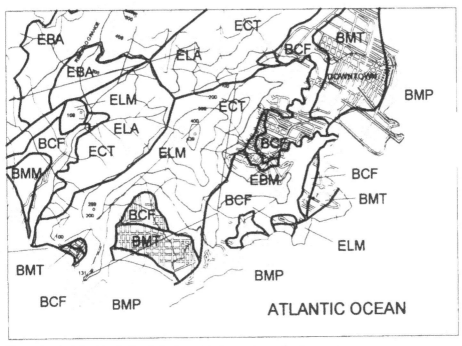

Figure 3. Detail of the Regional Physiographic Compartimentation Map depicting the main urban area and neighbourhood. It shows the bpus situated in the Escarpment (E) and Lowlands (B) sectors. Cristalline basement bedrock lithologies are leucogranitoid (L) and biotite gneiss (B). Continental quaternary sediments (C) are colluvionar/ fluvial deposits (F) and Colluvionar/ Talus/ Boulder (T). Marine sediments relate to the beach compartiment (P).

Therefore we can formulate an example about the significance of the *bpu* code. If the resulting *bpu* code is EBR2, it means that the bpu is located in the regional morphological sector called as escarpment (represented by letter E). The bedrock lithology involved is biotite gneiss grading to banded migmatites (letter B). Local landforms correspond to the scarp face in rectilinear section (letter R) with slope steepness gradients of 30% to 40%. The fourth character (number 2) represents particular features which differentiate the unit from others in similar hierarquical levels. Specifically to this unit they were recognised as follow: a) strong foliation dipping to NW in angles higher than slope steepness; b) unit boundaries originated by tectonics; c) high jointing density in E-W and N-S trends; d) rocky edges and restricted extensions of weathering veneer; e) frequent rockfall and block tilt phenomena

LAND SUSCEPTIBILITY AND RISK ASSESSMENT

The land susceptibility consisted in analysing and classifying the bpus in relation to the natural tendency of occurrence of hazard inducing phenomena. Current and potential situations are included in this anaylisis. The single characteristics of each level of compartimentation had to guide the evaluation of the bpus for susceptibility. It means that mass movement processes (which included landslides and gullying for example) were analysed with reference to Plateau (P) and to Escarpment (E) sectors (see Table 2). The susceptibility to flooding was investigated at the Lowlands/ Coastal Plain sector (B) whilst the coastal erosion was studied only at the beaches - a portion of the latter sector. Table 3 presents the standardized attributes related to the Land Susceptibility Class Map which represents a land-zoning in degrees of hazard. A detail of this map is shown in Figure 4. The attributes are quite different depending on the sector in appraisal as mentioned above. In general they correspond to types of reported and potential phenomena besides the bpus properties/characteristics applied to classification. These so-called classification factors (Table 3) were the following: predominant weathering materials, slope profile, bedrock structural features and steepness for Plateau and Escarpment; surface and sub-surface sediment type; drainage basin characteristics and steepness for Lowlands/ Coastal Plain (except beaches); finally steepness, wave energy, type of sediment and coastal transportation for beach compartiment.

Types of Reported and Potential Phenomena
The phenomena (geodynamic processes) numbered below from I to VI refer to the compartiments comprehended at Plateau and Escarpment sectors, VII at Lowlands/ Coastal Plain and VIII refers to beaches compartiments.
I. Soil Creep. Slow dislocation of several horizons observed by the inclination of trees, of walls, and of electric power stakes. The process is increased by water content in soil.
II. Landslides. These slope mass movements range in size as well as in type of transported material (soil, talus/boulder and colluvionar deposits). They are observed as different features corresponding to varied magnitudes and time of occurrence: landslide scars or accumulated mass, current active or recently formed, old or stabilized. Generally, the observed features indicated superficial movements involving predominantly the land cover with excessive water content.
III. Falls, Slab Failures and Topples. The evidences indicated that slab failure is a preliminary stage for rockfall and toppling generation. These processes usually occur where a steeply rock-face contains such a sort of natural well-structured bedrock (high jointing density, foliation, etc).
IV. Block Tilt and Rock Slide. Apparently they represent a second stage from precedent falls. Movements may occur when original support of either so-buried blocks or rockfall debris is removed.

Table 3. Attributes applied to the Land Susceptibility Class Map

PHYSIOGRAPHIC COMPARTIMENT	TYPE OF PHENOMENON	SUSCEPTIBILITY CLASS	FACTORS CONSIDERED TO CLASSIFICATION (CHARACTERISTICS AND PROPERTIES OF THE BPU's)			
PLATEAU AND ESCARP MENT	I - Soil creep		*Resulting weathering*	*Crest profile*	*Geological structures*	*Slope Stepness*
	II - Landslides	LOW	Clayey	Convex	Unconformable low fracturing	5 a 10%
	III - Slab failure and rockfall	MEDIUM	Clayey-sandy	Convex - concave	Conformable low fracturing	10 a 20%
	IV - Block tilt V - Mud and debris flow	HIGH	Sandy-clayey	Rectilinear	Unconformable high fracturing	20 a 30%
	VI - Erosion by rilling and gullying	VERY HIGH	Sandy	Concave - rectilinear	Conformable high fracturing	> 30%
LOWLAND OR COASTAL PLAIN (excepting beaches)	VII - Flood		*Type of Sediment, Slope Stepness, Surface Drainage*			
		LOW	It relates to **BMT** units on Compartimentation Map. Predominantly sandy sediments associated to sub-actual marine terraces.			
		MEDIUM	It relates to **BCF** units on Compartimentation Map. Actual fluvial sediments and lowland colluvionar deposits. Flood events generally associated to the main rivers.			
		HIGH	It relates to **BMM** units on Compartimentation Map. Mixed deposits (sandy to organic clayey sediments) associated to sub-actual lacustrine and actual fluvial deposits.			
BEACHES	VIII - Coastal erosion		*Slope stepness, Wave energy, Type of sediment*			
		LOW	Low energy - dissipative beaches (**BMP4**) Also low energy - reflexive beaches (**BMP5**)			
		MEDIUM	Reflexive beaches (**BMP1**)			
		HIGH	Transitional to intermediate beaches (**BMP2**) Also high energy - dissipative beaches (**BMP3**)			

V. Flows. These phenomena are always associated to high water contents within less cohesive soils. In the studied area, mud flows occur more frequently in the vicinity of the drainage basins.

VI. Soil erosion by water. It concerns surface erosional features as rills and gullies originated by waterflow at preferential paths. The increasing of erosion can generate ravines of some metres depth.

VII. Flooding. In the region, it could be associated to a combination of many natural factors such as basin characteristics, type of bedrock sediments, rainfall values, tidemarks. On the other hand, human influence on drainage network is a very important factor to be considered.

VII. Coastal erosion. It was qualitatively evaluated by using beach morphodynamical characteristics[13]. Some indicative features are: shoreline recession observed by the destruction of beach ridges and by the formation of sporadic washover fans; accelerated sediment movement.

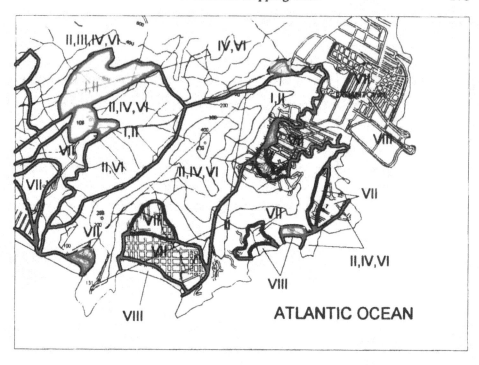

Figure 4. Detail of the Land Susceptibility Class Map depicting the main urban area and neighbourhood. It displays the types of potential and reported phenomena included in dashed zones which represent the classes of susceptibility.

Classes of Susceptibility

At the Land Susceptibility Class Map, the bpus analysed by the above described procedures are represented by color and graphic textures which indicate degrees of susceptibility. In this way, the degrees of susceptibility had been assessed and separated at intervals which meant the classes of susceptibility. The map representation also includes overwritten roman numbers (from I to VIII) which indicate the type of hazardous phenomenon. The classes of susceptibility are defined as follow:

Low susceptibility. It comprehends areas situated at the upland or escarpment sectors with nil or very slight degree of hazard. Those areas represent movement-free zones and zones where future mass movement is highly unlikely. They have low restrictions to excavations and man-made cuttings, besides they are more adequate to foundations or other engineering purposes. At the coastal plains, the low susceptibility areas correspond to those where the probability of occurrence of flooding phenomenon or stream erosion is very unlikely.

Mid-susceptibility. At the upland or escarpment sectors it corresponds to the areas where the degree of hazard is slight to moderate. Generally they conform to zones with no or few indication of current instability but potentially favourable to mass movement due to ground conditions. In other cases mid-susceptibility is characterized in some restricted areas affected by, or commonly subjected to small-scale erosional processes/ mass movements. At the lowland sector, zones of reported flooding events associated to the main drainage networks. In both kind of sectors, the units will present considerable restrictions for land-use and technical solutions and protection measures must be adopted to reduce or avoid potential risk. Slight level of coastal erosion at the beaches compartments.

High susceptibility. Areas with high degree of hazard in any sector (upland or escarpment, coastal plain and beaches). It corresponds to highly unstable zones affected by one or by a combination of the above listed phenomena-types. They are recognised as unfavourable zones for land-use wherein any engineering project implies accurate studies of suitability, and consequently higher costs. Such sort of zone which is presently in use normally demands some immediate remedial action or restoration.

Very high susceptibility. This class is only related to the areas belonging to the upland or escarpment sectors. It refers to those areas where geological hazards are imminent and possibilities of disasters are potential or reported. Engineering purposes as well as other kind of land-use should be avoided in those zones.

Selecting areas for risk assessment

Considering that the whole extension of 580 square kilometers (municipality territory) was not in need of risk assessment, only some portions should be selected to the detailed engineering geological work on a second stage. Some criteria and procedures should arise to select areas for such study. Thus, the units recognised within very high and high to mid-susceptibility class were chosen to be evaluated under land-use plus climatologic criteria. This analysis applied the information from Land-Use and Existing Services Inventory Chart, from Mineral Exploitation Inventory Chart and from Regional Rainfall Evaluation to the chosen units at the Land Susceptibility Class Map. The details concerning the contents of each inventory chart are no subject of this paper, although it is interesting to notice its socio-economic approach. The Land-Use and Existing Services Inventory Chart included a classification based on utility patterns and economical standards for building/housing, urban growth tendencies and aerial view vegetation analysis for non-building areas. The Mineral Exploitation Inventory Chart presented active and abandoned sites of mineral exploitation (quarries in general) including some considerations on geotechnical conditions and restoration.

By means of these procedures, five major areas were selected for risk assessment in relation to mass movements and twenty areas were studied in regard to flooding.

CONCLUSIONS

The experience of the IG/SMA-SP in performing earth-science studies for planning and environmental management in regions with distinct physiographic characteristics has indicated the necessity of using different approaches and methods. The way-of-working should be outlined as a function of the objectives and of the expected results. The type of action intended by the public authorities will be very important whenever choosing the methods and criteria for analysis. The elaboration of the São Sebastião Geohazards Prevention Chart implied two different research stages in both regional and detailed scales. This procedure represented a suitable response to divergent but integrated objectives of the work: regional/ local planning and immediate reduction of disasters.

The types of mapping units mentioned here refer to levels of land-system analysis in which certain predictable combinations of landscape components are likely to be observed and reported.

The use of remote sensing techniques for physiographic compartmentation has been revealed as a powerfull work-tool into the scope of the regional land evaluation. The method is particularly adequate to survey those regions where a highly contrasting topography is disclosed. It is just the example for the São Sebastião region and for the majority of brazilian southeastern coastal zone where the hazards are commonly originated by landslides and flooding.

The Land-Use and Existing Services Inventory Chart comprised a socio-economic building classifification. It also included a time-space analysis of the urban growth based on the interpretation of aerial photographs (sets of different ages). It is interisting to highlight the applicability of this inventory chart both for planning purposes and for risk assessment in a subsequent stage of detailed engineering geological mapping.

The results contained in each product (see Figure 1) enabled the local governmental authorities to obtain better understanding of the landscape before implementing public policies. The information provided by map record of landscape characteristics can be very helpful on decisions concerning the necessity of a certain specific remedial/mitigation action.

Acknowledgements

We are grateful to the field geologists and to the office staff of the IG/SMA-SP by their helpful collaboration during all the work. Special thanks to N.S.V. Moura Fujimoto, C.R.G. Souza, M.C. Holl and S.V. Cazzoli for permission to include their data on the land susceptibility analysis and on the topics of risk assessment, to O.C.B. Gandolfo for supporting all computer work, to C.M. Nunes and F. Caruso for their help to organize this text. The project was finnancially supported by the Mayoralty of São Sebastião and the IG/SMA-SP. We thank the convenor of the 30thIGC, Prof. Chengyou Ha, for his enthusiastic advising.

REFERENCES

1. G.A.C. Campanha & H.H. Ens. Estrutura Geológica na região de São Sebastião. In: *Simp. Geol. Sudeste*, 3th. Rio de Janeiro, *Bras. Soc. Geol. Abs.*, 51-52 (1993).
2. G.A.C. Campanha, H.H. Ens & W.L. Ponçano. Análise morfotectônica do Planalto do Juqueriquerê, São Sebastião (SP). *Rev.Bras. Geoc.*, **24**-1, 32-42 (1994).
3. J.C. Doornkamp, D. Brunsden, R.U. Cooke, D.K.C. Jones and J. F. Grifflths. Environmental Geology Mapping - an international review. In: *Planning and Engineering Geology* M.G. Culshaw, F.G. Bell, J.C. Cripps and M. O'Hara (Eds). Geol. Soc. Eng. Geol. Special Publication n.4, 215-219 (1987).
4. Geological Society of London. Land Surface Evaluation for Engineering Practice - Working Party Report. *Q.J.Eng. Geol. London*, 15, p.p. 265-316 (1982).
5. K. Grant & A.A. Finlayson. The Assessment and evaluation of geotechnical resources in urban or regional environments. *Eng. Geol.*, **12**, 219 - 293 (1978).
6. Y. Hasui *et al.*. Geologia da Região Administrativa 3 (Vale do Paraíba) e parte da Região Administrativa 2 do Estado de São Paulo, *Inst. Pesq. Tecn. Mon.* **1**, 1-78 (1978).
7. R.C.A. Hirata *et al.* Aplicação e discussão do método de unidades homogêneas para o planejamento territorial. In *Simp.Geol.Sudeste*, 2nd. São Paulo. *Bras. Soc. Geol. Abs.*, 373-382 (1992).
8. Instituto Geológico. Subsídios do meio físico ao planejamento regional e urbano da porção média da bacia hidrográfica do Rio Piracicaba, Estado de São Paulo. *IG/SMA-SP Unpubl. Report*, 3 volummes (1995).
9. C.Q.T. Maffra, P.C. Fernandes da Silva & G.A.C. Campanha. Estruturação do embasamento cristalino na região de São Sebastião: evidências de um domínio transpressivo. In: *Cong.Bras.Geol.*, 36th. Salvador. *Bras. Soc. Geol. Abs.*, **1**, 385-387 (1996).

10. A.M. Meijerink. Data acquisition and data capture through terrain maping units. *ITC Journal*, 23-44 (1988).
11. MEXE. Terrain classification and data storage. MEXE Part 1 Reporty No. **959**, Christchurch, Hants (1969).
12. A.T.F.S. Silva *et al.* Projeto Santos - Iguape. Relatório Final, geologia. São Pauo, *DNPM/CPRM*. v.1. (1977).
13. C.R.G Souza & K. Suguio. Coastal Erosion and Beach Morphodynamics along the State of São Paulo (SE Brazil). *An. Acad.Bras.Ci*, **68**-3, 405-424 (1996).
14. C.R.G Souza. Sedimentology applied to coastal management of the State of São Paulo, Brazil. In: *Bordomer 95 - Coastal Change*, Bordeaux, France. *Actes Proceedings*, **Tome I**, 217-228 (1995).
15. C.R.G Souza. *Considerações sobre os Processos Sedimentares Quaternários e Atuais na Região de Caraguatatuba, Litoral Norte do Estado de São Paulo*. [Unpublished M.Sc. thesis] University of São Paulo, Brazil (1990).
16. K. Suguio & L. Martin. Quaternary marine formations of the state of São Paulo and southern Rio de Janeiro, In: *Proc. International Symposium on Coastal Evolution in the Quaternary*, São Paulo. Special Publication, **1**, 1-55 (1978).
17. A. Thomas. Réflexion sur la cartographie géotechnique. In: *Proceedings of the 1st International Congress of Intl. Ass. Eng. Geol.* Paris, (1970).
18. U.S. ARMY. Application of remote sensors to army facility management and the use of remote sensing systems for acquiring data for environmental management purposes. *US Army Eng. Wat.Exp. Stn.Tech.* Report n. **74-2**. (1976.)
19. R. Vedovello. *Zoneamento geotécnico por sensoriamento remoto para estudos de planejamento do meio físico: aplicação em expansão urbana*. [Unpublished M.Sc. thesis] National Space Research Institute, São José dos Campos, Brazil (1993).
20. L.V. Zuquete. *Análise crítica da cartografia geotécnica e proposta metodológica para condições brasileiras*. [Unpublished PhD Thesis] São Carlos School of Engineering, Brazil (1987).

Proc.30th Int'l.Geol.Congr.,vol.24,pp.282-294
Yuan Daoxian (Ed)
© VSP 1997

Presentation of a methodology to produce geological hazard map to evaluate natural risks, using GIS: state of the art of the study in the Serchio-Gramolazzo river basin (Tuscany - Italy)

MARCO AMANTI, PIERLUIGI CARA, SILVANA FALCETTI, RENATO VENTURA
PCM - DSTN- National Geological Survey, Via Curtatone 3, 00185 Rome - ITALY

MASSIMO PECCI
ISPESL - Productive Settlements and Environmental Interaction Department, Via Urbana 167, 00184 Rome - ITALY

Abstract

Italian Geological Survey is producing the Geological Map of Italy at 1:50.000 scale and some related geothematic maps; among them the geological hazard map should represent one of the most important tool to evaluate natural risks, such as landslides, in infrastructured areas (typical ISPESL's issue). A preliminary draft of specific guidelines to make geological hazard maps has been published [4].

In these guidelines a new methodology was introduced; it is based on numerical process of multiple data-set using multivariate statistic, integrated by geological technical characterization in GIS environment.

According to the guidelines a pilot project in a test area (Serchio-Gramolazzo river basin -Tuscany, Italy) started. The aims of the project were the verification of the methodology and the development of software procedures and of related database. The final results of this project should represent an example of full application of the above described methodology in a whole Italian river basin.

The data-set includes cartographic information and descriptive and detailed landslide characterisation. The geo-lithological map derives from a new field survey at 1:5.000 scale; the landslide map was produced using aerophotograms and field controls. A detailed digital elevation model was generated and new information were obtained from numerical process (e.g.: watershed boundaries detection). The land units to which refer all the data were finally obtained by intersection of watershed boundaries and drainage network.

Automatic procedures using multivariate statistic on tabular data related to cartographic data in GIS database, allow to characterise each land unit, according to different key parameters in order to evaluate geological hazard.

Keywords : Geological hazard mapping, landslide, natural risks, GIS

INTRODUCTION

Classical landslides hazard map realization techniques are now supported by GIS technology and multivariate statistical models [11, 12, 13, 15, 17, 20]; these new approaches to the problem allow to use much more variables and to obtain more and more reliable models on large areas.

In this work the authors present the methodology [4] and the obtained and available products of a research carried out for the landslide hazard mapping in the upper part of the experimental river Serchio basin (Italian law n.183/89 about land defence), called in the following "Serchio di Gramolazzo" river basin (Fig. 1).

Figure 1. Geographic location of the study area.

The work started with the collection of available technical end scientific literature and data; successively the field activity followed up with the collection of lithological, geological-technical, geomorphologic and related to landslide and land use data, as well as proposed in [4]. Some papers have been already presented regarding specific landslide case studies in the study area [1, 2, 3].

The authors want here emphasise the importance of an "objective" field data collection; in fact the quality of the obtained results is directly linked to the quality and reliability of the used data and, of course, of the used model.

This purpose can be achieved using specific data collecting forms with coded tables and a specific vocabulary. In the present study two different specific forms have been studied, proposed and used: a *rock mass form*, to collect geological-technical data [6] according to [8], to classify rock masses and a *landslide form* to classify gravitational phenomena [3, 5]. A Land Information System has been designed [16] and implemented using GIS technology. The data-set includes both base and thematic data. A detailed digital terrain model (DTM) was generated and new information were obtained from numerical process (e.g.: watershed boundaries detection). Automatic procedures using multivariate statistic on tabular data related to cartographic data in GIS database, allow to characterise each land unit according to different key parameters, in order to evaluate geological hazard.

In this paper the state of the art in performing the proposed methodology is presented.

METHODOLOGY

The proposed methodology is composed of four main phases shown in Fig. 2.

They will be discussed in the following paragraphs.

The last one, regarding the actual landslide hazard zonation, is at present in development.

METHODOLOGY

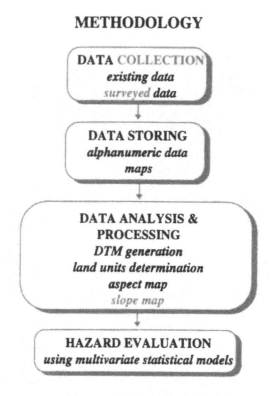

Figure 2. Flow diagram describing the used methodology for hazard evaluation.

Data collection

The field survey started with the identification of the main lithological units at a 1:5.000 scale coupled with the geological-technical data collection using the specific rock mass forms; a total number of 250 (for an average of about 3 per km²) field station has been surveyed, scattered on the whole area.

The survey took particularly into account superficial terrain (eluvial and alluvial deposits, debris, landslide deposits, moraines) widely distributed all over the basin.

During the field activity the geological units had been divided or joined, depending on their *supposed* mechanical behaviour; at the end of the survey some lithological units have been joined again, because of their similar behaviour, according to the *measured* parameters.

The frequency of the stations on each lithological unit is shown in Fig. 3; the frequency is directly proportional to the outcropping area for each lithological unit (Fig. 4).

The landslide distribution map (including also main geomorphologic features) was then realised by means of the following studies and activities:

1) aereophotogeological interpretation of available photographs and particularly Tuscany region flights, carried out between 1974 and 1984, with flight heights included between

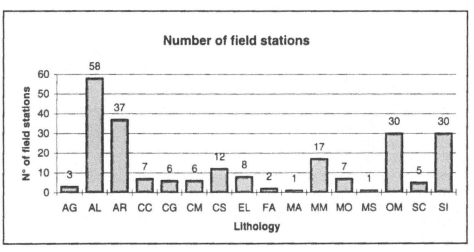

Figure 3. Histogram showing the number of field stations for each lithology (AG = clays, AL = alternances, AR = sandstone, BM = metamorphic breccia, CC = "Cavernoso" limestone, CG = conglomerate, CM = massive limestone, CS = layered limestone, DA = alluvial deposit, DI = waste disposal, EL = eluvial deposit, EN = eluvial cohesive deposit, FA = debris deposit, MA = marls, MM = massive marble, MO = morainic deposit, MS = layered marble, OM = ophiolite, RA = marble quarries disposal, SC = "Scaglia" marl, SI = schist, TR = travertine) .

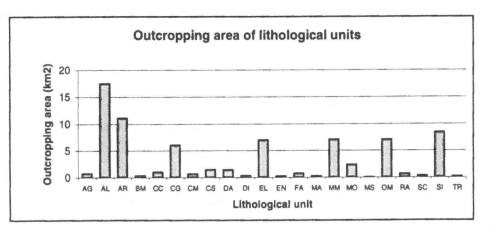

Figure 4. Histogram showing the extent of each lithology (see Fig. 3 caption for lithology code explanation) outcropping in the study area.

2.000 and 2.800 meters, integrated by "Italia flight" of 1954 at a scale of about 1 :33.000 and "High definition" flight of 1988-89;

2) field control and data collection using the specific landslide forms; the field survey was mainly aimed at the classification of each phenomenon, according to a modified landslide classification [3], and at its evolution in terms of velocity and state of activity; a total number of 219 landslides have been mapped and classified (Fig. 5).

Data storing

All the available data have been successively validated and introduced in a data base related to a GIS; a problem was represented by the data value and quality, depending both on the goodness of field control and on the experience and preparation of surveyors. So the higher is the level of field control and data validation, the better is the data quality and consequently of the results.

At the end of all the procedures the following set of data layers was obtained; it was partially collected during field surveys (*) and partially from available sources (mainly Tuscany Region) in digital form:

Lithological map (*) (1:5.000)
Simplified lithological map (*) (1:5.000)
Structural map (*) (1:5.000)
Landslides inventory map (*) (1:5.000)
Field stations distribution map (*) (1:5.000)
Simplified geomorphologic map (*) (1:5.000)
Land use map (1:25.000)
Hydrographic and related drainage/divide network map (1:5.000)
Topographic contour lines (1:5.000)
Spot heights map (1:5.000)

Data processing and analysis

The third step included data processing and analysis, with the production of the following derived maps and modified data sets:

Digital Terrain Model (DTM)
Aspect map
Slope map
Slope units map
Geological technical database

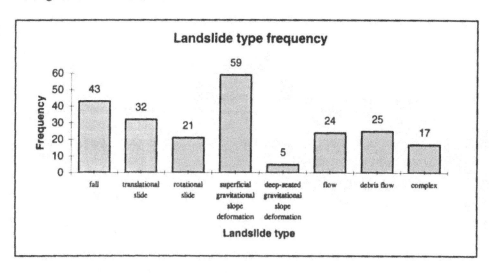

Figure 5. Histogram showing the frequency of each landslide type occurring in the study area.

A detailed DTM was generated using 10 meters contour lines, spot heights and drainage network.

The aspect map and the slope map were generated directly from the DTM.

The definition of a land unit to which refer all the data was a very important step of the process. Different approaches could have been used: a regular square grid, a geomorphologic unit and a slope unit. All of these units have their pros and cons; the square grid does not directly refer to the terrain, being only superimposed to the study area without a physical significance; the geomorphologic unit is directly linked to the terrain features but is not very objective, in fact different operators can divide the same territory in many ways, according to their own point of view.

Finally the slope unit is easily and objectively derived from the DTM and from the drainage/divide network and physically represents the land portion where the phenomenon takes place.

So, according to [12], the whole area was divided into about 500 slope units, whose dimensions are directly related to the landslides' ones.

Hazard evaluation

The fourth phase deals with hazard evaluation as shown in Fig. 6. Many difficulties can occur during this step of the work; in the following some critical features are described.

The information related to each slope unit, could be too much to be easily processed; furthermore also the results of the subsequent elaboration could be numerically unstable. It will be so necessary to reduce the variables' number and to use only the significant ones, both from the geological point of view and from the statistical one.

HAZARD EVALUATION

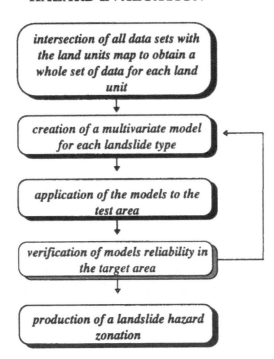

Figure 6. Logical procedure for hazard evaluation.

Multivariate statistical analysis can operate these simplifications using techniques that analyse the variables' relations; these techniques substitute single factors to subsets of related variables or eliminate from the set the ones showing a low contribution to the model. Since there are no exact rules to select a variable to eliminate, if the resulting model should not reach a good reliability, it is possible to insert again the excluded variables and to test their importance in the model.

The following step will be the slope units classification using the reduced variables set; according to [12] discriminant analysis and logistic regression techniques will be used.

Logistic regression will be performed in the case of dichotomous variables. We shall preliminary be able to classify all the slope units with actual landslides as unstable and the other ones as stable.

Using estimation procedures the analyst should never base one's confidence regarding the correct classification of future observations on the same data set from which the discriminant functions were derived; rather, if the aim is classifying cases predictively, it is necessary to collect new data to "try out" the utility of the discriminant functions.

Discriminant Analysis will automatically compute the classification functions. These must not be confused with the discriminant functions. The classification functions can be used to determine to which group each case most likely belongs. There are as many classification functions as there are groups. Each function allows to compute classification scores for each case for each group, by applying the formula:

$$S_i = c_i + w_{i1} * x_1 + w_{i2} * x_2 + ... + w_{im} * x_m \qquad (I)$$

In (I) the subscript i denotes the respective group; the subscripts $1, 2, ..., m$ denote the m variables; c_i is a constant for the i'th group, w_{im} is the weight for the m'th variable in the computation of the classification score for the i'th group; x_m is the observed value for the respective case for the m'th variable. S_i is the resultant classification score.

It is possible to use the classification functions to directly compute classification scores for some new observations.

Once we have computed the classification scores for a case, it is easy to decide how to classify the case: in general we classify the case as belonging to the group for which it has the highest classification score (unless the *a priori* classification probabilities are widely disparate) [21].

Table 1. An example of classification table; if possible the number of classes used to classify the area can be increased to 5, adding some intermediate classes such as *medium-low hazard* or *medium-high hazard*.

classification	% of occurrence coming from the discriminant analysis
high hazard	*100 - 60*
medium hazard	*59 - 40*
low hazard	*39 - 0*

In other words it is possible to use the simplified set of variables to classify each slope unit using the discriminant analysis; the result of the classification will be a number between 1 and 0 (or a percentage) assigned to each slope unit; each slope unit will be so classified, according to a specific table (Tab. 1).

The final product should be a map in which each slope unit is coloured with an increasing tone of red according to the predicted hazard.

PRELIMINARY PROCESSING RESULTS IN THE VITELLINO SUB-BASIN

In the previous paragraphs the general methodology of data collecting, storing and processing has been briefly explained and discussed; at present the full hazard evaluation has not yet completely performed. The case study of "Vitellino" sub-basin is here presented as a significant example in order to evaluate hazard condition performing statistical analysis on queried data from the geographical data base. The same sub-basin may also be considered representative of the whole area from the geological and geomorphologic point of view.

Geological, geomorphologic, hydrogeological and seismic outline of " Vitellino basin"

The "Vitellino" sub-basin (Fig. 7) is located in the upper part of the "Acqua Bianca" river, secondary right stream of the principal "Serchio di Gramolazzo". The mouth of "Vitellino" torrent into the "Acqua Bianca" river is located at an height of about 650 m a.s.l. close to the village of Gorfigliano. The plain is important in the area for the presence of relevant productive settlements, such as Apuanian marble cutting devices and treatment and factories for shoes production.

From the geological point of view [10, 19] the area represents just the boundary between the "Apuanian metamorphic core Auct." and the Tuscany thrust belt ("Falda toscana Auct."). The tectonic contact is represented in the area by an overthrust often in outcropping condition of reverse fault ("Giovetto" area) and marked by a strong thickness of coverage/weathered deposits. The units belonging to the "Falda Toscana" are composed, from the bottom to the top, by dolomitic limestone and "Cavernoso Auct." limestone (upper Trias), "Massiccio Auct." limestone and stratified "Angulati Auct." limestone (Lias) . The "Falda Toscana" domain is directly in contact with flysch deposits in clayey-arenaceous facies ("Tosco-Emiliano Complex Auct.") by means of a strike slip fault along the "Vitellino" torrent. The toes of the slopes are usually covered by moraines and fluvial-glacial deposits, while the bottom of "Acqua Bianca" valley is filled with fluvial deposits.

The main geomorphologic processes in evolution are the gravitational and fluvial ones, probably linked to a general uplift of the area and to an active tectonic.

The surveyed landslides in the sub-basin have been classified [3] as: a quick-active TRANSLATIONAL SLIDE; a slow-dormant Deep Seated Gravitational Slope Deformation (DSGSD); two quick-active ROCK FALLS; one quick-stabilised DEBRIS FLOW; two quick-dormant ROCK FALLS (Fig. 8).

The rapid evolution along the slopes of "Vitellino" torrent is mainly due to the great erosive action of the stream.

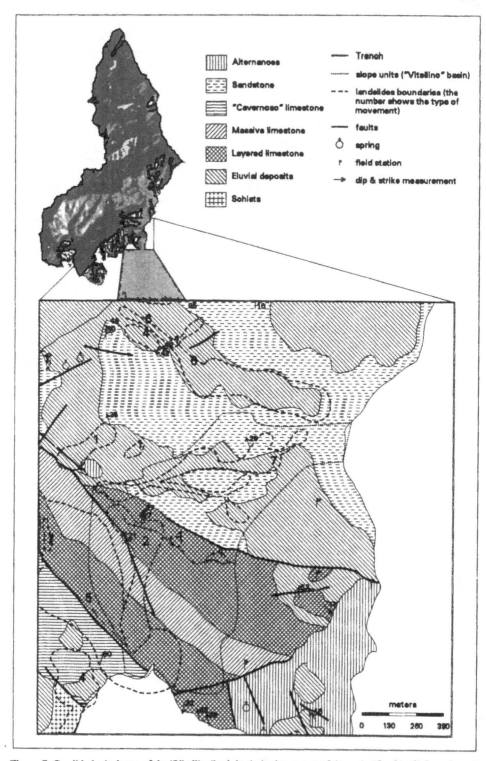

Figure 7. Geo-lithological map of the "Vitellino" sub-basin in the context of the main "Serchio di Gramolazzo" basin. In the upper left the 20 m DTM is shown ("Gramolazzo" lake and drainage network in black). The numbers in the map indicate the landslide type (1 = fall, 2 = translational slide, 4 = superficial gravitational slope deformation, 5 = deep-seated gravitational slope deformation, 6 = flow, 7 = debris flow).

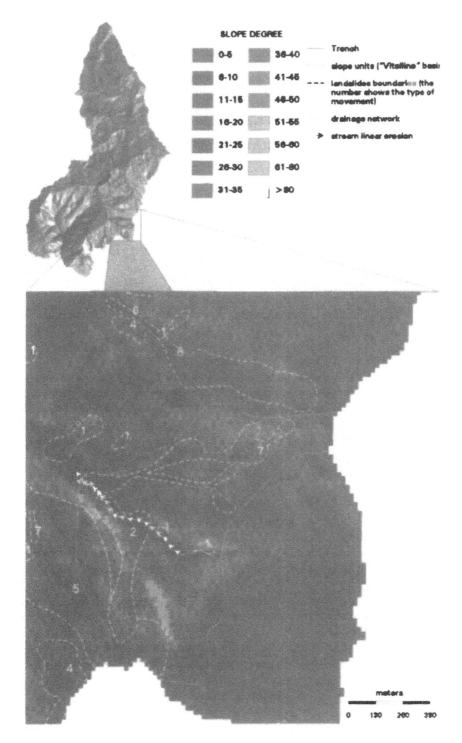

Figure 8. Slope map (cell-size 20 m) of the "Vitellino" sub-basin in the context of the main "Serchio di Gramolazzo" basin. In the upper left the 20 m DTM is shown ("Gramolazzo" lake and drainage network in black). To facilitate the comparison with Fig.7 some features (i.e slope units) have been duplicated.

From the hydrogeological point of view the full knowledge of the setting of the studied basin is very difficult, owing to the complexity of stratigraphic and tectonic relationships of many different units, characterised by extremely varying permeability features [7].

Nevertheless it is possible to simplify the hydrogeological condition of the slopes using some schemes and to classify them using the permeability and dip conditions of each slope. These information are available in the geological technical data base.

So, according to [12] we could identify: top aquifer slope, intermediate aquifer slope, bottom aquifer slope, full aquiclude slope and full aquifer slope coupled with overdip slope, reverse dip slope, underdip slope and chaotic slope.

Consequently the left Vitellino sub basin may be classified as a top aquifer - underdip slope, with the bedrock in layered/massive limestone covered by some meters of debris; the right one as a top aquifer - reverse dip slope in sandstone often covered by eluvial deposits.

From the structural and seismic point of view the whole area is classified [14] as a 2nd category, S = 9 (high seismic activity), with the last severe earthquake occurrence on the 2nd of April, 1994, two days before the fall of the large Vitellino landslide, according to local witnesses.

The effects of a strong tectonic, still active [18] are also strictly linked to geomorphologic and structural evidences, mainly consisting in [9] :

- presence of fluvial Pleistocene deposits at an height of 250-400 m above the actual river streams;
- probably fluvial catch of the northern (upper) side of Serchio di Gramolazzo river by Aulella river, at present some hundred metres lower than Serchio river;
- presence of faults in Pleistocene lacustrine deposits in the whole Serchio river basin;
- a general uplift of "Alpi Apuane", causing eastward dipping in "Serie Toscana Auct." units, different from main Apennine and Antiappennine tectonic orientations;
- presence only of Wurm traces deposits, mainly due to the low height of uplifting chain, allowing the presence of glaciers only during last glacial phase;
- many severe earthquakes registered in the area during historic period.

Mechanical features of rock masses

In the "Vitellino" basin area there are outcrops of several different lithological units (Fig. 7): Massive limestone, Layered limestone, Sandstone, "Cavernoso" limestone, Eluvial deposits, Alternances and Schist.

The slope unit in which the great translational slide took place is interested mainly by Limestone, both layered and massive; their modified RMR indexes [6] are shown in table 2.

The related rock mass quality values (ranging from 3 to 4) indicate that the mechanical behaviour and properties are medium to good and probably do not directly affect by themselves the landslide failure.

On the opposite slope the outcropping Sandstone shows a slightly worse behaviour (ranging from 2 - poor to 3 - medium) generating a thicker eluvial/detritic coverage and related localised landslides.

The authors believe that the factors playing a principal role in inducing the studied landslides are the hydrogeological conditions of slopes and the seismic activity widely distributed all over the whole basin. So special maps containing hydraulic condition of slopes (derived from lithology and geological technical database) and buffers of seismic activity areas, surrounding main faults and discontinuities, will be designed and processed in the hazard evaluation analysis.

Table 2. Details about field stations (see Fig.7) information in the "Vitellino" sub-basin. Some parameters (e.g. the Rock Mass Quality) are derived [4, 6] by several items in the data base. About lithology code see Fig. 3 caption.

Field station name	Lithology	Modified RMR index	Rock Mass Quality
λ	AR	53	3
ν	AR	38	2
ι	CC	65	4
δ	CC	71	4
β	CM	77	4
γ	CM	52	3
ε	CM	46	3
α	CS	62	4
409	CS	47	3
602	EL	high permeability	thickness about 3 m
408	EL	very high permeability	thickness about 4 m

PERSPECTIVES

Starting from the available data sets and cartography the development of the methodology is shown in the logical diagram of Fig. 6.

The "Vitellino" sub-basin has been chosen for the variety and distribution of geological and geomorphologic phenomena into a single slope units. The training carried out in the studied area builds the "knowledge base", in the authors' thought, to be applied in the hazard model proposed (Fig. 6).

Consequently the landslide hazard analysis and zonation on the whole area is now ready to be achieved.

REFERENCES

1. M. Amanti, P. Cara, G. Di Filippo and M. Pecci. Information Integration about Deep-Seated Gravitational Slope Deformations with GIS Technologies Aimed at The Knowledge of Natural Risks: a Case Study in the Upper Part of River Serchio Basin (Tuscany, Italy). Proceedings of : *II International Meeting of Young Researchers in Applied Geology*, pp. 7 - 12, Politecnico di Torino, Peveragno (1995).
2. M. Amanti, P. Cara and M. Pecci. Landslide Hazard Study in the Serchio River Basin (Tuscany - Italy): Presentation of Preliminary Results in the Gramolazzo Sub Basin. Proceedings of : *First European Regional Cartography Congress and Geographic System*, pp. 300-301, Bologna, (1994).
3. M. Amanti, P. Cara, M. Pecci and R. Ventura. Preliminary Results in the Field Control of a New Proposal of Landslide Classification for the Prevention of Natural Risks in the Experimental

Serchio River Basin (Tuscany, Italy). Proceedings of : *Prevention of Hydrogeological Hazard : the Role of Scientific Research*, Alba (in press).

4. M. Amanti, A. Carrara, G. Castaldo, P. Colosimo, G. Gisotti, M. Govi G. Marchionna, R. Nardi, M. Panizza, M. Pecci and G. Vianello. Linee guida per la realizzazione di una cartografia della pericolosità geologica connessa ai fenomeni di instabilità dei versanti, alla scala 1:50.000 (preliminary draft), pp. 1-53, Servizio Geologico d'Italia, Roma (1992).

5. M. Amanti, N. Casagli, F. Catani, M. D'Orefice and G. Motteran. Guida al censimento dei fenomeni franosi ed alla loro archiviazione. *Miscell. Serv. Geol. d'It.*, VII, pp. 1-109, Roma (1996).

6. M. Amanti and M. Pecci. Proposta di una scheda per la raccolta e l'informatizzazione dei dati utili alla classificazione e caratterizzazione degli ammassi rocciosi. *Quad. Geol. Appl.*, 1-95, 1-8, Bologna (1995).

7. F. Baldacci, S. Cecchini, G. Lopane and G. Raggi. Le risorse idriche del fiume Serchio ed il loro contributo all'alimentazione dei bacini idrografici adiacenti. *Mem. Soc. Geol. It.*, 49, 365-391, (1993).

8. Z. T. Bieniawski, Geomechanics Classification of Rock Masses and its Application in Tunnelling. Proceedings of : *IV Congress on Rock Mech.*, vol. A (1979).

9. V. Bortolotti. Geologia dell'alta Garfagnana tra Poggio, Dalli e Gramolazzo. *Mem. Soc. Geol. It.*, 83, 25 - 154, (1964).

10. Carmignani. Carta geologico-strutturale del complesso metamorfico delle Alpi Apuane, (1984).

11. Carrara. Multivariate models for landslide hazard evaluation. *Math. Geol.*, 15, 403-426, (1983).

12. Carrara, M. Cardinali, F. Guzzetti and P. Reichenbach. Gis Technology in Mapping Landslide Hazard. In : *Geographic Information Systems in Assessing Natural Hazard*. A. Carrara and F. Guzzetti (Eds.). pp. 135-176. Dordrecht : Kluwer Academic Publishers, The Netherlands, (1995).

13. Carrara, M. Cardinali, F. Guzzetti and P. Reichenbach. Gis Based Techniques for Mapping Landslide Hazard (on-line version). http ://deis158.deis.unibo.it/gis/ chapt1.html (1996).

14. Consiglio Superiore del Lavori Pubblici - Servizio Sismico. Atlante della classificazione sismica nazionale, Roma, (1984).

15. Jäger, G. and F. Wieczorek. Landslide Susceptibility in the Tulley Valley Area, Finger Lakes region, New York, USGS Open-File-Report 94-615 (On-line version). http :// www.geog.uni-heidelberg.de/~stefan/tully.html (1996).

16. Marchionna, R. Ventura. Schema logico della base informativa "Carta della pericolosità geologica" connessa ai fenomeni di instabilità dei versanti (scala 1 :50.000). Unpublished report. Presidenza del Consiglio dei Ministri, Servizio Geologico - (1993).

17. Mark and S. D. Ellen. Statistical and simulation models for mapping debris-flow hazard. In : *Geographical Information Systems in Assessing Natural Hazard*. A. Carrara and F. Guzzetti (Eds.). pp. 93-106. Dordrecht : Kluwer Academic Publishers, The Netherlands (1995).

18. Moretti. Evoluzione tettonica della Toscana settentrionale tra il Pliocene e l'Olocene. *Boll. Soc. Geol. It.*, 111, 459-492, Roma (1992).

19. Nardi, A. Pochini, A. Puccinelli, G. D'Amato Avanzi and M. Trivellini. Valutazione del rischio da frana in Garfagnana e nella media valle del fiume Serchio (Lucca). *Boll. Soc. Geol. It.*, 104, 585-599, Roma (1984).

20. C. J. van Westen. GIS in landslide hazard zonation : a review, with examples from Andes of Colombia. In : *Mountain environment and Geographic Information System*. M.F. Price and D.I. Heywood (Eds.). pp. 135-165. Taylor & Francis, UK (1994).

21. SAS Institute Inc., SAS/STAT User's Guide, Version 6, Fourth edition, Volumes 1 & 2, Cary, NC :SAS Institute Inc. (1989).

Proc. 30th Int'l. Geol. Congr., vol. 24, pp. 295-299
Yuan Daoxian (Ed)
© VSP 1997

Geological Hazards and their Provincial Atlas in China

DUAN YONGHOU

China Institute of Hydrogeology and Engineering Geology Exploration, Dahuisi 20, Haidian, Beijing 100081, China

Abstract

China is one of the countries suffered most serious geological hazards in the world due to her complex geographic and geological conditions. The atlas of provincial geological hazards in China reflects the main types and subtypes of geological hazards and their distribution, harmfulness in each province, while the geological background of geological hazards, such as geomorphologic condition, landform, active faults, type of soil and rock, etc., are shown on it as its base map. The zonation of risk assessment of geological hazards is also presented on the atlas. A systematic classification of geological hazards is built by the atlas compilation. The atlas is the contribution of geologist in China to the sustainable development of China and IDNDR.

Keywords: Geological hazards, Atlas, Distribution, Classification

INTRODUCTION

China is a country of a vast territory with a complex geological condition and frequent crustal movement, and is one of countries suffered serious geological hazards in the world. The definition and classification of geological hazards have been paid more attention by researchers. We define geological hazards as such geological process that is caused by natural and artificial activities, and endanger to human lives and property loss and make environment degraded.

PURPOSE AND SIGNIFICANCE OF THE MAP COMPILATION

The atlas of provincial geological hazards in China was compiled on the basis of geological hazards investigation, in which the main type, regional distribution, damage and their trends of geological hazards were reflected. It also reflected the environmental background and geological conditions, such as topography, geological structure (including active faults), rock-soil type of engineering geology and hydrogeology. Their types and subtypes have been divided and evaluated on the basis of development (density, frequency and scale), seriousness and harmfulness of geological hazards. The purpose of the atlas compilation was to serve for preventing and controlling of geological hazards and reforming the geological environment of human-living in favor of social and economical sustainable development of China and IDNDR.

Geological hazard is one kind of the most serious hazards in China, which produces a huge economical loss and endangers more people's lives than other natural hazards and has characteristics of sudden, frequent and group occurrence in a long run. However, currently research and investigation on geological hazards are not so deep to meet the requirement of preventing and controlling hazards and developing economy. The atlas is compiled on the basis of investigation (conducted before 1992) data from more than ten thousands of typical geological hazards spots. This is first time to compile the provincial atlas of geological hazards in China and it provides us an opportunity to understand the current situation and developing trends of the hazards in China.

The research on the distribution and development of geological hazards and the work on monitoring, preventing, controlling of the hazards are important part for environment protection. It is an important step to implement the sustainable development strategies of environment and development listed in "China Agenda for 21st Century", and also benefits to the later generation.

PRINCIPLE AND METHOD OF MAP COMPILATION

To compile the " Provincial Atlas of Geological Hazards in China", an unified scheme of map compilation, legend and color was designed.

1). The geographic base map was provided by China Map Press. The elements of topography and landform concerned geological hazards were remained according to the specific requirement.

2). The soil and rock type in engineering geology were taken as basic unit for mapping and basic materials at which geological hazards developed. According to current standards, first grade is rock type including magma rock, metamorphic rock, clastic rock and carbonate rock. The second grade is rock group which is determined by lithology, strength and structure type. Soil types were divided into pebble, gravel,sand, clay and special soil. Soil structure was divided into single, dual and multi-layer structure.

3). The active faults since Quaternary, playing a controlling role to earthquake and other types of geological hazards, were expressed as lines on the atlas. The earthquake center with the intensity of more than VII and division of earthquake intensity were obviously reflected as major geological hazards type and a controlling factor to induce other geological hazards.

4). The different geological hazards or their groups were expressed as a special legend. Geological hazards developed in the form of surface were shown as surface legend. Measurable geological hazards were presented as a contour line.

5). Geological hazards, divided and evaluated according to their types, developing density and velocity, frequency, deformational scale, trends of increasing or decreasing and damage extent, were expressed as different colors.

6). For better understanding to non-geologists and decision makers, the atlas would be more clear and easier to see and use. The important contents were expressed as inlay or profile map.

7). About 2000 words were attached to each map to explain hazards type, harmfulness and prevention strategy.

MAIN TYPES OF GEOLOGICAL HAZARDS IN CHINA

According to noteworthy case of geological hazards and their composition, dynamic process, destructive form and rate, the geological hazards in China are classified into ten types and thirty-eight subtypes.

Type	Subtype
Earthquake and volcano	Structural earthquake Volcanic earthquake Human activity induced earthquake
Slope soil and rock movement	Landslide Rockfall and avalanche Debris flow
Ground deformation	Land subsidence Ground collapse Ground fissure Liquefaction
Land degeneration	Soil erosion Desertization and desertification Salinization Cold soaked farmland
Ocean (coast) dynamic hazards	Sea level rising Sea water intrusion Coastal erosion Port deposition Storm tide
Hazards related to mining and underground engineering	Water bursting in mine pit Mine pit collapse Gas bursting and explosion Coalfield self-combustion Rock blasting
Hazards related to special soil	Loess collapse Expansion soil Soft soil Frozen soil Laterite
Hazards related to geo-environment	Endemic disease Polluted soil by pesticides & herbicides
Hazards related to groundwater change	Rising and down of groundwater table Groundwater pollution
Hazards around rivers, lakes and reservoirs	Bank failure Deposition Seepage Flood Bank bursting

THE CHARACTERISTICS AND DEVELOPING TRENDS OF GEOLOGICAL HAZARDS IN CHINA

1). The distribution of geological hazards in China is controlled by geomorphic and geological conditions, and latitude position. For example:
Landslide, rockfall and debris flow concentrate in west mountain area;
Land subsidence, ground fissure distribute in east plain area;
Karst collapse occur in southwest mountain area and part of east hill area;
Sea water intrusion, coastal erosion and deposition occur along the coastal zone;
Land desertization, desertification, salinization happen in northern China, northeastern China and northwestern China.

This regularity of geological hazards' distribution is a basis of decision making in geological hazards prevention and environment protection.

2). With the rapid economic development, the impact of geological hazards to society and economy is getting severer. Almost every province has suffered different kinds of geological hazards. Since no systematic planning before economic development, more and more geological hazards have been and are being induced by human activities, such as over exploiting natural resources, waste disposal, mining, etc. It has been investigated that the harmfulness and frequency of geological hazards is linearly related to the increase of population and economic development. It is sure that geological hazards will become more and more if there is no development planning and measure to protect environment and coordinate the relationship between economic development and resources utilization.

3). Geological hazards occur as a group and interact each other, which consist of a geological hazard system or a hazard chain. It is quite common that some geological hazards cause other kinds geological hazards, for example: a debris flow happen after a landslide which is a result of soil erosion, then the debris flowing into a river or a lake or reservoir make them silted, which may effect the river's navigation or reservoir's working. The suspend river bed of the Yellow River in its lower reach is the result of soilerosion in its upper and middle reach. Due to over pumping groundwater, land subsidence, sea water intrusion occur as a hazards group in coastal zone, as their result,

the harmfulness of jetty bursting, storm tide would become more serious. The geological hazards chain or group present that the hazards prevention should be a comprehensive work under a systematic planning.

4). The impacts of geological hazards last a long term and their process is periodic and not reversible. The periodicity of slope movement follows the one of rainfall. The earthquake in China has been in fifth active periodicity since 1980's, which is predicted that the period would last until 2000. The height lost by land subsidence have not recovered by artificial recharging groundwater since it occurred in several coastal cities and east plain of China. The impact of environment from groundwater pollution and decreasing of deep groundwater table which is about 70 meters lower than sea level in some coastal area is restricting and will restrict the regional sustainable development.

5). The types of geological hazards are getting more and more, their damage serve, owing to the rapid growth of population and economic development. Land subsidence, most of karst collapse, ground fissure are caused by over extracting groundwater. Over 20 provinces have suffered from the land subsidence and ground fissure by 1990's instead of a few of cities and provinces in 1950's. Most of slope movement hazards along the railways of Baoji - Chengdu, Baiji - Lanzhou, Chengdu - Kunming have resulted from artificial slope cutting, loading. It is estimated that the amount of landslide, rockfall, debris flow in Sichuan in 1990's is as same as three times in 1950's, while the population in Sichuan has increased two times during the period. Before 1950's, debris flow in eastern Liaoning was one time, two times during 1960's and 1970's, six times in 1980's, while strength and harmfulness of debris flow are getting higher and higher.

6). Slow on-site geological hazards, such as land subsidence, soil erosion, land desertization, groundwater pollution, etc. is becoming more serious. As its result, ecological environment is degrading and natural resources exhausting. Soil erosion and land desertization are much severer than other slow on-site geological hazards, which greatly change ecological environment and cause a large mount of economic loss. The statistic result shows that the deserted land in Inner Mongolia has increased 106,600 Km^2 with a rate of 3000 Km^2 per year for thirty year. If there is no measure to mitigate the hazard, the deserted area will be 800,000 Km^2, which is about 70% of total area in Inner Mongolia. Soil erosion in loess plateau is 430,000 Km^2 with a erosion mode of 8000 Ton/Km^2*year, while it is also getting servere in laterite area, granite area and limestone area, etc., especially in Sichuan, Yunan, Jiangxi, Hunan, etc. Slow on-site geological hazards prevention should thus be paid attention too for improving human's living environment.

7). It is quite hard that geological hazards prevention and control which rely on not only a great deal of fund, but also a systematic research and monitoring on geological hazards. If environment protection is regarded as same important as economic development, a lots of geological hazards may be prevented.

REFERENCES

1. Academy of Geosciences of China. Atlas of Geology of China, Geological Publishing House, 78-80 (1973).
2. 562 Team of Academy of Geoscience of China. Tectonic system and its explanation of China, Map Press of China, 7 (1979).
3. Huang Zhentai. Investigation on slope hazards and its control, Bulletin of Institute of Energy and Resource of Taiwan (1992).
4. Huan Zhentai. Land subsidence in Taiwan, Bulletin of Institute of Energy and Resource of Taiwan (1992).
5. Geological Survey of Japan. Atlas of Geological hazards in Japan (1:3,000,000) (1992).
6. Duan Yonghou, Luo Yuanhua, Liu Yuan, et al. Geological hazards of China, Architecture Press of China (1993).
7. Ministry of Geology and Mineral Resource of P.R.China, State science and Technology Committee of PR China, State Planning Committee of P. R. China, Geological hazards and prevention. Geological Publishing House (1991).
8. China Institute of Hydrogeology and Engineering Geology Exploration. Map series of environmental geology, Map Press of China (1992).

Milton Keynes UK
Ingram Content Group UK Ltd.
UKHW040448071024
449327UK00020B/1081